Applications of Antenna Technology in Sensors

Applications of Antenna Technology in Sensors

Editor

Pedro Pinho

MDPI • Basel • Beijing • Wuhan • Barcelona • Belgrade • Manchester • Tokyo • Cluj • Tianjin

Editor
Pedro Pinho
Departament of Electronics,
Telecommunications and
Informatics
Universidade de Aveiro
Aveiro
Portugal

Editorial Office
MDPI
St. Alban-Anlage 66
4052 Basel, Switzerland

This is a reprint of articles from the Special Issue published online in the open access journal *Sensors* (ISSN 1424-8220) (available at: www.mdpi.com/journal/sensors/special_issues/AATS).

For citation purposes, cite each article independently as indicated on the article page online and as indicated below:

LastName, A.A.; LastName, B.B.; LastName, C.C. Article Title. *Journal Name* **Year**, *Volume Number*, Page Range.

ISBN 978-3-0365-4488-5 (Hbk)
ISBN 978-3-0365-4487-8 (PDF)

Contents

About the Editor

Pedro Pinho

Pedro Pinho was born in Vale de Cambra, Portugal, in 1974. He received Licenciado and Master's degrees in Electrical and Telecommunications Engineering in 1997 and 2000, respectively, and a PhD degree in Electrical Engineering from the University of Aveiro in 2004. He is currently an Assistant Professor in Electronics, Telecommunications and Informatics Engineering Department at the University of Aveiro (UA) and a Senior Member of the research staff at the Instituto de Telecomunicações (IT), Aveiro, Portugal.

Dr. Pinho is also a senior member of the IEEE and serves as the Technical Program Committee in several conferences and reviewer of several IEEE journals and is a member of the IEEE APS. He has authored or co-authored one book, edited four books, ten book chapters, and more than 200 papers for conferences and international journals. He participated as the principal investigator or coordinator in projects with scientific and/or industry focus, both at a national and international level. To date, he has led and leads 8 PhD students and 60 MSc students. His current research interest is in antennas and propagation.

Preface to "Applications of Antenna Technology in Sensors"

During the past few decades, information technologies have been evolving at a tremendous rate, causing profound changes to our world and to our ways of living. This reprint aims to introduce and treat a series of advanced and emerging topics in the field of antenna sensors.

Antenna sensors are a topic of increasing interest for industry and the scientific community and for wearable applications. Essentially, they are electronics devices with two functions, communication and sensing, that can be obtained by an appropriate change in the antenna structure in order to change its behavior, typically by changing the resonance frequency of the antenna. Nowadays, these sensors are developed with an emphasis on flexible solutions, for integration with textiles or in the body.

The first chapter presents a novel approach to direction-of-arrival (DoA) estimation, using a two-row electronically steerable parasitic array radiator (ESPAR) antenna, which has 12 passive elements and allows for the elevation and azimuth beam switching using a simple microcontroller. Thus, this approach relies solely on the received signal strength (RSS) values measured at the antenna output port. The results obtained for different signal-to-noise ratio levels indicate that a two-row ESPAR antenna can produce, even for low SNR values, accurate DoA estimation in the horizontal plane, without prior knowledge of the elevation direction of the unknown radio frequency (RF) signals, by using appropriate combinations of only twelve 3D antenna radiation patterns.

The second chapter presents a simplified method for the calculation of mutual inductance of the planar spiral coil, inspired by the Archimedean spiral. The accuracy of the calculation results in terms of the simulation and the measurement results demonstrates that it is a good candidate for wireless power applications.

The third chapter presents a circularly polarized flexible and transparent circular patch antenna, suitable for body-worn wireless communications. The performance of the explored flexible-transparent antenna is also compared with its non-transparent counterpart manufactured with a non-transparent conductive fabric. The compatibility of the antenna in wearable applications is evaluated by testing the performance on a forearm phantom and calculating the specific absorption rate (SAR).

Antennas in wireless sensor networks (WSNs) are characterized by enhancing the capacity of the network, providing a longer range of transmission, better spatial reuse, and lower interference. The fourth chapter presents a planar patch antenna for mobile communication applications, operating at 1.8, 3.5, and 5.4 GHz. The results indicate that the proposed planar patch antenna can be utilized for mobile applications such as digital communication systems (DCS), worldwide interoperability for microwave access (WiMAX), and wireless local area networks (WLAN).

Sparse arrays have grating lobes in the far-field pattern due to the large spacing of elements, residing in a rectangular or triangular grid. Random element spacing removes the grating lobes, but it also produces large variations in element density across the aperture. The fifth chapter introduces a low discrepancy sequence (LDS) to generate the element locations in sparse planar arrays without grating lobes. The author presents the mathematical formulation for implementing an LDS, to generate an element lattice for sparse planar arrays, and presents numerical results of their performance. Multiple array configurations are studied, and it is shown that these LDS techniques are not affected by the type/shape of the planar array.

The sixth chapter presents the effect of superstrates and metamaterials and proposes the idea of dielectric fill-in between the antenna and the superstrate in order to enhance the transmission between two antennas. Preliminary studies show that the proposed method can significantly increase transmission between a pair of antennas. Afterward, an analysis of a representative passive neurosensing system with realistic biological tissues shows very low transmission loss, as well as considerably better performance than state-of-the-art systems.

In the seventh chapter, the authors present a compact, low-cost, and low-power prototype of a pulsed Ultra Wide Band (UWB) oscillator and a UWB elliptical dipole antenna integrated on the same RF Printed Circuit Board (PCB) and its digital control board for Real-Time Locating System (RTLS) applications. The PCB has been manufactured, and the entire system has been assembled and measured. Simulated and measured results are in excellent agreement with respect to the radiation performances, as well as the power consumption.

Bearing in mind the need for communication with distributed sensors/items, the eighth chapter presents the design of a single-port antenna with dual-mode operation, representing the front end of a future generation tag acting as a position sensor, with identification and energy harvesting capabilities. A conformal design, supported by 3D-printing technology, is pursued to check the versatility of the proposed architecture, considering any application involving its deformation and tracking/powering operations.

The ninth chapter presents the development and experimental validation of an ISA100.11a simulation model for industrial wireless sensor networks (IWSN). Several metrics related to the timing of events and communication statistics are used to evaluate the behavior and performance of the tested IWSNs. The analysis of the results demonstrates the potential of the proposed model to accurately predict IWSN behavior.

The tenth chapter provides an integrated process route for smelting gallium-based liquid metal (GBLM) in a high vacuum and injecting GBLM into the antenna channel in high-pressure protective gas, which avoids the oxidation of GBLM during smelting and filling. Then, a frequency-reconfigurable antenna, utilizing the thermal expansion characteristic of GBLM, is proposed. The designed antenna also provides a new approach to the fabrication of a temperature sensor, which detects temperature in challenging situations for conventional temperature sensing technology.

The eleventh chapter presents the design and fabrication of two multi-element structurally embedded vascular antennas (SEVAs). These are achieved through advances in additively manufactured sacrificial materials and demonstrate the ability to embed vascular microchannels in both planar and complex-curved epoxy-filled quartz fiber structural composite panels. Experimental and predicted results demonstrate the operation of antennas in canonical states. Additional results for the array topology demonstrate the ability for beam steering and contiguous operation of interconnected elements in the multi-element structure.

The last chapter covers a new class of Radio Frequency Identification (RFID) tags, namely the three-dimensional (3D)-printed chipless RFID. These tags can be realized with low-cost materials and inexpensive manufacturing processes and can be mounted on metallic surfaces. The performance of such a class of chipless RFID tags is finally assessed by measurements on real prototypes.

The editor and the authors would like to express their gratitude to the publisher for the assigned time, invaluable experience, efforts, and staff that successfully contribute to enriching the final overall quality of the reprint.

When I needed you the most, you were there. Thank you, Carla Silva!
To Íris Pinho and Petra Pinho

Pedro Pinho
Editor

Article

Direction of Arrival Estimation Based on Received Signal Strength Using Two-Row Electronically Steerable Parasitic Array Radiator Antenna

Mateusz Rzymowski *, Krzysztof Nyka and Lukasz Kulas

Department of Microwave and Antenna Engineering, Faculty of Electronics, Telecommunications and Informatics, Gdansk University of Technology, Narutowicza 11/12, 80-233 Gdansk, Poland; krzysztof.nyka@pg.edu.pl (K.N.); lukasz.kulas@pg.edu.pl (L.K.)

* Correspondence: mateusz.rzymowski@pg.edu.pl

Abstract: In this paper, we present a novel approach to direction-of-arrival (DoA) estimation using two-row electronically steerable parasitic array radiator (ESPAR) antenna which has 12 passive elements and allows for elevation and azimuth beam switching using a simple microcontroller, relying solely on received signal strength (RSS) values measured at the antenna output port. To this end, we thoroughly investigate all 18 available 3D antenna radiation patterns of the antenna measured in an anechoic chamber with respect to radiation coverage in the horizontal and vertical direction and propose a generalization of the power-pattern cross-correlation (PPCC) algorithm involving a high number of multiple calibration planes (MCP) as well as specific combinations of radiation pattern sets. Additionally, a new way of RSS-based DoA estimation accuracy assessment, which involves thorough testing conducted along the elevation direction when RF signals impinging on the antenna arrive from arbitrary θ angles, has been reported in this paper to verify the overall algorithm's performance. The results obtained for different signal-to-noise ratio (SNR) levels indicate that two-row ESPAR antenna can produce, even for low SNR values, accurate DoA estimation in the horizontal plane without prior knowledge about the elevation direction of the unknown RF signals by using appropriate combinations of only 12 3D antenna radiation patterns.

Keywords: Internet of Things (IoT); wireless sensor network (WSN); switched-beam antenna; electronically steerable parasitic array radiator (ESPAR) antenna; received signal strength (RSS); direction-of-arrival (DoA) estimation

Citation: Rzymowski, M.; Nyka, K.; Kulas, L. Direction of Arrival Estimation Based on Received Signal Strength Using Two-Row Electronically Steerable Parasitic Array Radiator Antenna. *Sensors* **2022**, *22*, 2034. https://doi.org/10.3390/s22052034

Academic Editor: Christian Vollaire

Received: 14 February 2022
Accepted: 2 March 2022
Published: 5 March 2022

1. Introduction

Wireless sensor networks (WSNs), especially in the Internet of Things (IoT) applications, depend on low-cost wireless transceivers, which are usually integrated with simple microcontrollers to provide the functionality required by different IoT applications [1–3]. To increase WSNs capabilities, especially when they are installed in challenging environments, in which connectivity problems may be present due to multipath propagation or presence of interfering radio frequency (RF) signals [3,4], WSN nodes can be integrated with energy-efficient switched-beam antennas (SBAs) [5–9] providing a number of directional radiation patterns. Such patterns can easily be switched electronically by a WSN node's integrated microcontroller [9,10] in order to improve the overall network performance, e.g., by focusing antenna beams of WSN nodes towards specific directions, and increasing its energy efficiency [9,11–15]. Moreover, for a simple low-cost WSN node, it can enable a capability of direction-of-arrival (DoA) estimation for received RF signals incoming from different WSN nodes belonging to the same network [10].

Electronically steerable parasitic array radiator (ESPAR) antenna is one of the promising SBA designs that can successfully be integrated with a WSN node [10] or a WSN gateway [16]. It relies on a simple, yet very effective, a concept introduced originally by

Harrington [17], in which a centrally placed active element is surrounded by 6 passive elements that are connected to variable reactances. In this concept, a directional radiation pattern can be formed by setting the correct values of the reactances, which is well-suited for systems having a single RF transceiver port that could benefit from beamforming capabilities. The first ESPAR antenna designs and prototypes, based on the original Harrington proposal, relied on monopoles connected to varactor diodes [18]. A standard RF transceiver can be connected to the proposed ESPAR antenna output port and, additionally, DoA estimation of impinging RF signals with 2° precision can be performed in the horizontal plane relying solely on received-signal strength (RSS) measurements. However, the necessary bias voltages needed to provide correct values of the reactances at the terminals of 6 passive elements in [18] were produced by 12-bit digital-to-analog converters (DAC) of a digital signal processing (DSP) device. In consequence, this antenna system could not be practically used in energy-efficient wireless sensor systems.

To further adapt the ESPAR antenna proposed in [18] to the applications in WSNs, especially integrated within low-cost battery-powered WSN nodes, a simplified beam steering concept has been proposed [19]. In that approach, 12 passive elements surrounding the active one were used and the varactor diodes were replaced by simple low-cost and low-current integrated single-pole double-throw (SPDT) FET switches to provide load impedances at the terminals of passive elements close to an open or short circuit. Such ESPAR antenna can produce 12 directional radiation patterns which can be formed and rotated in the horizontal plane with 30° discreet steps using digital I/O ports of a simple microcontroller integrated within an RF transceiver. Thus, it can also be used to estimate DoA with 2° precision in an anechoic chamber using power-pattern cross-correlation (PPCC) algorithm, originally proposed in [18], relying solely on RSS measurements at the antenna output port [20]. Therefore, by integrating energy-efficient ESPAR antennas within WSN nodes employing simple and inexpensive RF transceivers that can measure RSS of incoming packets, it was possible to develop WSN nodes capable of performing DoA estimation [10]. Together with accompanying beam focusing functionality it can improve coverage, connectivity and energy efficiency as well as reduce latencies in the whole network [13–15,21,22].

One of the main benefits of ESPAR antennas is that, according to existing literature, they can easily be integrated with a WSN node and provide DoA estimation functionality to the node relying solely on RSS measurements gathered for 12 ESPAR antenna radiation patterns [10]. The PPCC algorithm used in [10] for DoA estimation has been originally introduced in [18]. It relies on strong horizontal diversity of ESPAR antenna radiation patterns which are measured beforehand in the horizontal direction in an anechoic chamber and correlates these measurements with RSS values gathered by the WSN node. The original PPCC algorithm has been extended to involve multiple calibration planes (MCP) [23,24] being the radiation patterns measured in an anechoic chamber for different θ directions. It allows 1D DoA estimation of RF signals impinging on the antenna without prior knowledge about their vertical direction, which may vary as the elevation of WSN nodes, may be different in practical implementations [25,26]. Unfortunately, due to the conical shape of ESPAR antenna directional radiation patterns [27], DoA estimation results do not keep the low accuracy levels available for $\theta = 90°$ when impinging RF signals have low θ angles [23,24,27]. This effect is especially well pronounced when a reduced number of directional beams are considered. Experiments conducted for a low-profile ESPAR antenna having 8 passive elements in [28] indicate that DoA estimation not only loses accuracy for lower θ angles when 6 or 8 directional radiation patterns are used but also creates ambiguous results making the PPCC-MCP algorithm results unreliable for lower signal-to-noise ratio (SNR) values [28].

To address WSN connectivity challenges in practical industrial installations [25,26] and also possible drone applications [29,30], in which the elevation of WSN nodes may significantly vary, a two-row ESPAR antenna concept has been proposed recently that has 12 passive elements arranged in two rows around the active element. It is capable to create

18 directional radiation patterns covering 3 different elevation directions [31]. The new ESPAR antenna has been designed to provide better connectivity in IoT applications relying on WSN nodes, in which the nodes positions are not restricted to the horizontal plane only, and to provide DoA functionality in such setups. It relies on simple low-power SPDT integrated switches, and therefore, it can easily be incorporated in a WSN gateway or a WSN node. As a consequence, simple elevation and azimuth beam switching allows to create 6 directional beams in the horizontal plane for each of 3 vertical direction.

Combined connectivity and DoA estimation capabilities can highly improve IoT systems involving WSN nodes and gateways installed on the ground in smart factories, buildings and cities [1–3,5,6,25,26]. Moreover, it can be used in modern IoT applications involving long-range communication to such objects as unmanned aerial vehicles (UAVs), high-altitude pseudo-satellites (HAPS) or satellite platforms [29,30,32,33], in which the elevation between transceiver and receiver may change in a fast manner. In this regard particularly interesting and growing application area, in which beamforming plays significant role, is providing multiple access and increased security in satellite and aerial integrated networks for IoT communication [34–37]. In such applications, the proposed two-row ESPAR antenna can successfully provide beamforming capabilities as low weight, and high efficiency have to be considered for practical space applications [38].

Connectivity performance and RSS-based DoA estimation approaches, which are crucial to develop new WSN gateways and nodes integrated with two-row ESPAR antenna, have not been investigated in the original publication [31]. According to the authors' knowledge, they are also not currently available in the literature. Therefore, the main contributions of this paper are:

- In-depth analysis of the antenna from the connectivity perspective with respect to possible beam steering in horizontal and elevation directions;
- Presentation of an approach for DoA estimation relying solely on RSS values gathered at the antenna output, which is a prerequisite for energy-efficient WSN nodes having DoA functionality [9,10], that is suitable for the antenna and can provide acceptable DoA estimation results also for low θ angles and are free from ambiguities for lower SNR values;
- Proposal of a detailed DoA algorithm performance testing method for more accurate DoA estimation accuracy assessment that involves all θ angles to address strong error variation at low θ angles.

The rest of this paper is organized as follows: in Section 2 the two-row ESPAR antenna is described together with its simulated and measured radiation patterns as well as their influence on the overall connectivity in WSN-based IoT setups, in which the positions of nodes are not restricted to the horizontal plane only. Section 3 describes a PPCC-MCP algorithm and its proposed generalization for the usage with the two-row ESPAR antenna. Results obtained using a new RSS-based DoA estimation accuracy assessment involving thorough testing conducted along the elevation direction are presented in Section 4, while Section 5 presents concluding remarks.

2. Two-Row ESPAR Antenna

2.1. Antenna Design

The proposed antenna concept has been discussed in [31] and is presented in Figure 1. It was aimed to achieve a low-cost reconfigurable antenna that is able to provide maximal angular radiation coverage by steering the beam in elevation and azimuth. The active quarter-wave monopole is located in the center of the metallic ground plane realized on the top of a dielectric substrate and surrounded by 12 passive radiators symmetrically arranged in two circular rows. The active radiator is fed coaxially with RF signal, while the passive radiators can be open or shorted to the ground plane by changing the state of the SPDT switches. Forming the radiation pattern is realized by setting proper configuration of the passive elements that become reflectors when shorted to the ground or directors when opened. Such an approach simplifies the beam steering control that can be realized with external microcontroller or transceiver having its general purpose input/output (GPIO) lines connected to the RF switches. As a result, each radiation pattern corresponds to a steering vector $V = [v_1, v_2, \ldots, v_s, \ldots, v_{12}]$, in which v_s denotes the state of sth passive element: $v_s = 1$ for open and $v_s = 0$ for the shorted one. The antenna was optimized to provide three sets of directional radiation patterns with different inclination angles $\theta_{max\_up} = 46°$, $\theta_{max\_down} = 56°$, and $\theta_{max\_mid} = 52°$ obtained for three different steering vectors sets. The symmetrical design allows rotating each beam by 360°, which means that it is possible to steer the radiation pattern in both, azimuth and elevation. Therefore, for each elevation direction, 6 directional radiation patterns having maximum in azimuth at 30°, 90°, 150°, 210°, 270°, 330° are available for steering vectors $V_{UP}^{\varphi_{max}}$, $V_{DOWN}^{\varphi_{max}}$, and $V_{MID}^{\varphi_{max}}$, where $\varphi_{max} = \{30, 90, 150, 210, 270, 330\}$ degrees. The antenna was designed and optimized in Altair FEKO simulation tool to operate at center frequency 2.44 GHz and the detailed design procedure including optimization goals was described in [31].

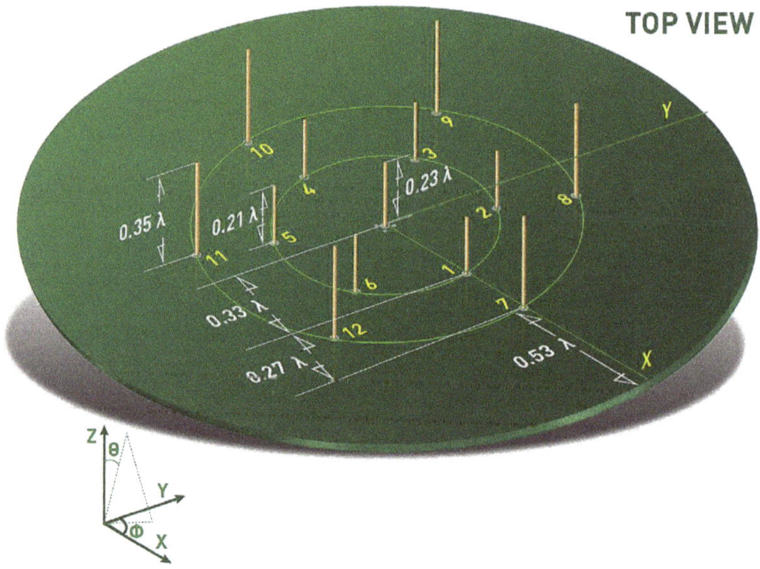

Figure 1. Two-row ESPAR antenna model.

2.2. Realized Antenna

The fabricated antenna is presented in Figure 2. 1.55 mm thick FR4 substrate with metalized top layer has been used as a PCB for the switching circuitry and the ground plane of the antenna. The RF connector and switching circuits, as well as LED indicators have been located on the bottom layer as illustrated in Figure 2. The switching circuits employ NJG1681MD7 SPDT switches providing a trade-off between high isolation, low insertion losses and power consumption in comparison to PIN diodes or varactors [28,39,40]. To provide proper performance and avoid eventual RF signal degradation, a decoupling capacitor and ESD protection coil have been placed in close proximity of the switch pins. An LED indicator has been located near each switching circuit to indicate the actual state of the switch (lights when the circuit is in open state). The steering is controlled by an external microcontroller connected to the board. The fabricated antenna has been measured and verified with the simulation model. The assumed three groups of similar beam configurations with different angle tilt in elevation have been simulated, measured and gathered in Figures 3 and 4, while the main antenna parameters for each configuration at the center frequency have been summarized in Table 1. The simulated results have been confirmed by the measurements and only small discrepancies in the backward radiation level can be observed. Measured gain for all characteristics is on a similar level around 8 dB. The input impedance matching results for all configurations are presented in Figure 5 which shows that the achieved |S11| values are close or below −10 dB. The large difference of impedance matching between configurations is a trade-off to achieve the highest possible difference of maximal direction of the radiation pattern in elevation.

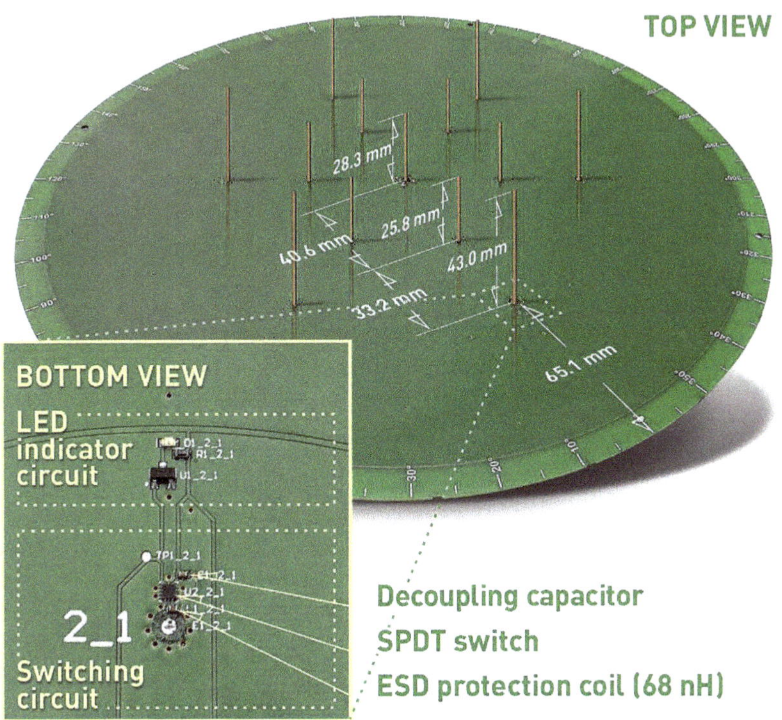

Figure 2. Fabricated two-row ESPAR antenna.

Table 1. Measured parameters of the realized two-row ESPAR antenna.

Configuration	θ_{max}	G_{max}	$HPBW_\theta$	$HPBW_\varphi$	SLL_θ	SLL_φ	|S11|
UP	46°	7.7 dB	45°	88°	10 dB	14 dB	−20.9 dB
MID	52°	7.8 dB	43°	100°	10 dB	28 dB	−10.5 dB
DOWN	56°	8.1 dB	46°	93°	12 dB	25 dB	−8.8 dB

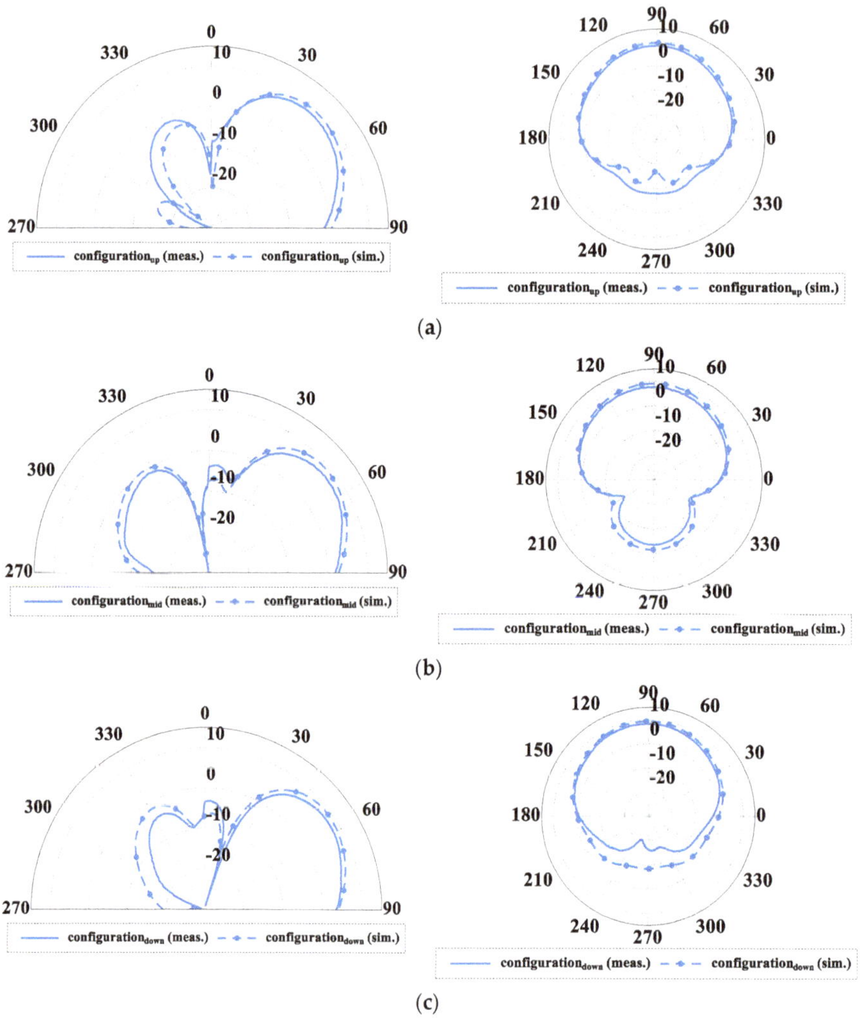

(a)

(b)

(c)

Figure 3. Simulated and measured antenna radiation patterns at 2.44 GHz in the elevation plane (for $\varphi = 90°$) and in the horizontal plane for three configurations (see text for explanations): (**a**) V_{UP}^{90} (maximum at $\theta_{max\_up} = 46°$), (**b**) V_{MID}^{90} (maximum at $\theta_{max\_mid} = 52°$), and (**c**) V_{DOWN}^{90} (maximum at $\theta_{max\_down} = 56°$).

Figure 4. Measured 3D radiation patterns for each considered configuration with different inclination angles: θ_{max_up}, θ_{max_mid}, θ_{max_down} (from left to right).

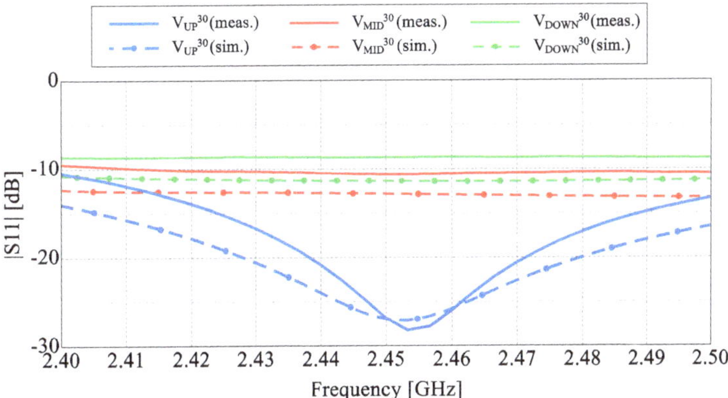

Figure 5. Values of |S11| for all three considered configurations.

2.3. Antenna Radiation Performance Analysis

The main goal of the proposed antenna was oriented towards connectivity improvement by increased angular coverage of the directional radiation pattern that can be switched in both, azimuth and elevation. For this reason, a more detailed analysis of the antenna radiation capabilities has to be considered. In Figure 4, measured 3D radiation patterns have been presented which proves the similarity of the directional beams for all three sets. For detailed analysis and comparison, the measured radiation patterns for all configurations in their maximal directions in both, horizontal and elevation plane have been presented in Figures 6 and 7. The similarity of the radiation pattern sets in horizontal plane have been confirmed for all directions of the antenna. The slight differences can be noticed in half power beam width (HPBW) where the values vary from 88° to 100° and in case of the sidelobes level (SLL), 10 dB vs 25–28 dB when comparing UP configuration to MID and DOWN configurations. In Figure 7, one can see the measured radiation patterns in elevation plane. The tilt difference and angular coverage for each set is visible. HPBW in the elevation plane is almost the same for all radiation pattern sets being around 45 degrees. It can be observed that the results are repeatable for all measured configurations.

To assess the angular coverage of the antenna that include all possible beam configurations, aggregated radiation patterns need to be considered. This form of presentation is known in the context of time-modulated arrays and 3D coverage analysis for 5G applications where the aggregated characteristic consists of all available antenna radiation patterns to show the complete spatial coverage providable by the antenna [41]. Aggregation is selecting the highest gain values for each angle from the available radiation pattern sets. In the discussed construction the aggregated radiation pattern consists of 18 characteristics: 6 for UP configuration, 6 for MID configuration and 6 for DOWN configuration. In Figure 8, it can be seen that the angular coverage range in elevation for aggregated gain above 0 dB starts from $\theta \approx 10°$. It means that only limited space under the antenna can be considered as a blind zone. To emphasize the difference in the antenna spherical coverage dependent on the selected beam configurations, a cumulative distribution function (CDF) of aggregated gain can be used [42]. In Figure 9, a CDF for the discussed antenna is presented and aggregated beams for each configuration are compared with aggregated function of all 18 radiation patterns. For example, it can be seen that for 50% of the hemisphere the aggregated gain value is 2 dB higher when considering all 18 radiation patterns instead of only one of the selected configurations. The results clearly indicate higher angular coverage of the antenna when all configurations are used.

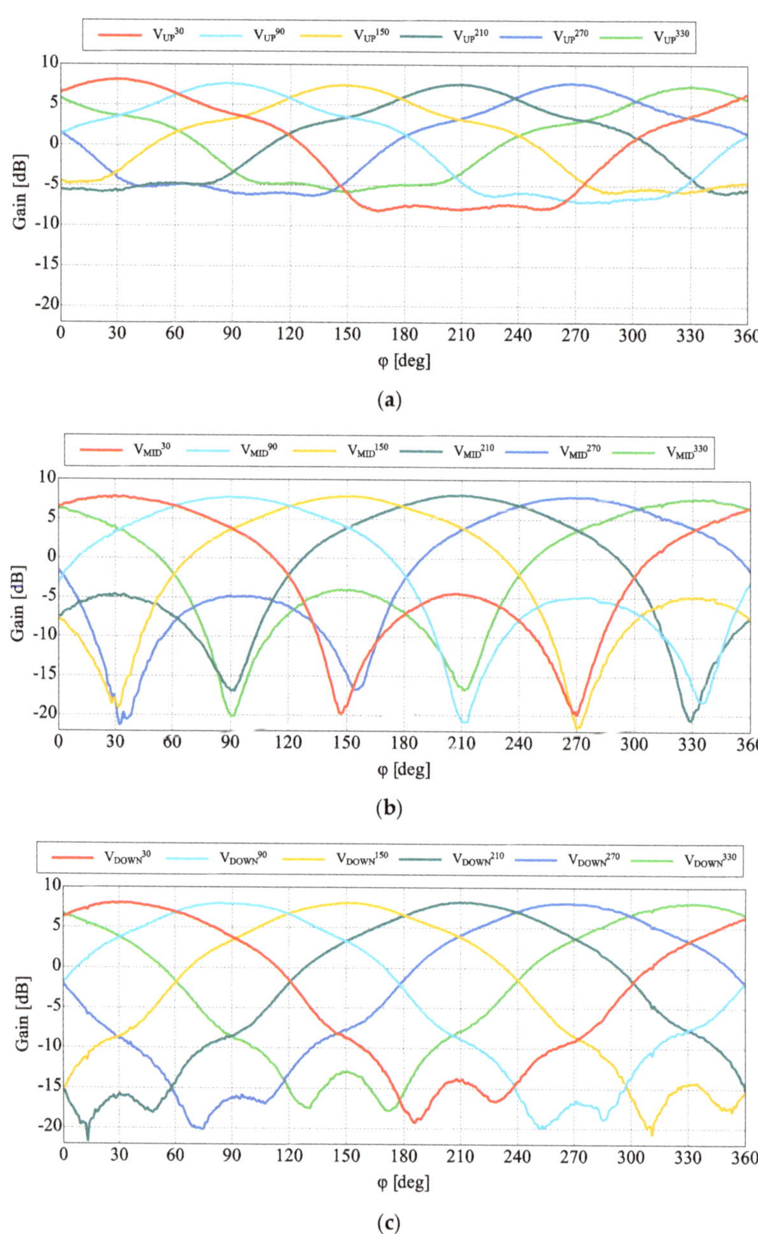

Figure 6. Measured radiation patterns for each configuration in all directions for steering vectors: (**a**) $V_{UP}^{\varphi_{max}}$, (**b**) $V_{MID}^{\varphi_{max}}$, and (**c**) $V_{DOWN}^{\varphi_{max}}$.

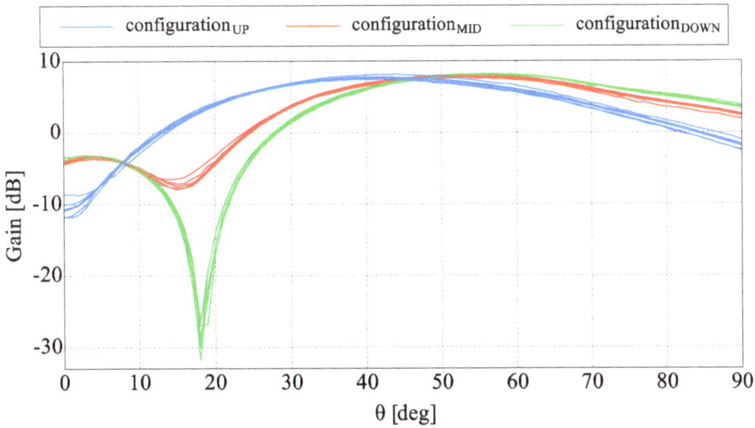

Figure 7. Measured radiation patterns in elevation plane for all directions.

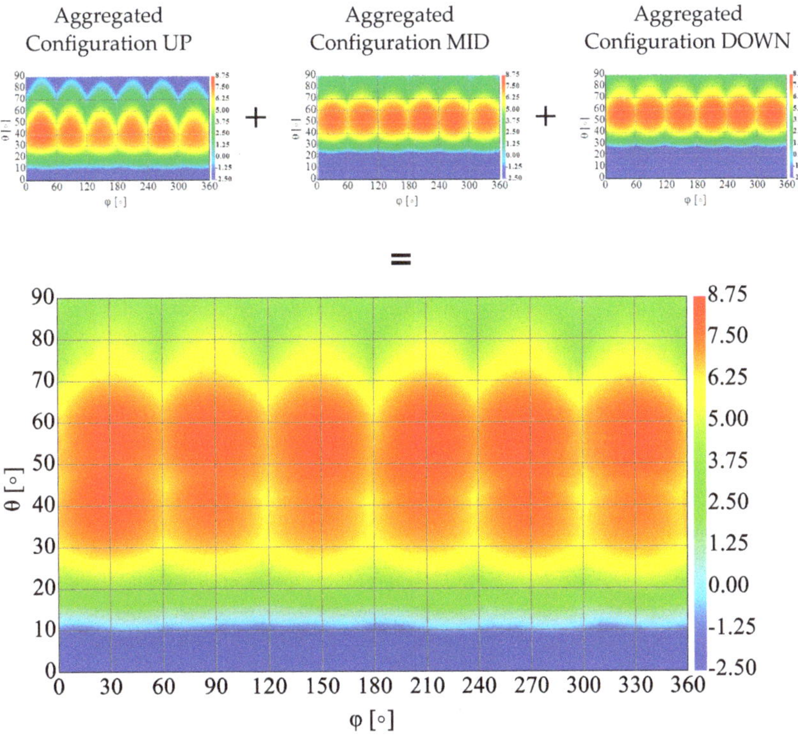

Figure 8. Aggregated gain for all measured configurations.

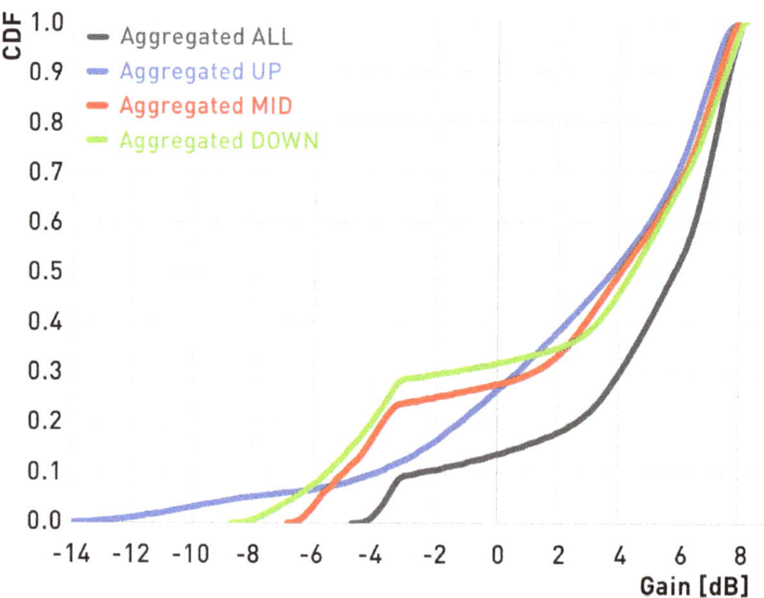

Figure 9. Cumulative Distribution Function of the aggregated gain for different sets of characteristics.

3. RSS-Based DoA Estimation for Two-Row ESPAR Antenna

Power-pattern cross-correlation algorithm is one of the most practical methods to determine unknown directions of RF signals impinging on an ESPAR antenna [10,18–20]. It relies solely on RSS measurements gathered at the antenna output port and uses simple correlation coefficient calculation [18]. When written as matrix-vector operations, they can easily be integrated with a simple microcontroller and provide DoA estimation functionality to the WSN node [10]. In its original implementation, the algorithm relies on ESPAR antenna radiation patterns measured in an anechoic chamber only in the horizontal direction [18]. In consequence, it makes DoA estimation results accurate in the $\theta = 90^\circ$ direction (i.e., the horizontal plane) with 2–4° precision being the maximum DoA estimation error [18–20]. When the directions of RF signals impinging on an ESPAR antenna are different, one may expect an accuracy drop. It has been reported that for low values of θ angle of RF signals impinging on the antenna the maximal error, referred to as precision, of 1D DoA estimation performed in the horizontal direction can easily reach 13° due to the fact that the shape of ESPAR antenna radiation patterns with respect to φ angles gradually changes for $\theta > 90^\circ$ [23]. To improve the overall accuracy for the θ angles different than 90°, one can employ multiple calibration planes (MCP) [23,24] in the original PPCC formulation, which relies on a single $\theta = 90^\circ$ calibration plane [18].

PPCC-MCP algorithm uses ESPAR antenna radiation patterns measured in chosen vertical directions. In the first implementation, all ESPAR antenna radiation patterns were measured in an anechoic chamber at $\theta = 90^\circ$ and $\theta = 45^\circ$ and, based on such two calibration planes, it was possible to maintain 4° precision for RF signals impinging from angles between $\theta = 50^\circ$ and $\theta = 90^\circ$ [23]. Involvement of 9 calibration planes, namely $\{\theta_1 = 10^\circ, \theta_2 = 20^\circ, \ldots, \theta_9 = 90^\circ\}$, provides 6° 1D DoA precision for RF signals impinging from 9 test vertical angles spanning equally between $\theta = 10^\circ$ and $\theta = 90^\circ$. However, it has been shown that PPCC-MCP algorithm accuracy is sensitive to correct placement of calibration planes positions [24], while obtained results depend on testing setup [28] parameters, especially the choice of testing directions in horizontal and vertical planes as well as SNR values set for testing signals [24,28]. Moreover, the existing PPCC formulation was created to provide DoA results based on a number of directional radiation patterns, while new

ESPAR antennas can have a number of possible radiation patterns that can be used in DoA estimation [32,33] having different performance, especially when testing signals may come from multiple horizontal and vertical directions. Therefore, to be used for DoA estimation together with new ESPAR antennas, including the two-row ESPAR antenna proposed in [31], PPCC-MCP algorithm and associated DoA testing methods have to be generalized.

3.1. PPCC-MCP Algorithm

PPCC algorithm which allows DoA estimation based on ESPAR antenna radiation patterns measured in an anechoic chamber relies on a cross-correlation coefficient between the measured radiation patterns and RSS values recorded for each directional antenna radiation pattern [18]. The cross-correlation coefficient for the ESPAR antenna having 12 directional beams has the following form:

$$\Gamma(\varphi) = \frac{\sum_{n=1}^{12} (P(V_{max}^n, \varphi)\, Y(V_{max}^n))}{\sqrt{\sum_{n=1}^{12} P(V_{max}^n, \varphi)^2} \sqrt{\sum_{n=1}^{12} Y(V_{max}^n)^2}} \tag{1}$$

where V_{max}^n is ESPAR antenna's steering vector that allows to create n-th directional radiation pattern, $P(V_{max}^n, \varphi)$ are directional ESPAR antenna radiation patterns measured during the calibration phase in an anechoic chamber in the azimuth plane for $0^\circ \leq \varphi < 360^\circ$, $Y(V_{max}^n)$ are RSS values measured at the antenna output port for all 12 directional radiation patterns during the DoA estimation phase for an unknown RF signal, and $\Gamma(\varphi)$ is the cross-correlation coefficient, which have its highest value associated with the estimated direction of arrival angle $\hat{\varphi}$ [18].

In practical implementations, the angular step precision $\Delta\varphi = 1^\circ$ is commonly used during the calibration phase, and therefore, the cross-correlation coefficient can be written in a convenient vector form as [19]:

$$g = \frac{\sum_{n=1}^{12} (p^n Y(V_{max}^n))}{\sqrt{\sum_{n=1}^{12} (p^n \circ p^n)} \sqrt{\sum_{n=1}^{12} Y(V_{max}^n)^2}} \tag{2}$$

where vector $p^n = \left[p_1^n,\ p_2^n, \cdots,\ p_I^n\right]^T$, where the superscript ^{T} is the vector transpose operator, contains $I = 360$ measured discreet values of $P(V_{max}^n, \varphi)$ and '$\circ$' denotes the element-wise product of vectors. In result, the cross-correlation coefficient $g = [\Gamma(\varphi_1),\ \Gamma(\varphi_2),\ \ldots, \Gamma(\varphi_I)]^T$ is also a vector with $I = 360$ entries being discretized values of the correlation coefficient $\Gamma(\varphi)$ for every considered value of φ in $\varphi = [\varphi_1,\ \varphi_2,\ \ldots,\ \varphi_I]^T = [0^\circ,\ 1^\circ,\ \ldots,\ 359^\circ]^T$ and $\hat{\varphi}$ corresponds now to the maximum value of g. One should note, however, that as only a single calibration plane at $\theta = 90^\circ$ is used to measure directional ESPAR antenna radiation patterns in an anechoic chamber during the calibration phase, cross-correlation operation Equation (2) will produce the most accurate results when RF signals impinging on the antenna arrive from $\theta = 90^\circ$ vertical angle, which is aligned with the calibration plane. The results will significantly deteriorate for lower θ angles as the shape of ESPAR antenna radiation patterns gradually changes for $\theta > 90^\circ$ [23,24].

To further extend PPCC algorithm and improve its accuracy in situations when θ angles of RF signals impinging on the antenna are different than 90°, PPCC-MCP algorithm, which allows to incorporate higher number of calibration planes, has been introduced [24]. It is based on the following cross-correlation coefficient:

$$g_\theta = \frac{\sum_{n=1}^{12} (p_\theta^n\, Y(V_{max}^n))}{\sqrt{\sum_{n=1}^{12} (p_\theta^n \circ p_\theta^n)}\, \sqrt{\sum_{n=1}^{12} Y(V_{max}^n)^2}} \tag{3}$$

where, for the considered number of M calibration planes Equation (3), vector $p_\theta^n = \left[\left(p_{\theta_1}^n\right)^T,\ \left(p_{\theta_2}^n\right)^T,\ \cdots,\ \left(p_{\theta_M}^n\right)^T\right]^T$ of length $I{\cdot}M$ contains ESPAR antenna's radiation pattern values measured at $\{\theta_1, \theta_2, \ldots, \theta_M\}$ angles in elevation direction and vector g_θ of length $I \cdot M$ contains PPCC-MCP cross-correlation coefficient in a vector form. As every discretized radiation pattern $p^n_{\theta_m}$, where $m = \{1, 2, \ldots, M\}$, corresponds to discretized values of φ in the same vector φ, the values in p^n_θ correspond to those in the vector $\varphi_\theta = \left[\varphi^T_{\theta_1}, \varphi^T_{\theta_2}, \cdots, \varphi^T_{\theta_M}\right]^T = \left[\varphi^T, \varphi^T, \cdots, \varphi^T\right]^T$ [24,28]. In consequence, the estimated DoA angle $\hat{\varphi}$ is now a value in φ_θ that corresponds to the highest value in g_θ [28]. It should be underlined, however, that both the number of calibration planes and choice of their θ angles highly influence the overall DoA estimation accuracy when RF signals impinging on the antenna arrive from arbitrary θ angles [23,24,28,34].

3.2. Generalized PPCC-MCP Algorithm for Two-Row ESPAR Antenna

The vector form of the PPCC algorithms in Equations (2) and (3) is appropriate for implementation in a WSN node's microcontroller to compute DoA estimation based on RSS measured for incoming packets. It is also straightforward to use such a node with ESPAR antennas having different number of directional radiation patterns or other switched beam antennas [28]. However, it is possible to create many different directional radiation patterns in ESPAR antennas and also patterns that do not have a clearly shaped directional beam. For an ESPAR antenna having 12 passive elements designed for vehicle-to-everything (V2X) applications in 802.11p frequency band, due to performance of microwave switches used at the terminals of passive elements in 5.9 GHz frequency band, it was possible to form 5 different directional radiation patterns types [32] having different beam parameters in the horizontal plane, e.g., gain, HPBW and sidelobe level (SLL), as well in the vertical plane. As every radiation pattern type can be rotated in the horizontal plane to form 12 directional beams, PPCC and PPCC-MCP algorithms can successfully be used in RSS-based DoA estimation [33]. Unfortunately, although such estimation could involve different radiation pattern types in a single estimation, to further increase the overall accuracy, appropriate and convenient PPCC-MCP algorithm formulation has not been created so far. Similarly, among 18 available two-row ESPAR antenna radiation patterns, there exist 6 groups having similar main beam direction in the horizontal plane with differences radiation pattern shape in elevation.

To perform RSS-based DoA estimation based on radiation patterns available for two-row ESPAR antenna, one has to create a generalized PPCC-MCP algorithm that can easily be implemented within simple WSN nodes and involves all possible radiation patterns. Due to their spatial characteristics, it is possible to increase the overall accuracy of the estimation while mitigating the necessity of correct placement of calibration planes. As the two-row ESPAR antenna radiation patterns, which are created using associated steering vectors V^n, exhibit certain spatial performance, including gain, SLL, direction of maximal radiation and HPBW, in both horizontal and vertical directions, we propose rewriting Equation (3) in the following general form:

$$g_{\theta_{ALL}} = \frac{\sum_{n=1}^{N}\left(p^n_{\theta_{ALL}}\, Y(V^n)\right)}{\sqrt{\sum_{n=1}^{N}\left(p^n_{\theta_{ALL}} \circ p^n_{\theta_{ALL}}\right)}\,\sqrt{\sum_{n=1}^{N} Y(V^n)^2}} \tag{4}$$

Total number of radiation patterns used in the DoA estimation and $n = \{n_1, n_2, \ldots, n_N\}$ are numbers of chosen steering vectors V^n associated with specific two-row ESPAR antenna radiation patterns $p^n_{\theta_{ALL}}$. Since θ_{ALL} means that all M = 90 possible angles in elevation direction, namely $\{\theta_1 = 90°, \theta_2 = 89°, \ldots, \theta_{90} = 1°\}$, are used as calibration planes, a high-precision turntable in the anechoic chamber is required. In consequence, vectors $p^n_{\theta_{ALL}}$, $g_{\theta_{ALL}}$, $\varphi_{\theta_{ALL}}$ will be much longer with the total length $I \cdot M = 32400$, which will make PPCC-MCP calculation more time consuming. However, the overall DoA estimation accuracy of the proposed generalized PPCC-MCP algorithm is no longer dependent on the choice of number and placement of calibration planes. Therefore, the overall DoA estimation accuracy will not be affected when RF signals impinging on the antenna arrive from arbitrary θ angles.

4. Results and Discussion

4.1. Mesurement Setup

To verify DoA estimation performance of the two-row ESPAR antenna using the generalized PPCC-MCP method 18 available antenna radiation patterns were measured in our 11.9 m × 5.6 m × 6.0 m anechoic chamber, shown in Figure 10, at 2.44 GHz using P9374A Keysight Streamline USB Vector Network Analyzer. The antenna was mounted at H = 4.1 m on a precise and equipped with a digital encoder turntable, which is able to set any 2D position with 0.1° precision in both horizontal and vertical directions. The transmitting antenna was placed on a pole stand at the same height H = 4.1 m, 8 m from the two-row ESPAR antenna. Every radiation pattern was measured with the angular step $\Delta\varphi = 1^\circ$ in the horizontal plane for $0^\circ \leq \varphi < 360^\circ$ and with the angular step $\Delta\theta = 1^\circ$ in the vertical plane for $0^\circ < \theta \leq 90^\circ$. As a result, I = 360 calibration points were produced for every elevation plane and M = 90 calibration planes were created, namely $\{\theta_1 = 90^\circ,\ \theta_2 = 89^\circ,\ \ldots, \theta_{90} = 1^\circ\}$.

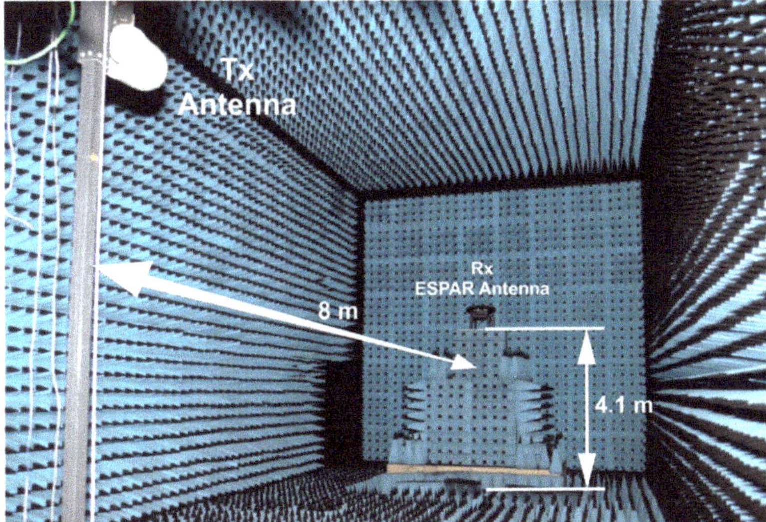

Figure 10. Anechoic chamber measurement setup used for the calibration phase of the proposed generalized PPCC-MCP algorithm for two-row ESPAR antenna.

4.2. Deatiled DoA Testing Method

To examine DoA estimation accuracy, measured calibration planes were imported to Matlab, where 10 snapshots of sinusoidal test signal were generated for all considered test directions (φ_t, θ_t) of impinging RF signals. During each test, as snapshots have to be received at the antenna's output [19,20,23,33], all sinusoidal test signals were multiplied by two-row ESPAR antenna radiation patterns values measured at appropriate test directions. Then, additive white Gaussian noise (AWGN) was added to obtain a required signal-to-noise ratio (SNR) and RSS value was calculated. Thus, the SNR value has not to be measured as it is determined by setting the appropriate value of spectral power density of numerically added AWGN. This reflects the measurement procedure presented in [20]. In result, for every considered test direction (φ_t, θ_t) an RSS value has been obtained for all considered two-row ESPAR antenna radiation patterns.

To compare the results with those already available in the literature, the test signal directions have to be set with discrete angular steps equal to $\Delta\varphi_t = 10^\circ$ and $\Delta\theta_t = 10^\circ$ in horizontal and elevation directions, respectively [24,28,33]. In consequence, one obtains 36 test horizontal directions $\varphi_t \in \{0^\circ, 10^\circ, \ldots, 350^\circ\}$ for 9 elevation angles $\theta_t \in \{90^\circ, 80^\circ, \ldots, 10^\circ\}$, which results in 324 testing points. Then, for every elevation angle, PPCC-MCP algorithm is used and, based on obtained 36 estimated direction of arrival angle $\hat{\varphi}$ values, root-mean-square error (RMSE) is calculated together with precision being the maximal DoA estimation error value within the plane. However, it has been observed in [43] that when $\Delta\varphi_t = 5^\circ$ and $\Delta\theta_t = 5^\circ$ is used, more accurate test results can be provided. Therefore, additional new testing directions, especially in elevation, reveal lack of accuracy of DoA estimation using ESPAR antenna together with PPCC-MCP algorithm for low θ angles when $M = 9$ calibration planes are used. As it was overlooked in [24], to verify DoA estimation performance in a detailed way for the proposed two-row ESPAR antenna and generalized PPCC-MCP algorithm, the number of testing points has to be further increased. To this end, we propose to use detailed DoA testing method that involve $\Delta\varphi_t = 1^\circ$ and $\Delta\theta_t = 1^\circ$ in horizontal and elevation directions correspondingly. In result, 360 test horizontal directions $\varphi_t \in \{0^\circ, 1^\circ, \ldots, 359^\circ\}$ for 90 elevation angles $\theta_t \in \{90^\circ, 89^\circ, \ldots, 1^\circ\}$ are created and, consequently, 90 RMSE and 90 precision values will be available for each plane.

4.3. DoA Estimation Results

To verify DoA estimation performance of the two-row ESPAR antenna using the generalized PPCC-MCP algorithm, which uses all 90 available calibration planes, $N = 18$ available antenna radiation patterns were divided into 3 separate groups having different angle tilt in elevation, as shown in Tables 1 and 2. In result, $n_{up} = \{1, 2, 3, 4, 5, 6\}$, $n_{down} = \{7, 8, 9, 10, 11, 12\}$, and $n_{mid} = \{13, 14, 15, 16, 17, 18\}$ were created as sets of steering vectors numbers associated with the specific two-row ESPAR antenna radiation patterns that can be created when particular steering vectors sets, namely $V^{n_{up}} = \{V^1, V^2, V^3, V^4, V^5, V^6\}$, $V^{n_{down}} = \{V^7, V^8, V^9, V^{10}, V^{11}, V^{12}\}$, and $V^{n_{mid}} = \{V^{13}, V^{14}, V^{15}, V^{16}, V^{17}, V^{18}\}$, are used. It allows verification of DoA estimation performance when $g_{\theta_{ALL}}$ is calculated using each of steering vectors set separately but also when they are combined in order to find the most appropriate combination for the two-row ESPAR antenna.

Table 2. Two-row ESPAR antenna steering vectors used in RSS-based DoA estimation.

n	V^n	Steering Vector Short Name	φ_{max}	θ_{max}
1	[001101 010100]	V^{30}_{UP}	30°	46°
2	[100110 001010]	V^{90}_{UP}	90°	46°
3	[010011 000101]	V^{150}_{UP}	150°	46°
4	[101001 100010]	V^{210}_{UP}	210°	46°
5	[110100 010001]	V^{270}_{UP}	270°	46°
6	[011010 101000]	V^{330}_{UP}	330°	46°
7	[001010 100100]	V^{30}_{DOWN}	30°	56°
8	[000101 010010]	V^{90}_{DOWN}	90°	56°
9	[100010 001001]	V^{150}_{DOWN}	150°	56°
10	[010001 100100]	V^{210}_{DOWN}	210°	56°
11	[101000 010010]	V^{270}_{DOWN}	270°	56°
12	[010100 001001]	V^{330}_{DOWN}	330°	56°

Table 2. *Cont.*

n	V^n	Steering Vector Short Name	φ_{max}	θ_{max}
13	[000100 001000]	V^{30}_{MID}	30°	52°
14	[000010 000100]	V^{90}_{MID}	90°	52°
15	[000001 000010]	V^{150}_{MID}	150°	52°
16	[100000 000001]	V^{210}_{MID}	210°	52°
17	[010000 100000]	V^{270}_{MID}	270°	52°
18	[001000 010000]	V^{330}_{MID}	330°	52°

To test DoA estimation accuracy in a detailed way, 360 test horizontal directions $\varphi_t \in \{0°, 1°, \ldots, 359°\}$ for 90 elevation angles $\theta_t \in \{90°, 89°, \ldots, 1°\}$ were created and then corresponding RMSE and precision values were calculated for all 7 possible combinations of steering vector sets which are gathered in Table 3, at SNR = 10 dB and the obtained results are presented in Figures 11 and 12.

Table 3. Combinations of steering vector sets used in verification of DoA estimation performance using generalized PPCC-MCP algorithm (see text for explanations).

Combination of Steering Vector Sets	Steering Vectors Numbers	Total Number of Steering Vectors
$\{V^{n_{up}}\}$	$\{1, 2, 3, 4, 5, 6\}$	6
$\{V^{n_{mid}}\}$	$\{13, 14, 15, 16, 17, 18\}$	6
$\{V^{n_{down}}\}$	$\{7, 8, 9, 10, 11, 12\}$	6
$\{V^{n_{up}}, V^{n_{mid}}\}$	$\{1, 2, 3, 4, 5, 6,$ $13, 14, 15, 16, 17, 18\}$	12
$\{V^{n_{up}}, V^{n_{down}}\}$	$\{1, 2, 3, 4, 5, 6,$ $7, 8, 9, 10, 11, 12\}$	12
$\{V^{n_{mid}}, V^{n_{down}}\}$	$\{13, 14, 15, 16, 17, 18,$ $7, 8, 9, 10, 11, 12\}$	12
$\{V^{n_{up}}, V^{n_{mid}}, V^{n_{down}}\}$	$\{1, 2, 3, 4, 5, 6,$ $13, 14, 15, 16, 17, 18,$ $7, 8, 9, 10, 11, 12\}$	18

As it can easily be observed in Figure 11, when a single set of steering vectors is used precision values deteriorate for low θ values and the best results can be produced for $\{V^{n_{mid}}\}$. For steering vector combinations involving at least 12 steering vectors, shown in Figures 11 and 12, one can acquire precision lower than 19° for all possible θ angles, i.e., $0° < \theta \leq 90°$. Surprisingly, the most accurate results with precision lower than 11° for all θ angles can be obtained for the steering vector set $\{V^{n_{mid}}, V^{n_{down}}\}$ containing 12 vectors. When the complete set $\{V^{n_{up}}, V^{n_{mid}}, V^{n_{down}}\}$ containing all 18 available steering vectors is used, the performance of generalized PPCC-MCP algorithm drops for $\theta \in (9°, 33°)$.

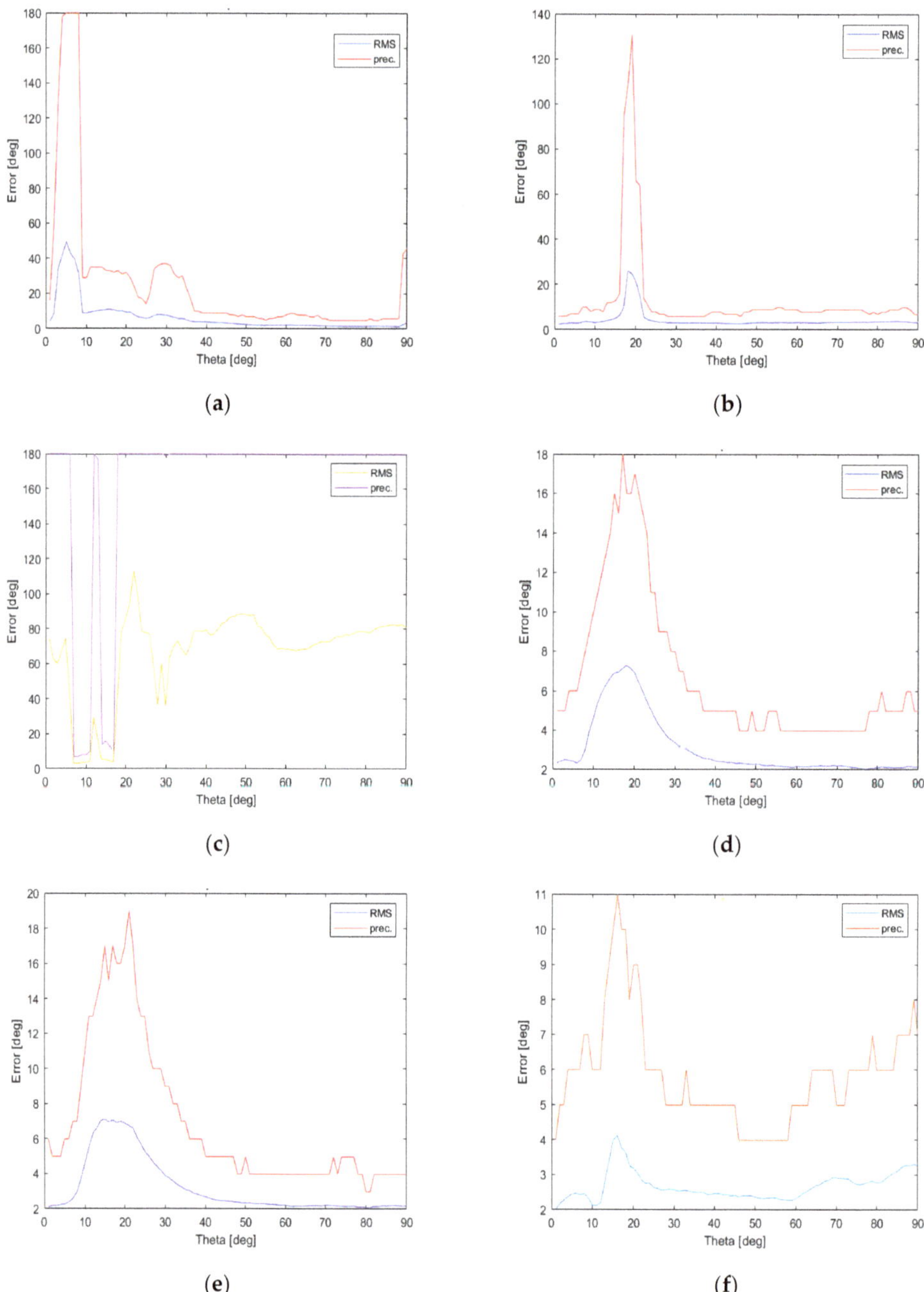

Figure 11. RMSE and precision values calculated using the proposed detailed DoA testing method, which involves all elevation angles during testing process, for the generalized PPCC-MCP algorithm relying on combinations of specific steering vector sets: (**a**) $\{V^{n_{up}}\}$; (**b**) $\{V^{n_{mid}}\}$; (**c**) $\{V^{n_{down}}\}$; (**d**) $\{V^{n_{up}}, V^{n_{mid}}\}$; (**e**) $\{V^{n_{up}}, V^{n_{down}}\}$; (**f**) $\{V^{n_{mid}}, V^{n_{down}}\}$.

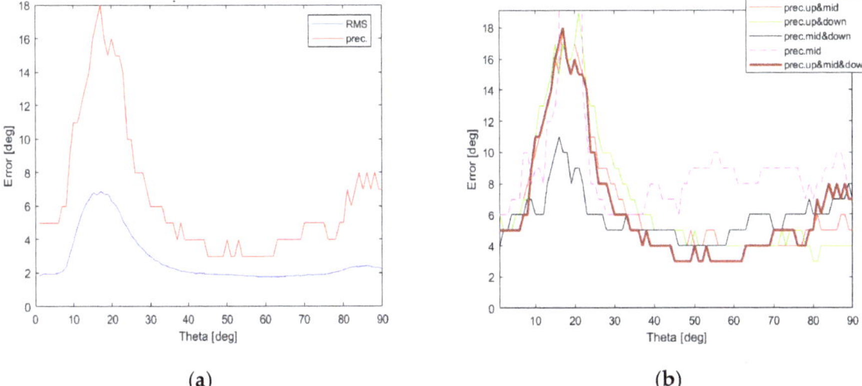

(a) (b)

Figure 12. RMSE and precision values calculated using the proposed detailed DoA testing method, which involves all elevation angles during testing process, for the generalized PPCC-MCP algorithm relying on all possible steering vectors $\{V^{n_{up}}, V^{n_{mid}}, V^{n_{down}}\}$ (**a**) together with precision value comparison for combinations of sets giving the most accurate results (**b**).

To verify the performance of generalized PPCC-MCP algorithm for different SNR levels, two additional DoA tests, with SNR = 5 dB and SNR = 0 dB, were performed. As it can easily be noticed in Figure 13, the most accurate results in both cases can be obtained for only 12 steering vectors when $\{V^{n_{mid}}, V^{n_{down}}\}$ steering vectors set is used. Therefore, two-row ESPAR antenna together with the proposed generalized PPCC-MCP algorithm can successfully be used for RSS-based DoA estimation in WSN nodes and gateways, even in noisy environments.

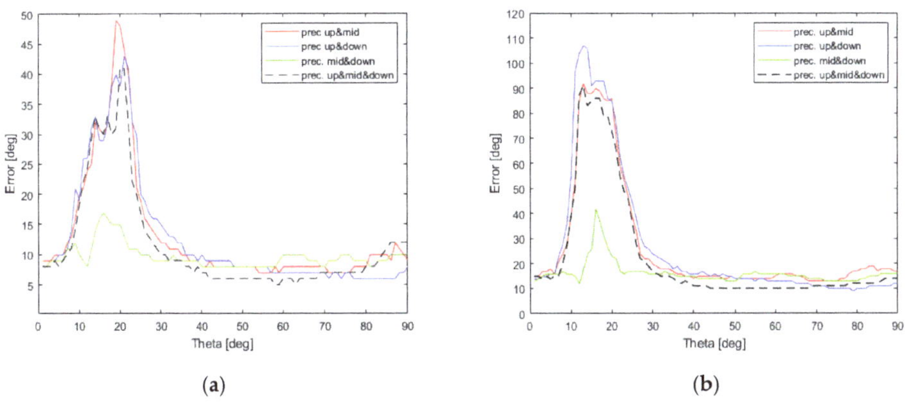

(a) (b)

Figure 13. Comparison of precision values obtained using the proposed detailed DoA testing method and the combinations of steering vector sets giving the most accurate results for different SNR levels: (**a**) SNR = 5 dB (**b**) SNR = 0 dB.

5. Conclusions

In this paper, we show how a two-row ESPAR antenna can be used for RSS-based DoA estimation. To this end, we have presented an analysis of all 18 available 3D antenna radiation patterns, being measured in an anechoic chamber, with respect to radiation coverage in the horizontal and vertical direction as well as their directional performance important in the estimation process. Moreover, we have proposed PPCC-MCP algorithm generalization that involves the usage of a very high number of calibration planes and can handle specific combinations of two-row ESPAR antenna radiation pattern sets. Detailed DoA estimation accuracy assessment, which involves thorough testing conducted along the elevation direction when RF signals impinging on the antenna arrive from arbitrary θ angles, shows that accurate results can be produced for only 12 steering vectors used. Precision lower than 11°, 17°, and 42° have been obtained for all θ angles when SNR was equal to 10 dB, 5 dB, and 0 dB, respectively. These results indicate that a two-row ESPAR antenna can produce accurate DoA estimation in the horizontal plane without prior knowledge about the elevation direction of the unknown RF signals even for low SNR values. Therefore, as the antenna beam switching in elevation and azimuth can be realized using a simple microcontroller and the DoA estimation relies solely on RSS values measured at the antenna output port, the antenna can easily be integrated with a WSN node or a WSN gateway. Moreover, 2D beamforming capabilities and accurate DoA estimation opens up new frontiers for ESPAR antenna applications involving new portable low-cost ground penetrating radars [44] and noninvasive microwave imaging devices supported by machine learning (ML) algorithms [45].

Author Contributions: Conceptualization, M.R. and L.K.; methodology, M.R.; software, L.K.; validation, M.R. and L.K.; writing—original draft preparation, M.R.; writing—review and editing, K.N. and L.K.; visualization, M.R.; supervision, L.K.; project administration, L.K.; funding acquisition, L.K. All authors have read and agreed to the published version of the manuscript.

Funding: This research was funded by InSecTT (www.insectt.eu, accessed on 26 February 2022) project that has received funding from the ECSEL Joint Undertaking (JU) under grant agreement No 876038. The JU receives support from the European Union's Horizon 2020 research and innovation programme and Austria, Sweden, Spain, Italy, France, Portugal, Ireland, Finland, Slovenia, Poland, Netherlands, Turkey. The document reflects only the author's view and the Commission is not responsible for any use that may be made of the information it contains.

Institutional Review Board Statement: Not applicable.

Informed Consent Statement: Not applicable.

Data Availability Statement: Not applicable.

Conflicts of Interest: The authors declare no conflict of interest. The funders had no role in the design of the study; in the collection, analyses, or interpretation of data; in the writing of the manuscript, or in the decision to publish the results.

References

1. Zanella, A.; Bui, N.; Castellani, A.; Vangelista, L.; Zorzi, M. Internet of things for smart cities. *IEEE Internet Things J.* **2014**, *1*, 22–32. [CrossRef]
2. Sotres, P.; Santana, J.R.; Sanchez, L.; Lanza, J.; Munoz, L. Practical lessons from the deployment and management of a smart city internet-of-things infrastructure: The SmartSantander Testbed Case. *IEEE Access* **2017**, *5*, 14309–14322. [CrossRef]
3. Low, K.S.; Win, W.N.N.; Er, M.J. Wireless Sensor Networks for industrial environments. In Proceedings of the International Conference on Computational Intelligence for Modelling, Control and Automation and International Conference on Intelligent Agents, Web Technologies and Internet Commerce (CIMCA-IAWTIC'06), Vienna, Austria, 28–30 November 2005; pp. 271–276. [CrossRef]
4. Joshi, G.P.; Nam, S.Y.; Kim, S.W. Cognitive radio Wireless Sensor Networks: Applications, challenges and research trends. *Sensors* **2013**, *13*, 11196–11228. [CrossRef] [PubMed]
5. Al-Karaki, J.N.; Gawanmeh, A. The optimal deployment, coverage, and connectivity problems in Wireless Sensor Networks: Revisited. *IEEE Access* **2017**, *5*, 18051–18065. [CrossRef]
6. Tran, T.; An, M.K.; Huynh, D.T. Symmetric connectivity in WSNs equipped with multiple directional antennas. In Proceedings of the 2017 International Conference on Computing, Networking and Communications (ICNC), Silicon Valley, CA, USA, 26–29 January 2017; pp. 609–614.
7. Brás, L.; Carvalho, N.B.; Pinho, P.; Kulas, L.; Nyka, K. A review of antennas for indoor positioning systems. *Int. J. Antennas Propag.* **2012**, *2012*, 1–14. [CrossRef]
8. Curiac, D.-I. Wireless Sensor Network security enhancement using directional antennas: State of the art and research challenges. *Sensors* **2016**, *16*, 488. [CrossRef]
9. Catarinucci, L.; Guglielmi, S.; Colella, R.; Tarricone, L. Compact switched-beam antennas enabling novel power-efficient Wireless Sensor Networks. *IEEE Sens. J.* **2014**, *14*, 3252–3259. [CrossRef]

10. Groth, M.; Rzymowski, M.; Nyka, K.; Kulas, L. ESPAR antenna-based WSN node with DoA estimation capability. *IEEE Access* **2020**, *8*, 91435–91447. [CrossRef]
11. Lysko, A.A. Towards an ultra-low-power electronically controllable array antenna for WSN. In Proceedings of the 2012 IEEE-APS Topical Conference on Antennas and Propagation in Wireless Communications (APWC), Cape Town, South Africa, 2–7 September 2012; pp. 642–645.
12. Loh, T.; Liu, K.; Qin, F.; Liu, H. Assessment of the adaptive routing performance of a Wireless Sensor Network using smart antennas. *IET Wirel. Sens. Syst.* **2014**, *4*, 196–205. [CrossRef]
13. Viani, F.; Lizzi, L.; Donelli, M.; Pregnolato, D.; Oliveri, G.; Massa, A. Exploitation of parasitic smart antennas in Wireless Sensor Networks. *J. Electromagn. Waves Appl.* **2010**, *24*, 993–1003. [CrossRef]
14. Skiani, E.D.; Mitilineos, S.A.; Thomopoulos, S.C.A. A study of the performance of Wireless Sensor Networks operating with smart antennas. *IEEE Antennas Propag. Mag.* **2012**, *54*, 50–67. [CrossRef]
15. Ademaj, F.; Rzymowski, M.; Bernhard, H.-P.; Nyka, K.; Kulas, L. Relay-aided Wireless Sensor Network discovery algorithm for dense industrial IoT utilizing ESPAR antennas. *IEEE Internet Things J.* **2021**, *8*, 16653–16665. [CrossRef]
16. Groth, M.; Nyka, K.; Kulas, L. Calibration-free single-anchor indoor localization using an ESPAR antenna. *Sensors* **2021**, *21*, 3431. [CrossRef] [PubMed]
17. Harrington, R. Reactively controlled directive arrays. *IEEE Trans. Antennas Propag.* **1978**, *26*, 390–395. [CrossRef]
18. Taillefer, E.; Hirata, A.; Ohira, T. Direction-of-arrival estimation using radiation power pattern with an ESPAR antenna. *IEEE Trans. Antennas Propag.* **2005**, *53*, 678–684. [CrossRef]
19. Kulas, L. RSS-based DoA estimation using ESPAR antennas and interpolated radiation patterns. *IEEE Antennas Wirel. Propag. Lett.* **2018**, *17*, 25–28. [CrossRef]
20. Plotka, M.; Tarkowski, M.; Nyka, K.; Kulas, L. A novel calibration method for RSS-based DoA estimation using ESPAR antennas. In Proceedings of the 22nd International Microwave and Radar Conference (MIKON), Poznan, Poland, 14–17 May 2018; pp. 65–68.
21. Asif, M.; Khan, S.; Ahmad, R.; Sohail, M.; Singh, D. Quality of service of routing protocols in Wireless Sensor Networks: A review. *IEEE Access* **2017**, *5*, 1846–1871. [CrossRef]
22. Shen, J.; Wang, A.; Wang, C.; Hung, P.C.K.; Lai, C.-F. An efficient centroid-based routing protocol for energy management in WSN-assisted IoT. *IEEE Access* **2017**, *5*, 18469–18479. [CrossRef]
23. Rzymowski, M.; Kulas, L. RSS-based direction-of-arrival estimation with increased accuracy for arbitrary elevation angles using ESPAR antennas. In Proceedings of the 12th European Conference on Antennas and Propagation (EuCAP 2018), London, UK, 9–13 April 2018; pp. 1–4. [CrossRef]
24. Groth, M.; Kulas, L. Accurate PPCC-based DoA estimation using multiple calibration planes for WSN nodes equipped with ESPAR antennas. In Proceedings of the 2018 48th European Microwave Conference (EuMC), Madrid, Spain, 23–27 September 2018; pp. 1565–1568. [CrossRef]
25. Cremer, M.; Dettmar, U.; Hudasch, C.; Kronberger, R.; Lerche, R.; Pervez, A. Localization of passive UHF RFID tags using the AoAct transmitter beamforming technique. *IEEE Sens. J.* **2015**, *16*, 1762–1771. [CrossRef]
26. Huang, S.; Gan, O.P.; Jose, S.; Li, M. Localization for industrial warehouse storage rack using passive UHF RFID system. In Proceedings of the 2017 22nd IEEE International Conference on Emerging Technologies and Factory Automation (ETFA), Limassol, Cyprus, 12–15 September 2017; pp. 1–8. [CrossRef]
27. Kulas, L. Simple 2-D direction-of-arrival estimation using an ESPAR antenna. *IEEE Antennas Wirel. Propag. Lett.* **2017**, *16*, 2513–2516. [CrossRef]
28. Burtowy, M.; Rzymowski, M.; Kulas, L. Low-profile ESPAR antenna for RSS-based DoA estimation in IoT applications. *IEEE Access* **2019**, *7*, 17403–17411. [CrossRef]
29. Buffi, A.; Nepa, P.; Cioni, R. SARFID on drone: Drone-based UHF-RFID tag localization. In Proceedings of the 2017 IEEE International Conference on RFID Technology & Application (RFID-TA), Warsaw, Poland, 20–22 September 2017; pp. 40–44. [CrossRef]
30. De Oliveira, M.T.; Miranda, R.K.; Da Costa, J.P.C.L.; De Almeida, A.L.F.; De Sousa, R.T. Low cost antenna array based drone tracking device for outdoor environments. *Wirel. Commun. Mob. Comput.* **2019**, *2019*, 5437908. [CrossRef]
31. Rzymowski, M.; Kulas, L. Two-row ESPAR antenna with simple elevation and Azimuth Beam Switching. *IEEE Antennas Wirel. Propag. Lett.* **2021**, *20*, 1745–1749. [CrossRef]
32. Duraj, D.; Rzymowski, M.; Nyka, K.; Kulas, L. ESPAR antenna for V2X applications in 802.11p frequency band. In Proceedings of the 2019 13th European Conference on Antennas and Propagation (EuCAP), Krakow, Poland, 31 March–5 April 2019; pp. 1–4.
33. Duraj, D.; Tarkowski, M.; Rzymowski, M.; Kulas, L.; Nyka, K. RSS-based DoA estimation in 802.11p frequency band using ESPAR antenna and PPCC-MCP method. In Proceedings of the 2020 23rd International Microwave and Radar Conference (MIKON), Warsaw, Poland, 5–8 October 2020; pp. 15–156. [CrossRef]
34. Lin, Z.; Lin, M.; de Cola, T.; Wang, J.-B.; Zhu, W.-P.; Cheng, J. Supporting IoT with rate-splitting multiple access in satellite and aerial-integrated networks. *IEEE Internet Things J.* **2021**, *8*, 11123–11134. [CrossRef]
35. Lin, Z.; Lin, M.; Zhu, W.-P.; Wang, J.-B.; Cheng, J. Robust secure beamforming for wireless powered cognitive satellite-terrestrial networks. *IEEE Trans. Cogn. Commun. Netw.* **2020**, *7*, 567–580. [CrossRef]

36. Lin, Z.; Lin, M.; Champagne, B.; Zhu, W.-P.; Al-Dhahir, N. Secure beamforming for cognitive satellite terrestrial networks with unknown eavesdroppers. *IEEE Syst. J.* **2020**, *15*, 2186–2189. [CrossRef]
37. Lin, Z.; Lin, M.; Wang, J.-B.; De Cola, T.; Wang, J. Joint beamforming and power allocation for satellite-terrestrial integrated networks with non-orthogonal multiple access. *IEEE J. Sel. Top. Signal Process.* **2019**, *13*, 657–670. [CrossRef]
38. Rahmat-Samii, Y.; Manohar, V.; Kovitz, J.M. For Satellites, Think Small, Dream Big: A review of recent antenna developments for CubeSats. *IEEE Antennas Propag. Mag.* **2017**, *59*, 22–30. [CrossRef]
39. Sugiura, S.; Iizuka, H. Reactively steered ring antenna array for automotive application. *IEEE Trans. Antennas Propag.* **2007**, *55*, 1902–1908. [CrossRef]
40. Zhang, L.; Gao, S.; Luo, Q.; Young, P.R.; Li, Q. Planar ultrathin small beam-switching antenna. *IEEE Trans. Antennas Propag.* **2016**, *64*, 5054–5063. [CrossRef]
41. Xu, B.; Ying, Z.; Scialacqua, L.; Scannavini, A.; Foged, L.J.; Bolin, T.; Zhao, K.; He, S.; Gustafsson, M. Radiation performance analysis of 28 GHz antennas integrated in 5G mobile terminal housing. *IEEE Access* **2018**, *6*, 48088–48101. [CrossRef]
42. Hazmi, A.; Tian, R.; Rintamaki, S.; Milosavljevic, Z.; Ilvonen, J.; Van Wonterghem, J.; Khripkov, A.; Kamyshev, T. Spherical coverage characterization of millimeter wave antenna arrays in 5G mobile terminals. In Proceedings of the 2019 13th European Conference on Antennas and Propagation (EuCAP), Krakow, Poland, 31 March–5 April 2019; pp. 1–5.
43. Groth, M.; Leszkowska, L.; Kulas, L. Efficient RSS-based DoA estimation for ESPAR antennas using multiplane SDR calibration approach. In Proceedings of the 2018 IEEE-APS Topical Conference on Antennas and Propagation in Wireless Communications (APWC), Cartagena, Colombia, 10–14 September 2018; pp. 850–853. [CrossRef]
44. Das, U.; Boer, H.J.; Van Ardenne, A. Phased array technology for GPR antenna design for near subsurface exploration. In Proceedings of the 2nd International Workshop onAdvanced Ground Penetrating Radar, Delft, The Netherlands, 14–16 May 2003; pp. 30–35. [CrossRef]
45. Salucci, M.; Gelmini, A.; Vrba, J.; Merunka, I.; Oliveri, G.; Rocca, P. Instantaneous brain stroke classification and localization from real scattering data. *Microw. Opt. Technol. Lett.* **2018**, *61*, 805–808. [CrossRef]

Article

Simplified Mutual Inductance Calculation of Planar Spiral Coil for Wireless Power Applications

Iftikhar Hussain and **Dong-Kyun Woo** *

Department of Electrical Engineering, Yeungnam University, Gyeongsan 38541, Korea; iftikhar.razai@gmail.com
* Correspondence: wdkyun@yu.ac.kr

Abstract: In this paper, a simplified method for the calculation of a mutual inductance of the planar spiral coil, motivated from the Archimedean spiral, is presented. This method is derived by solving Neumann's integral formula in a cylindrical coordinate system, and a numerical tool is used to determine the value of mutual inductance. This approach can calculate the mutual inductances accurately at various coaxial and non-coaxial distances for different coil geometries. The calculation result is compared with the 3D finite element analyses to verify its accuracy, which shows good consistency. Furthermore, to confirm it experimentally, Litz wire is used to fabricate the sample spiral coils. Finally, the comparison of a simplified method is also studied relative to the coupling coefficient. The accuracy of the calculation results with the simulation and the measurement results makes it a good candidate to apply it in wireless power applications.

Keywords: litz wire; lateral misalignment; magnetic field distribution; mutual inductance; Neumann integral formula; planar spiral coil

Citation: Hussain, I.; Woo, D.-K. Simplified Mutual Inductance Calculation of Planar Spiral Coil for Wireless Power Applications. *Sensors* **2022**, *22*, 1537. https://doi.org/10.3390/s22041537

Academic Editor: Pedro Pinho

Received: 18 January 2022
Accepted: 14 February 2022
Published: 16 February 2022

1. Introduction

Planar spiral coils have been widely adopted in many electromagnetic applications ranging from low power such as mobile phones, electric toothbrushes, and biomedical implants [1–10] to the high-power drone charging systems and electric vehicles. The wireless power transfer (WPT) method was employed to transfer power between these coils [11–15]. WPT consists of a primary coil and a secondary coil. Power is transferred between them by magnetic coupling. To design a coil or calculate the performance related to power, such as voltage gain, efficiency, output power, and transfer distance, determining mutual inductance (M) is necessary. In the last few decades, much work has been done to estimate the M between coils. Maxwell derived the M equation between two coaxial circular filamentary coils from the perspective of energy using a complete elliptic integral [16,17]. Another equation is obtained using the magnetic vector potential approach [18]. However, many applications require a spiral coil instead of a circular coil for higher power transfer and coupling coefficient. The Neumann integral formula was adopted to estimate M between two circular filaments carrying a constant current. M between spiral coils was predicted by assuming each spiral coil as a group of concentric circular coils. The self-inductances of each circular coil and mutual inductances between them are calculated to obtain the self-inductance of the spiral coil. Finally, the mutual inductance is computed by calculating the inductance between the two spiral coils [19,20]. However, this method needs a lot of calculations. Furthermore, there is always an inconsistency between the calculated result of mutual inductance and simulation and measurement results due to the supposition of the spiral coils as a collection of the circular coils. This problem is solved by employing the Archimedean spiral coil equation considering coil helicity [21]. This method has used a rectangular coordinate system to find the parameters of the Neumann integral formula for mutual inductance. As a result, the derivation steps and the number of parameters in the final M equation increase. Furthermore, a separate calculation of each lower and upper limit of a double integral is required. Thus, the M equation becomes more complex and needs more time to calculate the M. Therefore, a more simplified and fast method is necessary to estimate the M accurately.

The finite element method (FEM) using Maxwell 3-D electromagnetic software is one of the most suitable methods to compute mutual inductances; however, it requires a long computational time and a high-speed computer [22].

In this paper, a more simplified form of mutual inductance equation between two circular planar spiral coils is calculated in a cylindrical coordinate using Neumann' integral formula. Solving M in a cylindrical coordinate system significantly reduces the derivation steps and the calculation complexity.

In addition, compared to the conventional method, which required the calculation of each upper and lower limit of the double integral of the final M equation, our approach simplified it, starting from 0 to $2\pi N$, for all types of the circular spiral coil. Hence, a more simplified M equation is obtained, which can be solved by a numerical tool, such as Matlab. Moreover, a simplified M equation under lateral misalignment is also calculated. The simplified mutual inductance result is verified by simulation and experiment results. Thus, the comparison of the experiment result with the simulation and calculation result confirms the accuracy of our equation.

The rest of the paper is organized as below: Section 2 describes the derivation of the mutual inductance of planar spiral coil under perfectly aligned and lateral misalignment. The verification of the calculation result with the simulation result using ANSYS Maxwell 15 is discussed in Section 3. Section 4 is the detailed study of experimental verification of calculation and simulation results and is divided into many subsections and sub-subsections. The selection of the wire for the spiral coil and its measurement are determined in Section 4.1 while mutual inductance measurement method is described in Section 4.2. Section 4.3 express the comparison of calculation, simulation, and measurement results for different coil configurations under the coaxial case, and the errors between them are calculated. The effect of mutual inductance as a function of lateral misalignment for different coil configurations is represented in Section 4.4. The variation of calculation results relative to measurement and simulation are also computed in this subsection. Section 4.5 verify the calculation result with the coupling coefficient formula.

2. Mutual Inductance Equation Derivation

2.1. Two Perfect Aligned the Planar Spiral Coil

The planar spiral and its structural parameters are expressed in Figure 1a. To overcome the skin effect, assume a constant current flows into the coil. Mutual inductance depends on the geometric factors of the coil and the respective distance. Initially, we supposed that a spiral coil C_1 lies on the x–y plane with the center at the origin, and coil C_2 is placed above C_1 at a distance h apart, as shown in Figure 1b. The direction of the y-axis is into the page, not shown here. The primary coil's C_1 radius is R_A, the secondary coil C_2 radius is R_B, the distance between them is h, P and Q are the tangential elements on the C_1 and C_2, respectively, the gap between P and Q is represented by R, θ_1 is the angle of R_A relative to x-axis of C_1, and the angle of R_B is θ_2.

Mutual inductance of the planar spiral coil can be presented by Neumann's equation as follows:

$$M = \frac{\mu_0}{4\pi} \oint_{C_1} \oint_{C_2} \frac{dl_1 dl_2}{R} \quad (1)$$

where μ_0 is the vacuum permeability, dl_1 and dl_2 are line elements, and R is the separation between it. Eventually, the mutual inductance of the planar circular spiral coil can be determined by finding dl_1, dl_2, and R in Equation (1).

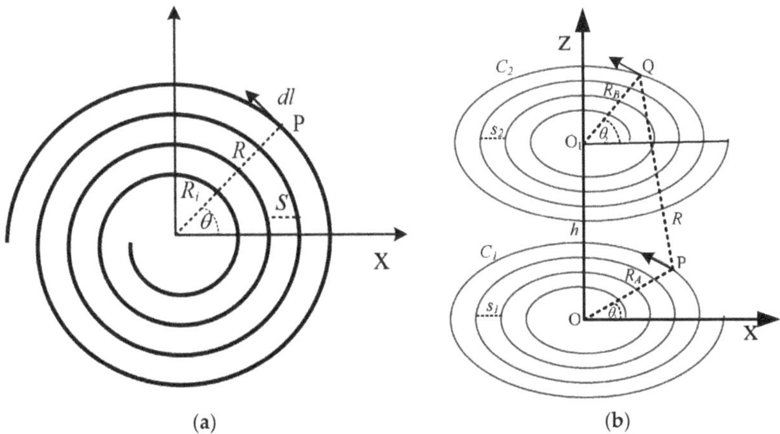

Figure 1. (**a**) Planar spiral coil. (**b**) Aligned circular spiral coil.

To estimate the mutual inductance, when the gap between the two coils is larger than wire diameter, the thick coil can be denoted by equation of planar spiral coil [23]. The equation of the circular spiral coil, motivated by the Archimedean spiral coil can be described by

$$R = R_i + a\theta, \theta_i \leq \theta \leq \theta_o \tag{2}$$

$$\theta = 2\pi N \tag{3}$$

$$a = \frac{s}{2\pi} \tag{4}$$

where R_i is an initial radius, N is the number of turns, s is the gap between turns, a is pitch factor, and θ_i and θ_o are initial and final angle of Archimedean spiral coil. The pitch factor a affects the gap between turns.

The values of the coil's parameters, such as inner radius R_i, outer radius R_o, number of turns N, and the gap between them s, are calculated from the spiral coil Equation (5).

$$R_o = R_i + N \times s \tag{5}$$

The pitch factor a can also be determined by substituting the value of s in Equation (2). Using Equation (2), the equation of C_1 and C_2 can be expressed as Equations (6) and (7).

$$R_A = R_{i1} + a_1\theta_1 \tag{6}$$

$$R_B = R_{i2} + a_2\theta_2 \tag{7}$$

Under the cylindrical coordinate system, the tangential elements dl_1 and dl_2 can be described as

$$dl_1 = -R_A \sin\theta_1 d\theta_1 \hat{x} + R_A \cos\theta_1 d\theta_1 \hat{y} \tag{8}$$

$$dl_2 = -R_B \sin\theta_2 d\theta_2 \hat{x} + R_B \cos\theta_2 d\theta_2 \hat{y} \tag{9}$$

where R_A and R_B represent the distance from the origin to the tangential elements C_1 and C_2, respectively.

$$dl_1.dl_2 = R_A R_B \cos(\theta_2 - \theta_1) d\theta_1 d\theta_2 \tag{10}$$

The distance R between dl_1 and dl_2 can be denoted by using cosine law.

$$\begin{aligned} R^2 = \ & (R_{i1} + a_1\theta_1)^2 + (R_{i2} + a_2\theta_2)^2 \\ & -2(R_{i1} + a_1\theta_1)(R_{i2} + a_2\theta_2)\cos(\theta_2 - \theta_1) + h^2 \end{aligned} \tag{11}$$

Substituting Equations (10) and (11) in Equation (1), the mutual inductance of perfectly aligned planar spiral coil is obtained in (12).

$$M = \frac{\mu_0}{4\pi} \iint_{2\pi N} \frac{j d\theta_1 d\theta_2}{\sqrt{(R_{i1}+a_1\theta_1)^2+(R_{i2}+a_2\theta_2)^2-2j+h^2}}$$
$$j = (R_{i1} + a_1\theta_1)(R_{i2} + a_2\theta_2)\cos(\theta_2 - \theta_1) \tag{12}$$

Equation (12) is a double integral from, therefore, it is hard to be represented with an elementary formula and it can be solved by numerical tool. This is a disadvantage of this method. However, maintaining the accuracy in various cases is a powerful advantage.

2.2. Lateral Misalignment of the Planar Spiral Coil

In most wireless power applications, the secondary coils are often laterally misaligned. The mutual inductance depends on the combined link of the magnetic field between two coils. Displacement of one coil relative to another significantly reduces the shared magnetic field and thus the mutual inductance. Therefore, it is important to consider the misalignment.

The arrangement of the planar spiral coil for lateral misalignment is shown in Figure 2. The secondary coil is laterally displaced towards the positive x-axis and denoted by the variable x. So, in this case, the distance between points P and Q can be obtained as Equation (13).

$$\begin{aligned} R^2 = \; & R_A^2 + R_B^2 + h^2 + x^2 - 2R_A R_B \cos(\theta_2 - \theta_1) \\ & +2xR_A \cos\theta_1 - 2xR_B \cos\theta_2 \end{aligned} \tag{13}$$

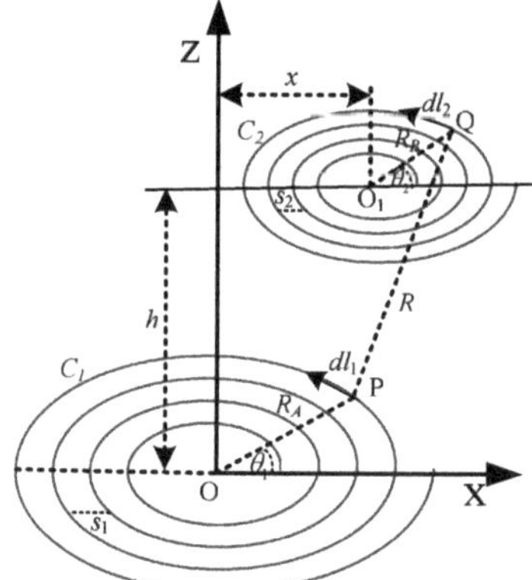

Figure 2. Planar spiral coil with lateral misalignment.

The final equation of mutual inductance with lateral misalignment can be achieved by replacing the denominator of Equation (12) with the Equation (13).

$$M = \frac{\mu_0}{4\pi} \iint_{2\pi N} \frac{j d\theta_1 d\theta_2}{\sqrt{\begin{array}{c} R_A^2 + R_B^2 + h^2 + x^2 - 2j \\ +2xR_A\cos\theta_1 - 2xR_B\cos\theta_2 \end{array}}} \tag{14}$$

Equation (14) can be applied to calculate the mutual inductance at non-axial lateral displacement. It can be used for axial distances as well when the $x = 0$. In this case, it will be simplified to Equation (12).

3. Simulation Verification

To verify the simplified equation, many couples of spiral coils with different sizes, gap distance, and number of turn, are simulated on 3-D Maxwell FEM. The calculation results are compared to the simulation results and errors are calculated relative to the simulation result as well. The type of solution is Magneto static. Mesh is assigned as a length based on the coil and its boundary. The element length is the default value. The current is assigned uniformly across the coil's cross-section for simplicity's purpose.

In the following Tables, the mutual inductances of a number of a circular spiral coils are calculated with different gap distances s between their turns, outer radiuses Ro, and number of turns, and the results are compared to the simulation result. Ri and h represent the inner radius and distance between the primary and secondary coil respectively. Both these parameters are considered constant in the simulation comparison. FEM shows the simulation result.

The above Tables validate the simplified mutual inductance equation relative to simulation results for all cases. Table 1 shows, smaller the gap between the turn s, lesser will be the error. When the outer radius R_o and the number of turns N increased, the error increased slightly, as indicated in Tables 2 and 3, respectively. However, for all cases the errors are below 4% which is in acceptable range. Thus, it proves the accuracy of the simplified equation.

Table 1. Comparison result with variation in gap between the turn.

Parameters				Results		
h (mm)	s (mm)	R_i (mm)	R_o (mm)	FEM (μH)	(11) (μH)	Error (%)
10	7.5	10	85	5.85	5.67	3.07
10	6.25	10	85	8.25	8.03	2.66
10	5.35	10	85	11.24	11.00	2.13
10	4.68	10	85	14.56	14.29	1.85

Table 2. Comparison result with variation in outer radius.

Parameters				Results		
h (mm)	N (mm)	R_i (mm)	R_o (mm)	FEM (μH)	(11) (μH)	Error (%)
10	10	10	85	5.85	5.67	3.07
10	10	10	95	6.61	6.40	3.17
10	10	10	105	7.39	7.15	3.24
10	10	10	115	8.19	7.91	3.41

Table 3. Comparison result with variation in number of turns.

Parameters				Results		
h (mm)	N (mm)	R_i (mm)	s (mm)	FEM (μH)	(11) (μH)	Error (%)
10	10	10	7.5	5.85	5.67	3.07
10	12	10	7.5	9.87	9.56	3.14
10	14	10	7.5	15.86	15.29	3.59
10	16	10	7.5	23.48	22.67	3.44

4. Experimental Verification

Wireless power technique has been employed in wide power applications, ranging from low power electric shaver, smartphone charger, and biomedical implant devices to medium power drone chargers and large power electric cars. Therefore, it is important to validate the simplified mutual equations for different geometry of the spiral coils. Depending on the application, we have constructed two small sample coils with turn number 10 and two large coils with the number of turn 16. The larger size of the coil is limited to

170 mm in diameter due to the capacity of a 3D printer which is used to make a bobbin for winding the wire. The PLA (polylactic acid) filament is adopted for constructing bobbins. These coils are sample examples for different wireless applications. The arrangement for determining the mutual inductance of the spiral coils with different axial and non-axial distances are as follows. Firstly, inductances are calculated between two large primary and large secondary coils; secondly, between small primary and small secondary; and finally, between large primary and large secondary. The calculation results are compared both by simulation and experiment.

Figure 3a,b show the bobbins and bobbins with spiral winding and Figure 3c describes the experimental setup for three different coil configurations. Special blocks with each 5 mm thickness are designed in the 3D printer for measuring the gap between the primary and secondary coil in the axial and non-axial cases. The stand is used to hold the secondary coil with the help of a pencil to show the coupling distance. A steel ruler and Vernier caliper are utilized to maintaining the accuracy of axial and non-axial distances.

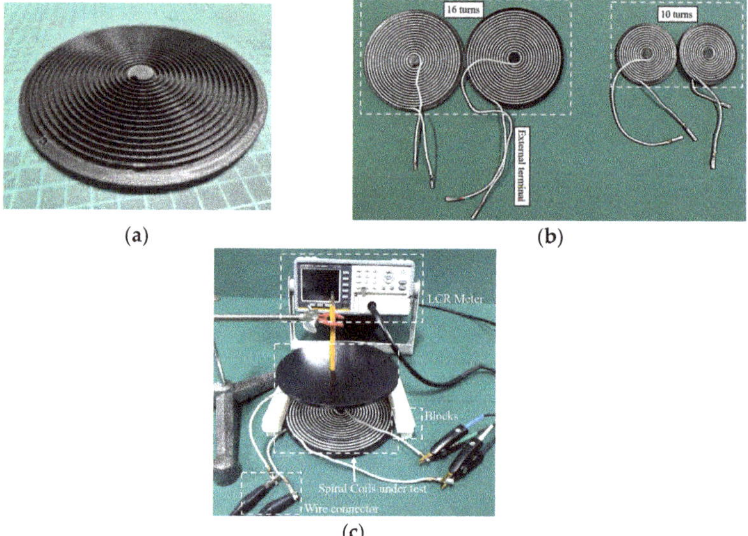

Figure 3. Experimental setup for measuring mutual inductance. (**a**) Bobbin. (**b**) Spiral coil with bobbin. (**c**) Measuring components.

The structural parameters of the sample coils are shown in Table 4 and categorized into the small and the larger coil. Where N is the turn number, s is the gap between turns, R_i is the inner radius of the planar spiral coil, and R_o is the outer radius.

Table 4. Geometrical parameters of experimental coils.

Coil Size	N (mm)	R_i (mm)	s (mm)	R_o (mm)
Small Coil	10	10	4	50
Large Coil	16	10	4.68	85

4.1. Selection of Wire and Its Measurement

We have used Litz wire for fabricating the spiral coil due to its vast advantages over solid wires. Litz wire has lower AC resistance than a solid conductor at higher frequencies. In a solid conductor, the current density is higher at its surface and exponentially decreases towards the center due to an effect called the skin effect. As a result, it decreases the effective cross-sectional area of the wire and, hence, increases the resistance. On the other hand, Litz wire is made of thin strands insulated with each other and twisted in a manner that, at some point, each strand holds a position in the cross-sectional area [24]. The thinning and twisting of the strand help to mitigate the skin and proximity effect, respectively. The proximity effect is caused by the current of nearby strands in the bundle, by the leakage fluxes from air gaps and core, and by the external field from other conductors [25].

In this study, the Litz wire with 500 strands and 0.12 mm diameter is chosen. The total diameter of Litz wire is 3.6 mm, and it is decided empirically by Equation (15).

$$D_e = 1.6\left(\sqrt{\left(\frac{D_s}{2}\right)^2 N_s}\right)2 \tag{15}$$

where *Ns* is the total number of strands, *Ds* is the diameter of one strand, and *De* is the estimated diameter of Litz wire [26].

The length of the wire is also one of the important parameters which influence the inductance and the *DC* resistance. Therefore, the accurate length and the resistance can be determined from Equations (16) and (17).

$$L_{total} = \int_0^{2\pi N} \sqrt{R_o^2 + \left(\frac{dR_o}{d\theta}\right)^2}\, d\theta \tag{16}$$

$$R_{DC} = \frac{l(\text{mm})}{\sigma n \pi r^2} \tag{17}$$

where l the length of the wire, σ is constant and its values is 58,000, n is the number of strands, and r is the radius of one strand.

Soldering quality is another factor that affects *DC* resistance. Good quality lead and specific pot temperature are required for high-quality soldering. Thus, a proper soldering set and particular jig were prepared for this purpose.

4.2. Mutual Inductance Measurement Method

The measurement of mutual inductance is performed by a Gwinstek LCR-6200, and their results are compared with the calculated and simulated results. The following method is used to measure the mutual inductance. Firstly, the inductance is measured when the primary and secondary coil is connected in forwarding mode, as shown in Figure 4a. In this mode, the inductance would be equal to the summation of the primary and secondary inductances and twice the mutual inductance. Secondly, the primary and secondary coils are connected in a backward mode, as shown in Figure 4b. In this mode, the inductance would be the sum of the inductances of each primary and secondary coil minus twice the mutual inductance Finally, the actual mutual inductance is obtained by subtracting the backward inductance from the forward connection inductance and dividing it by 4, as shown in Equation (18).

$$M = \frac{L_f - L_b}{4} \tag{18}$$

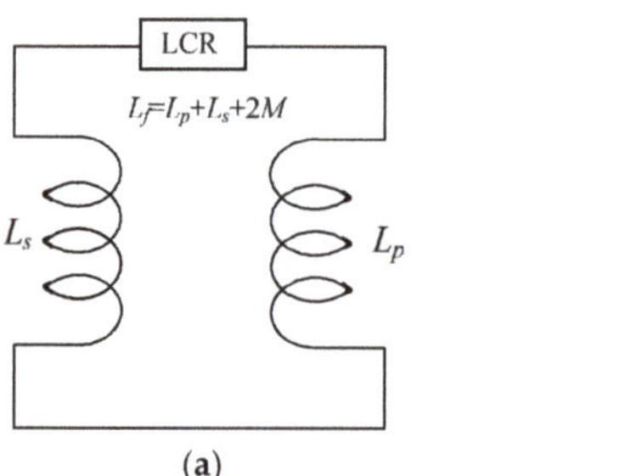

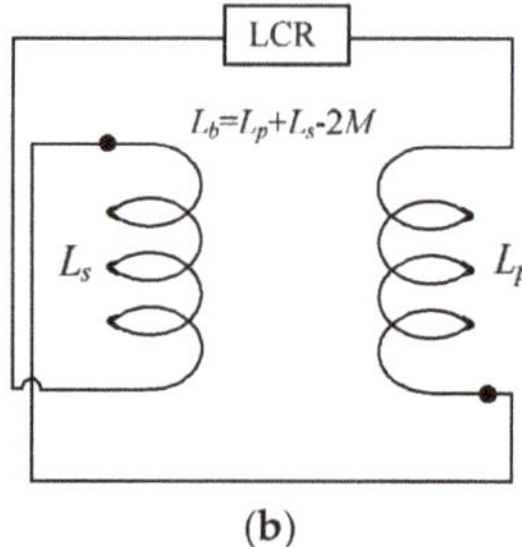

(**a**) (**b**)

Figure 4. Two different connection mode for measuring mutual inductance. (**a**) Forward mode. (**b**) Backward mode.

The operating frequency is selected 20 kHz to ignore the skin effect.

4.3. Mutual Inductances of Different Spiral Coil Configurations for Aligned Distances and Their Error Comparison

Considering the wireless charging applications, firstly, the mutual inductance between two large sample axial coils under the condition of increasing axial distance h is measured. The measured results are compared to the calculation and the simulation results. Some emerging applications, like transthermic wireless power transfer systems, require a small secondary coil compared to the primary coil [27].

To validate the simplified equation for these kinds of systems, mutual inductance between large primary and small secondary coils are also measured for various axial distances. Finally, inductances between two small primary and small secondary coils are measured. Their comparison results are shown in Figure 5 where N_p and N_s denote the number of turns of primary coil and secondary coil, respectively.

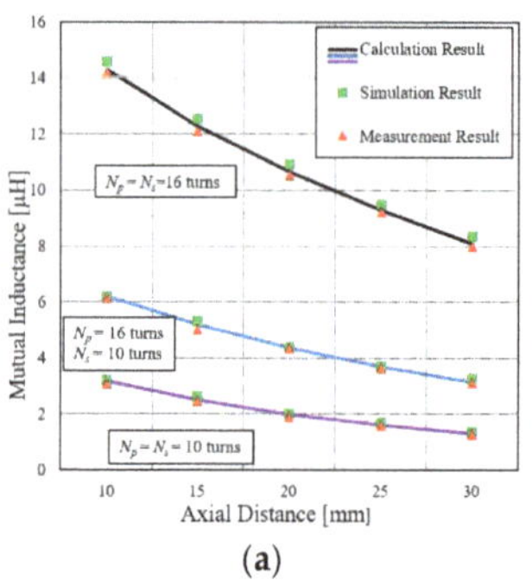

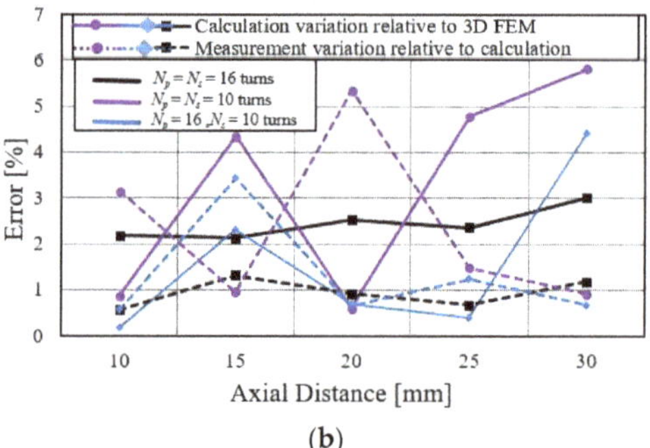

(**a**) (**b**)

Figure 5. Mutual inductance. (**a**) Comparison between different coils configuration at various axial distances. (**b**) Error comparison of coils relative to calculation and FEM.

The calculated results of the large, medium, and small configurations of coils are denoted by black, blue, and purple lines. The green square box shows the simulation results, and the red triangle box represents the measurement result. Figure 5a shows that the mutual inductance decreases with the increase in the axial distance, as can be described by the Faradays laws of magnetic induction, which state that the mutual inductance between two coils depends on the shared coupling of magnetic. The smaller the axial distance, the stronger the magnetic flux, and, thus, the higher the mutual inductance. The magnetic flux linkage would be weaker as the axial distance between coils increases. Therefore, this would cause the deceases in the mutual inductance.

If the axial distance keeps increasing, a point is reached where the mutual inductance would be zero, and all the lines will converge at the common point. Thus, to maintain receiving the power of WPT at the required level, it is necessary to keep the secondary coil at a certain optimized distance. Figure 5b is the error comparison of calculation results relative to simulation and measurement results. It shows that the error for the large coil is small, under 3%, compared to the small coil error, which is less than 9%. Thus, the comparison of simulation and measurement results with the calculated one certifies the accuracy of the simplified method.

4.4. Behavior of Mutual Inductance between Two Large Primary and Large Secondary Spiral Coils under Lateral Misalignment and Their Error Comparison

In some power transfer applications, such as biomedical implants and drone charging systems, the secondary coil is often misplaced. As a result, mutual inductance is drastically affected, and, therefore, so is the power link performance. This misaligned effect is important to be investigated. This case has shown, in Figure 2, that the secondary coil is laterally displaced. Regarding the applications perspective, using Equation (14), mutual inductance is calculated at discrete lateral misaligned distances for three different coil configurations: large primary and large secondary coil; small primary and small secondary; and the small secondary, large primary coil.

The calculated results are compared with simulation and measurement results to verify its correctness. The comparison details are represented in Figure 6a. In Figure 6a, the comparison between two large coils is shown. The calculated results are indicated with the black, blue, and purple lines. The green square box and a red triangle box represent the simulation and measurement results. Figure 6b shows the calculation variation relative to 3D FEM, while Figure 6c represents the variation of measurement result to the calculation result. In the configuration of this large coil, the variations are below 6% for all cases.

4.4.1. Behavior of Mutual Inductance between Two Small Primary and Small Secondary Spiral Coils under Lateral Misalignment and Their Error Comparison

Figure 7a shows the variation of mutual inductance for the lateral misalignment at various coupling distances for small primary and small secondary spiral coils. Their comparative error analysis is represented in Figure 7b,c. It is observed that the mutual inductance declines faster at small h compared to larger h. Moreover, it can be seen that the error for the small turn of the spiral coil ($N = 10$) increases gradually when the lateral displacement is more than the maximum radius of the coil. The variations of the calculation result to the FEM shows more than 20%, while measurement variation relative to the calculation is above 15%. Therefore, the working region is set as the outer radius of the coil for the lateral displacement.

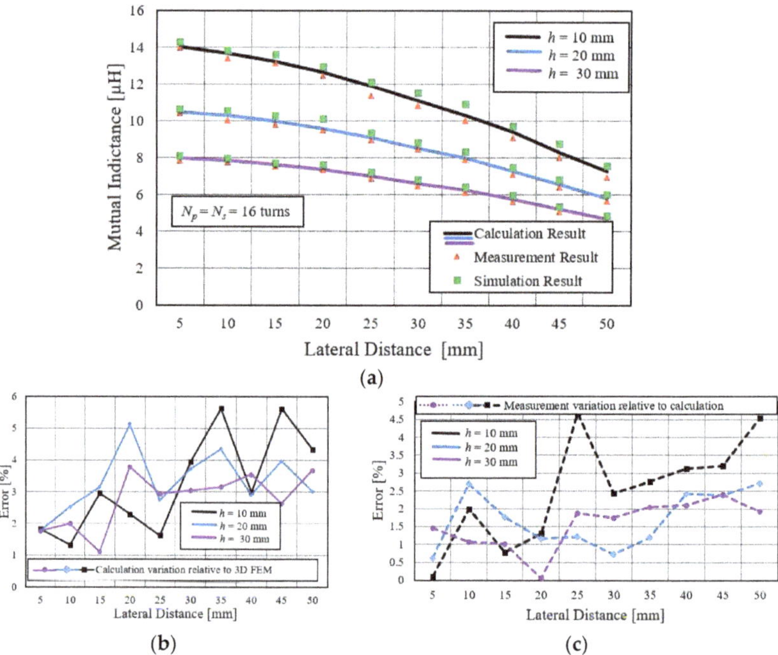

Figure 6. (**a**) Mutual inductance comparison between large primary and large secondary coils at various lateral misalignment distances. (**b**) Calculation variation relative to simulation (FEM) for $N = 16$. (**c**) Measurement variation relative to calculation.

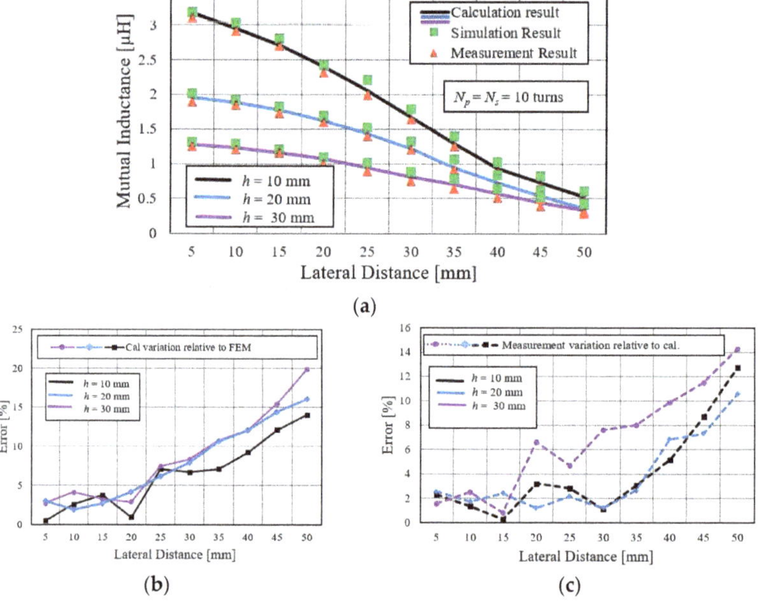

Figure 7. (**a**) Mutual inductance comparison between small primary and small secondary coils at various lateral misalignment distances. (**b**) Calculation variation relative to simulation (FEM) for $N = 10$. (**c**) Measurement variation relative to calculation.

Furthermore, such a big lateral displacement is not required in most wireless power applications because the mutual inductance also approaches zero in such large lateral displacement.

4.4.2. Behavior of Mutual Inductance between Large Primary and Small Secondary Spiral Coils under Lateral Misalignment and Their Error Comparison

Mutual inductances are calculated at three different axial distances: h = 10 mm, 20 mm, and 30 mm, and at various lateral distances, ranging from 0 to 50 mm with a 5 mm gap distance difference between them. The result of mutual inductance at each axial gap is differentiated with a unique color line.

Lateral misalignment of the coil also drastically affects the mutual inductance. This effect is higher at the larger misalignment than at the small one. As the misplaced distance increases, the mutual inductance reduces. This reduction of the M is due to the lower influence of the magnetic flux distribution of the primary coil on the displaced coil.

The detail of a large primary and a small secondary coil is illustrated in Figure 8a, while its error comparison is shown in Figure 8b,c. In this case, the calculation error relative to FEM and measurement result is less than 6%, which verifies the accuracy of the simplified method.

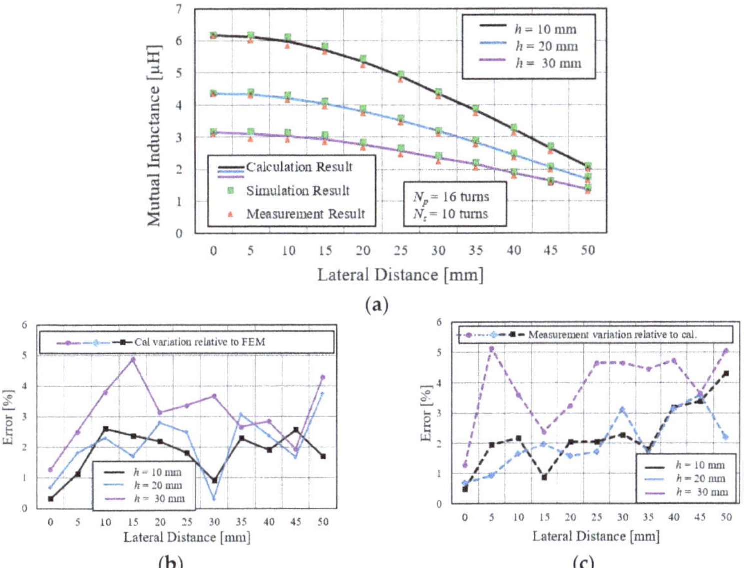

Figure 8. Mutual inductance comparison between large primary and small secondary coils at various lateral misalignment distances. (**a**) Primary large and secondary small coil. (**b**) Calculation variation relative to simulation (FEM) for N_p = 16, N_s = 10. (**c**) Measurement variation relative to calculation.

The mutual inductances of all coil configurations will meet at a common point if the lateral distance increases beyond a certain limit. The limit is the outer radius of the coil. The mutual inductance approaches zero when the lateral misaligned distance increases beyond the highest radius of the coil, and the simplified equation is not accurate under this condition. However, such a higher lateral distance may not require in practice. The maximum working region for lateral misalignment is less than the radial length of the coil.

Mutual inductance values in Figure 8a show that the curve decline rate is higher when the misplaced distance is increased compared to the smaller displacement, which indicates a flat curve. Thus, it can be concluded that the lateral misalignment should be little for obtaining consistent mutual inductance value. It can be described by the magnetic field theory, which states that the concentration of flux is higher at the center of the coil, and it gradually becomes weaker as it moves away from the center.

It is also observed from Figures 6a, 7a, and 8a that a decline in the rate of mutual inductance at higher axial distance is slightly lower than at the lower axial distance h. It can be easily justified by the field distribution of spiral coil. When the electric current flows into the coil, the field is generated and distributed mainly in two directions: radial direction, and vertical direction. Since the radial field has a negligible influence on mutual inductance, it can be ignored. The concentration of the vertical field is higher near the coil than at a distance away from it. Therefore, when the secondary coil moves away from the center of the primary coil at a lower axial distance, it cuts more flux line compared to coil at higher h where flux lines are fewer. It causes slightly more decline of M [28].

Thus, these comparisons for different coil configurations prove the accuracy of the simplified mutual inductance equation for all cases.

4.5. Coupling Coefficient of Spiral Coil

For wireless power applications, one of the parameters which determine the maximum output power is the coupling coefficient, and it can have a huge impact on the system efficiency [29].

The simplified mutual equation has also verified by the coupling coefficient. It is the fraction of the total flux flowing from one coil to another. Its value ranges from 0 to 1 and is represented by k. For perfectly coupled coil (whole flux produced by one coil link another coil), the value of $k = 1$, while $k > 0.5$ shows the tightly coupled and the loosely coupled coils can be described by $k < 0.5$. It depends on the physical configuration of the coil, such as orientation, windings, core, and the axial distance between them. k can defined by Equation (19).

$$k = \sqrt{\frac{M^2}{L_p L_s}} \tag{19}$$

where M is the mutual inductance, L_p is the self-inductance of primary coil, and L_s is the self-inductance of secondary coil. Self-inductance of the coil can be found by replacing the axial distance parameter h in the denominator of Equation (12) with the wire diameter w. Using a modified Equation (12), the self-inductance of the sampled 10 turns and 16 turns spiral coil were 4.584 μH and 18.734 μH, respectively. It can also be measured by an LCR meter. The measured self-inductance values are 4.712 μH and 19.310 μH, alternately. k relative of axial distance is depicted in Figure 9.

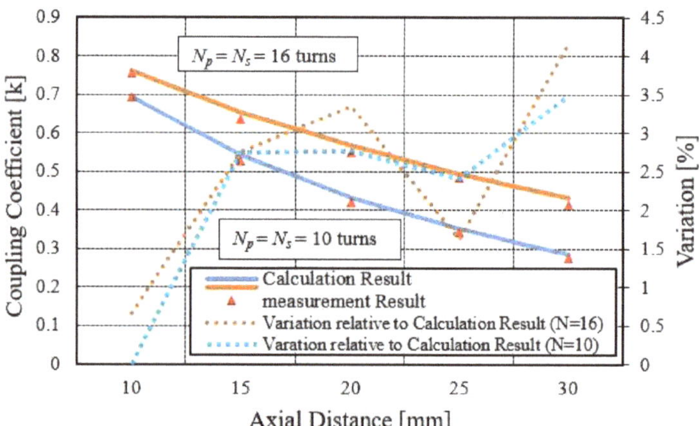

Figure 9. Coupling coefficient as a function of axial distance.

The minimum distance is 10 mm, and the maximum is 30 mm. The difference between successive h changes is 5 mm. The calculated coupling coefficient values is represented with the orange and blue lines, while the measured value is indicated by the small red triangular box.

The comparison result shows good agreement between measured and calculated values, which is below 5%. It can be seen that with the increase in the axial distance, the k reduces to zero, which is clear from the definition of the k that the magnetic flux linkage of one coil regarding another coil drops as the distance between them increases. Figure 9 also indicates that for the fixed axial distance, the value of k is a little smaller for the small coil compared to the large coil due to the higher magnetic flux linkage in the large coil than the small coil.

5. Conclusions

In this paper, a simplified and easy equation of mutual inductance for two coaxial and non-coaxial circular planar spiral coils has been presented. The Neumann formula has been adopted to derive this equation. Each step of mathematical derivation has been given. Compared to the conventional method, this simplified method reduces the calculation complexity and long numerical computations. The comparison results have shown that the error is larger for a spiral coil with a small number of turns. For the same axial distance, the error is below 9% for smaller coils while it is less than 3% for the larger spiral coil. And in the case of non-axial distance, the error increases when the lateral misaligned distance is greater than the maximum radius of the spiral coil. However, such a big misalignment is not required practically because the mutual inductance is approaching zero in such a case.

The calculation result has been compared by simulation for different geometry, either sparsely wounded or densely wounded. The comparison of the simulation result confirms the simplified equation's correctness with a difference of only 4%. Secondly, it has been compared with the simulation and the measurement result at various axial and non-axial distances for different coil categories, such as between small primary and small secondary coil, large primary and small secondary coil, and large primary and large secondary coil. This comparison also certifies the simplified equation accuracy. The final validity comparison has concluded between calculated and measured values of the coupling coefficient relative to axial distance. The error between them is under 5%. Thus, these comparisons have shown the correctness of the simplified mutual inductance equation, which makes it a suitable candidate to be adopted in different wireless power applications like biomedical implants, wireless charging systems, and contactless battery charging.

In this work, mutual inductance was derived by supposing the uniform current flow in the coil. However, the inductance value could be different for un-uniform current flow. It requires further research in this regard. Furthermore, the simplified mutual inductance equation is in double integral form. It can be simplified to a single integral form, which will be considered in future work.

Author Contributions: I.H. proposed the theoretical model and conducted the experiment, and D.-K.W. revised the draft and provided guidance. All authors have read and agreed to the published version of the manuscript.

Funding: This research received no external funding.

Institutional Review Board Statement: Not Applicable.

Informed Consent Statement: Not Applicable.

Data Availability Statement: Not Applicable.

Conflicts of Interest: The authors declare no conflict of interest.

References

1. Khan, S.R.; Choi, G. Analysis and Optimization of Four-Coil Planar Magnetically Coupled Printed Spiral Resonators. *Sensors* **2016**, *16*, 1219. [CrossRef]
2. Palagani, Y.; Mohanarangam, K.; Shim, J.H.; Choi, J.R. Wireless power transfer analysis of circular and spherical coils under misalignment conditions for biomedical implants. *Biosens. Bioelectron.* **2019**, *141*, 111283. [CrossRef]

3. Khan, S.R.; Sumanth, K.P.; Gerard, C.; Marc, P.Y.D. Wireless power transfer techniques for implantable medical devices: A review. *Sensors* **2020**, *20*, 3487. [CrossRef] [PubMed]
4. Krishnapriya, S.; Chandrakar, H.; Komaragiri, R.S.; Suja, K.J. Performance analysis of planar microcoils for biomedical wireless power transfer links. *Sādhanā* **2019**, *44*, 1–8. [CrossRef]
5. Haerinia, M.; Shadid, R. Wireless Power Transfer Approaches for Medical Implants: A Review. *Signals* **2020**, *1*, 209–229. [CrossRef]
6. Nguyen, M.Q.; Hughes, Z.; Woods, P.; Seo, Y.-S.; Rao, S.; Chiao, J.-C. Field Distribution Models of Spiral Coil for Misalignment Analysis in Wireless Power Transfer Systems. *IEEE Trans. Microw. Theory Tech.* **2014**, *62*, 920–930. [CrossRef]
7. Artan, N.S.; Amineh, R.K. Wireless Power Transfer to Implantable Medical Devices With Multi-Layer Planar Spiral Coils. In *Research Anthology on Emerging Technologies and Ethical Implications in Human Enhancement*; IGI Global: Hershey, PA, USA, 2021; pp. 457–481.
8. Waffenschmidt, E. Wireless power for mobile devices. In Proceedings of the 2011 IEEE 33rd International Telecommunications Energy Conference (INTELEC), Amsterdam, The Netherlands, 9–13 October 2011; IEEE: Piscataway, NJ, USA, 2011; pp. 1–9.
9. Lopez-Alcolea, F.J.; Del Real, J.V.; Roncero-Sanchez, P.; Torres, A.P. Modeling of a Magnetic Coupler Based on Single- and Double-Layered Rectangular Planar Coils with In-Plane Misalignment for Wireless Power Transfer. *IEEE Trans. Power Electron.* **2020**, *35*, 5102–5121. [CrossRef]
10. Pon, L.L.; Leow, C.Y.; Rahim, S.K.A.; Eteng, A.; Kamarudin, M.R. Printed Spiral Resonator for Displacement-Tolerant Near-Field Wireless Energy Transfer. *IEEE Access* **2019**, *7*, 172055–172064. [CrossRef]
11. Villa, J.L.; Jesús, S.; Andrés, L.; José, F.S. Design of a high frequency inductively coupled power transfer system for electric vehicle battery charge. *Appl. Energy* **2009**, *86*, 355–363. [CrossRef]
12. Sallán, J.; Juan, L.; Villa, A.L.; José, F.S. Optimal design of ICPT systems applied to electric vehicle battery charge. *IEEE Trans. Ind. Electron.* **2009**, *56*, 2140–2149. [CrossRef]
13. Miller, J.M.; Onar, O.C.; Chinthavali, M. Primary-Side Power Flow Control of Wireless Power Transfer for Electric Vehicle Charging. *IEEE J. Emerg. Sel. Top. Power Electron.* **2015**, *3*, 147–162. [CrossRef]
14. Nguyen, D.H. Electric Vehicle—Wireless Charging-Discharging Lane Decentralized Peer-to-Peer Energy Trading. *IEEE Access* **2020**, *8*, 179616–179625. [CrossRef]
15. Bouanou, T.; El Fadil, H.; Lassioui, A. Analysis and Design of Circular Coil Transformer in a Wireless Power Transfer System for Electric Vehicle Charging Application. In Proceedings of the 2020 International Conference on Electrical and Information Technologies (ICEIT), Rabat, Morocco, 4–7 March 2020; IEEE: Rabat, Morocco, 2020; pp. 1–6.
16. Good, R.H. Elliptic integrals, the forgotten functions. *Eur. J. Phys.* **2001**, *22*, 119–126. [CrossRef]
17. Maxwell, J.C. *A Treatise on Electricity and Magnetism*; Clarendon Press: Oxford, UK, 1873; Volume 1.
18. Ramo, S.; John, R.W.; Theodore, V.D. *Fields and Waves in Communication Electronics*; John Wiley & Sons: Hoboken, NJ, USA, 1994.
19. Ramrakhyani, A.K.; Mirabbasi, S.; Chiao, M. Design and Optimization of Resonance-Based Efficient Wireless Power Delivery Systems for Biomedical Implants. *IEEE Trans. Biomed. Circuits Syst.* **2011**, *5*, 48–63. [CrossRef]
20. Raju, S.; Wu, R.; Chan, M.; Yue, C.P. Modeling of Mutual Coupling between Planar Inductors in Wireless Power Applications. *IEEE Trans. Power Electron.* **2014**, *29*, 481–490. [CrossRef]
21. Liu, S.; Su, J.; Lai, J. Accurate Expressions of Mutual Inductance and Their Calculation of Archimedean Spiral Coils. *Energies* **2019**, *12*, 2017. [CrossRef]
22. *Maxwell-ANSYS*; ANSYS, Inc.: Canonsburg, PA, USA, 2014.
23. Kalantarov, P.L. *Inductance Calculations*; National Power Press: Moscow, Russia, 1955.
24. Vaisanen, V.; Hiltunen, J.; Nerg, J.; Silventoinen, P. AC resistance calculation methods and practical design considerations when using litz wire. In Proceedings of the IECON 2013—39th Annual Conference of the IEEE Industrial Electronics Society, Vienna, Austria, 10–13 November 2013; IEEE: Piscataway, NJ, USA, 2013; pp. 368–375.
25. Rossmanith, H.; Marc, D.; Manfred, A.; Dietmar, E. Measurement and characterization of high frequency losses in nonideal litz wires. *IEEE Trans. Power Electron.* **2011**, *26*, 3386–3394. [CrossRef]
26. Hussain, I.; Woo, D.-K. Self-Inductance Calculation of the Archimedean Spiral Coil. *Energies* **2022**, *15*, 253. [CrossRef]
27. Matsumoto, H.; Neba, Y.; Ishizaka, K.; Itoh, R. Model for a Three-Phase Contactless Power Transfer System. *IEEE Trans. Power Electron.* **2011**, *26*, 2676–2687. [CrossRef]
28. Liu, X.; Hui, S.Y.R. Optimal Design of a Hybrid Winding Structure for Planar Contactless Battery Charging Platform. *IEEE Trans. Power Electron.* **2008**, *23*, 455–463. [CrossRef]
29. Kim, M.; Park, H.; Jung, J.-H. Design Methodology of 500 W Wireless Power Transfer Converter for High Power Transfer Efficiency. *Trans. Korean Inst. Power Electron.* **2016**, *21*, 356–363. [CrossRef]

 sensors

MDPI

Article

Flexible and Transparent Circularly Polarized Patch Antenna for Reliable Unobtrusive Wearable Wireless Communications

Abu Sadat Md. Sayem [1,2,*], Roy B. V. B. Simorangkir [3], Karu P. Esselle [1], Ali Lalbakhsh [2], Dinesh R. Gawade [3], Brendan O'Flynn [3] and John L. Buckley [3]

1 School of Electrical & Data Engineering, University of Technology Sydney, Sydney, NSW 2007, Australia; karu.esselle@uts.edu.au
2 School of Engineering, Macquarie University, Ryde, NSW 2109, Australia; ali.lalbakhsh@mq.edu.au
3 Tyndall National Institute, T12 R5CP Cork, Ireland; roy.simorangkir@ieee.org (R.B.V.B.S.); dinesh.gawade@tyndall.ie (D.R.G.); brendan.oflynn@tyndall.ie (B.O.); john.buckley@tyndall.ie (J.L.B.)
* Correspondence: abusadatmd.sayem@uts.edu.au

Abstract: This paper presents a circularly polarized flexible and transparent circular patch antenna suitable for body-worn wireless-communications. Circular polarization is highly beneficial in wearable wireless communications, where antennas, as a key component of the RF front-end, operate in dynamic environments, such as the human body. The demonstrated antenna is realized with highly flexible, robust and transparent conductive-fabric-polymer composite. The performance of the explored flexible-transparent antenna is also compared with its non-transparent counterpart manufactured with non-transparent conductive fabric. This comparison further demonstrates the suitability of the proposed materials for the target unobtrusive wearable applications. Detailed numerical and experimental investigations are explored in this paper to verify the proposed design. Moreover, the compatibility of the antenna in wearable applications is evaluated by testing the performance on a forearm phantom and calculating the specific absorption rate (SAR).

Keywords: circular polarization; flexible; polymer; transparent; wearable

Citation: Sayem, A.S.M.; Simorangkir, R.B.V.B.; Esselle, K.P.; Lalbakhsh, A.; Gawade, D.R.; O'Flynn, B.; Buckley, J.L. Flexible and Transparent Circularly Polarized Patch Antenna for Reliable Unobtrusive Wearable Wireless Communications. *Sensors* **2022**, *22*, 1276. https://doi.org/10.3390/s22031276

Academic Editors: Pedro Pinho and Giovanni Andrea Casula

Received: 13 December 2021
Accepted: 1 February 2022
Published: 8 February 2022

1. Introduction

With the rapid advancement of flexible materials and manufacturing technologies, new opportunities for flexible electronics are emerging. The flexible electronics industry is worth a billion-dollar market value worldwide. A flexible antenna is considered as the most prominent invention in the flexible electronics industry. The applications of flexible antennas have spanned beyond the realm of traditional wireless communications. Nowadays, antennas have become an indispensable component in body centric wireless communications, remote sensing technology, national surveillance and security services, battle fields, firefighting and rescue missions, sports and fitness and vehicular communications. These applications have greatly benefited from flexible, robust, low-profile and small antennas owing to their flexible deployment on the system platforms, regardless of the curvature of the surface, thus maximizing the use of existing infrastructure, as well as their capability to withstand physical deformation and rigorous environmental circumstances. Apart from flexibility and robustness, unobtrusiveness is also a demanding characteristic of antennas in many applications not only for enhancing aesthetics but also for ensuring the reliability of operations. One example is wireless wearable technology [1,2], which is used in the healthcare and well-being sector, including remote sensing, monitoring and detection of patients suffering from chronic and fatal diseases, isolated communities, mental health patients, aged people and even contagious diseases like COVID-19. To ensure efficient sensing and monitoring performance of a wearable wireless healthcare system, antennas, as one of the enabling technologies, need to be comfortable, easy to wear, and visually imperceptible so that they impose minimum interference with daily activities.

Visual imperceptibility can be achieved in a number of different methods. For instance, the embroidery technique can be employed to realize an antenna embedded onto a patients' outfit, resulting in very little visual appearance. The major drawback is that, after repeated washing, the antenna performance is significantly degraded [3,4], thus affecting its long-term usage. Another approach is by hiding the antenna inside the wearer's outfit. This could possibly be the handiest way to make an antenna's appearance entirely invisible. This approach, however, implies that the antenna performance is significantly impacted by the dielectric properties of the outfit materials [5,6]. Such effect is severe, particularly upon the outfit's exposure to water (e.g., from the wearers' sweat or rain). As an alternative to these approaches, antennas' appearance can be made imperceptible by making the antennas optically transparent. In a wearable environment, using a robust, flexible and optically transparent antenna is the most effective solution for setting up an imperceptible communication network [7]. In addition to body centric communications, flexible-transparent antennas have a number of other potential applications, such as vehicular communications [8], concealed network terminals, smart cities [9] and solar cells, where antennas that are optically transparent and can be mounted on curved surfaces are desirable. However, the development of flexible and transparent antennas entirely depends on unconventional materials [10]. The unavailability of competent materials has obstructed the progress of flexible-transparent antennas' development.

To the best of our knowledge, there have been quite a few efforts that have succeeded in demonstrating antennas that are simultaneously flexible and transparent. The unavailability of perfect transparent conductors is one of the possible reasons of this scarcity. Transparent conductors used in antenna fabrication are broadly divided into two categories, i.e., thin films and meshed conductors; both groups are associated with some limitations. The group of thin film conductors includes indium-tin-oxide (ITO) [11], multi-layer ITO (IZTO/Ag/IZTO) film [12], fluorine-doped tin oxide (FTO) [13], gallium-doped zinc oxide (GZO) [14] and silver-coated polyester (AgHT-8) film [15]. In general, these conductors exhibit poor conductivity. Moreover, the common thin films, such as ITO [11], FTO [13] and GZO [14], are fragile, hence not compatible for conformal applications. Some approaches, however, have been reported to make them flexible. For example, in [12], Zn was mixed with an ITO film to increase its flexibility. Furthermore, a stacked topology of IZTO films comprising two layers of IZTO with Ag in between was introduced to increase the conductivity. Despite the flexibility of IZTO/Ag/IZTO film, it has 4.99 Ω/sq sheet resistance, which is quite high for an efficient antenna operation. The second group, meshed conductors, on the other hand, achieves transparency by incorporating perforations throughout the surface of traditional conductive sheets [16,17]. The transparency of meshed conductors depends on the ratio of opening to the total surface area. The larger the opening on the conductor surface, the higher the transparency that can be achieved. This unfortunately results in an increase of sheet resistance subsequently. On top of that, special treatment is required to integrate meshed conductors to flexible substrates to realize flexible transparent antennas [18]. It should also be noted that both groups of transparent conductors require strictly optimized fabrication processes to achieve the right density of the conductive structure for a balance between the conductivity performance and transparency [9], which are generally complex, cumbersome, and costly [19]. This might also result in reproducibility issues of an antenna with a desired performance. Using polymer and additive manufacturing, flexible, transparent and stretchable structures can be achieved, which is successfully demonstrated by [20–22]. Recently, a new technology has been reported that has combined commercially available transparent conductive fabric with solution-processable polydimethylsiloxane (PDMS) [7,23,24]. This technology is relatively simple, effective, cost-efficient and reproducible for realizing robust-flexible-transparent antennas. This technology is utilized in this work to fabricate the proposed circularly polarized (CP) antenna.

Circular polarization is an important property of antennas in many applications, especially when they are mounted on moving objects, such as in the case of wearable applications. CP antennas are less impacted by multi-path effects and polarization mismatch losses than linearly polarized (LP) antennas [8,25]; thus, CP antennas can enhance the reliability of wireless communication links. By avoiding the power loss due to polarization mismatch, CP antennas also play an essential role in extending the battery life, which is often a concern and limiting factor in low-power wireless communication systems [26]. In this paper, a flexible and transparent single-feed CP microstrip patch antenna is designed. The explored design incorporates a rectangular slot at the middle of the patch and a chamfer at the patch periphery to achieve CP radiation. While significant efforts in developing CP wearable antennas have been reported in recent years [27–38], to our knowledge, this is the first CP wearable antenna ever reported that is concurrently flexible and transparent.

The design process, fabrication methods and numerical and experimental investigations of the proposed flexible and transparent CP antenna are demonstrated in this paper. The proposed antenna operates at 2.4 GHz Industrial, Scientific and Medical (ISM) band that has a number of applications including wireless body area network (WBAN). The target application of the antenna is wearable technology. To evaluate its suitability in wearable applications, specific absorption rate (SAR) is calculated in a forearm phantom. Two antenna prototypes are fabricated, one with transparent fabric and another one with non-transparent fabric, as a reference. It is worth mentioning that a comparative study between flexible antennas based on transparent and non-transparent conductive fabric has never been reported before. This investigation will be an important source of information to understand the potential of the proposed materials for the development of robust, flexible, and unobtrusive wearable antennas.

2. Materials

The most crucial step in the development of flexible-transparent antennas is the selection of appropriate materials. The development of flexible antennas' entirety depends on unconventional materials where the selected conductive and dielectric materials should have high durability against physical deformations, endurance to environmental impacts, easy availability, low-cost, simple fabrication process, high optical transparency and low-loss. These properties are offered by the transparent conductive fabric-PDMS composite, for which reason it is used to fabricate the CP antenna proposed in this work.

The conductive parts of the demonstrated antenna (i.e., the patch, feed-line, and ground plane) are fabricated with transparent conductive fabric, VeilShield, developed by Less EMF Inc., Latham, NY, USA. VeilShield is an ultra lightweight, thin (57 μm), flexible and corrosion resistant mesh style conductive fabric. The mesh is formed by intertwining monofilament polyester threads, with 132/inch mesh accomplishing nearly 72% optical transparency. The polyester threads are coated with Nickel/Zinc blackened Copper, which makes the fabric conductive with 0.1 Ω/sq sheet resistance. This transparent conductive material benefits from its availability in the form of a fabric sheet with structural regularity. As a result, the electrical performance and optical transparency are more predictable and stable, which helps in solving the reproducibility issue encountered with other types of transparent conductive materials.

As the substrate and encapsulation of the antenna, PDMS, which is a mineral-organic elastomeric polymer constituting carbon and silicon, is used. PDMS exhibits some unique properties that make it emerging for flexible electronics fabrication. This includes high transparency (>94% [39]), high flexibility, waterproof, and biocompatibility. Another notable advantage of PDMS is its simple processing. The PDMS solution, a mixture between the base and cross-linking agent (curing agent), can be cured even in room temperature and does not necessitate clean room-based treatment. Due to its initial liquid form, once poured into customized mold and cured, nearly any shape and thickness of the flexible substrate can be achieved. Based on the measurement conducted with an Agilent 85070E Dielectric Measurement Kit, PDMS exhibits a relatively constant permittivity of 2.75 from 0.5–10.6 GHz with a frequency dependent loss tangent ranging from 0.008 to 0.07. In this work, the PDMS solution is made with Dow Corning Sylgard 184 silicone elastomer kit with base to cross-linking agent mixing ratio of 10:1. PDMS makes extremely strong integration with conductive fabric, thanks to the percolation of PDMS solution into the mesh, which leads to strong bonding and seals the fabric firmly upon curing [7,40]. To take into account the portion of PDMS penetration into the mesh, an effective sheet resistance of 0.7 Ω/sq, obtained through intensive investigations incorporating transmission lines and T-resonator samples [7], was used to model the conductive layer in the antenna simulation process.

A non-transparent version of the proposed CP antenna design was also fabricated as a reference. The utilized non-transparent conductive fabric is nickel–copper coated ripstop which has a thickness of 0.08 mm and sheet resistance of 0.03 Ω/sq. Similar to the transparent conductive fabric, the PDMS percolation into this fabric was taken into account during the antenna design process by using an adjusted sheet resistance of 0.23 Ω/sq [40] to model the antenna conductive parts.

3. Antenna Topology and Design

3.1. Antenna Configuration

The topology of the proposed flexible and transparent circularly-polarized antenna is shown in Figure 1. The demonstrated antenna is a microstrip circular patch antenna with a rectangular central slot and a chamfer at the edge. On the opposite side of the patch, there is a circular ground plane with a circular slot etched at the center. To ensure protection against humidity, dust, chemical and harsh bending scenarios, the antenna is encapsulated by PDMS. Very thin encapsulation is used to avoid RF performance degradation.

The numerical modelling and optimization of the proposed antenna were conducted in CST Microwave Studio 2020 that uses Finite Integration Technique (FIT) for solving Maxwell's electromagnetic equations. The optimized dimensions of the antenna are shown in Table 1.

3.2. Design Methodology

The CP characteristic of the antenna is achieved by exciting two orthogonal modes having the same magnitudes but with 90° phase difference [36]. The proposed circular-shaped antenna operates at its fundamental TM_{11} mode. The rectangular slot at the center of the patch and the chamfer at patch periphery concurrently degenerate the fundamental TM_{11} mode into two orthogonal modes that produce CP [36]. The dimensions of the slot and chamfer are optimized to maintain the same magnitudes and phase quadrature of the degenerated orthogonal modes (E_x and E_y). The slot and chamfer drive the surface current to rotate 90° at each time phase quadrature, which is the ideal characteristic of the CP wave. To observe the CP characteristic of the radiated wave of the antenna, its surface current distribution is analysed. Figure 2 illustrates the surface current distribution of the antenna at 2.4 GHz as the feeding phase progresses from 0° to 270° by 90° interval. For ease of observation, we added labels n_1 and n_2, each of which denotes the null location of the current distribution and marked accordingly the tendency of the current vector direction with an arrow. It can be seen from Figure 2 that the current vector of the antenna rotates clockwise as the phase progressing, which indicates a left-handed circular polarization in the direction of the +z-axis.

Due to the loss associated with the materials used, the antenna suffers from low efficiency and gain. A defected ground plane approach was therefore implemented by cutting a circular slot at the center of the ground to improve the gain and efficiency of the antenna. By cutting the slot on the ground, the overall transparency of the antenna also increases as the area where two layers of fabric overlay is reduced. However, the presence of slot in the ground plane increases the back radiation, which in turn decreases the front-to-back ratio of the radiation pattern. Further optimization of the size of the slot is, therefore, required.

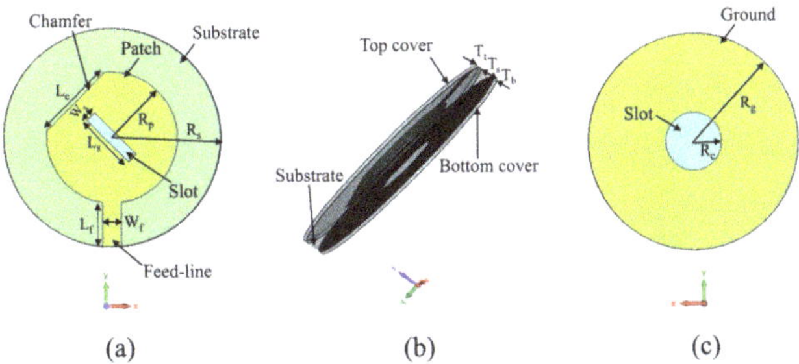

Figure 1. Geometry of the proposed flexible and transparent CP antenna: (**a**) top view, (**b**) side view, (**c**) bottom view.

Table 1. Dimensions of the flexible and transparent CP circular patch antenna.

Parameter	Description	Value (mm)
R_p	Radius of the patch	18
R_s	Radius of the substrate	30
R_g	Radius of the ground plane	30
R_c	Radius of the slot in the ground plane	8
L_s	Length of the slot in the patch	16
W_s	Width of the slot in the patch	3
L_c	length of the chamfer	22
L_f	Length of the feed-line	12.17
W_f	Width of the feed-line	5
T_t	Thickness of the top PDMS cover	0.2
T_b	Thickness of the bottom PDMS cover	0.2
T_s	Thickness of the substrate	3

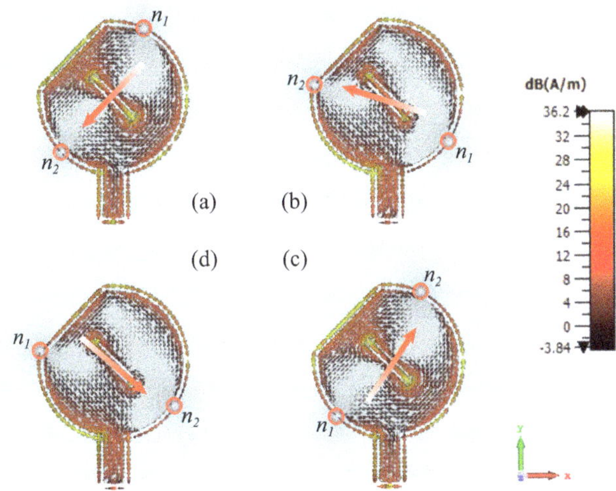

Figure 2. Surface current distribution of the antenna at 2.4 GHz at the time phases (**a**) 0°, (**b**) 90°, (**c**) 180° and (**d**) 270°.

The effect of the radius of the slot of the ground plane on the resonance characteristics of the antenna is illustrated in Figure 3, which shows that resonance frequency shifts with the change of the slot radius.

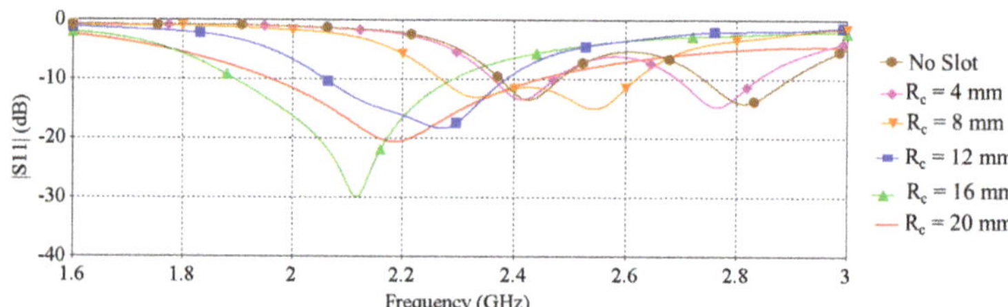

Figure 3. $|S_{11}|$ vs. frequency when varying the radius of the slot of the ground plane.

The influence of the slot radius on the peak gain, radiation efficiency and radiation patterns are illustrated in Figures 4–6, respectively. Figure 4 shows that the peak gain of the antenna improves significantly as the slot radius (R_c) extends up to 12 mm. Beyond that point, expanding the slot radius further does not substantially enhance the gain. It can also be noticed from Figures 5 and 6 that, with the increase of the slot radius, the radiation efficiency of the antenna increases as well as the back radiation. Considering the target wearable applications, the latter is not preferred as it means more radiation is directed towards the lossy human body tissue. Giving consideration to the antenna peak gain, efficiency and back radiation, the slot radius of 8 mm was selected in this design, which provides a maximum peak gain of 2.67 dBi, maximum radiation efficiency of nearly 43.2% and front-to-back ratio of 8.6 dB. The findings of this investigation are summarized in Table 2.

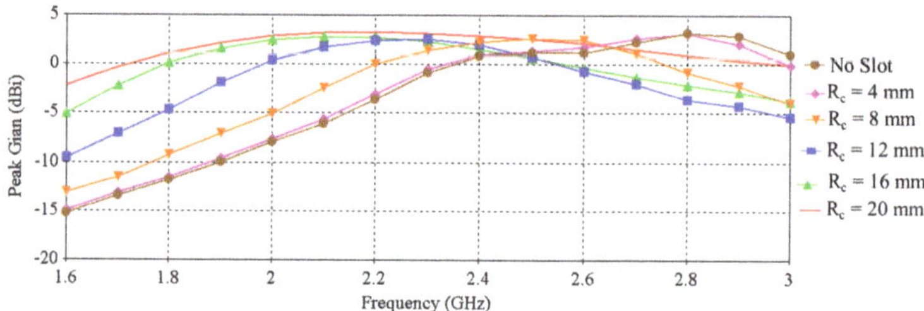

Figure 4. Peak gain vs. frequency when varying the radius of the slot of the ground plane.

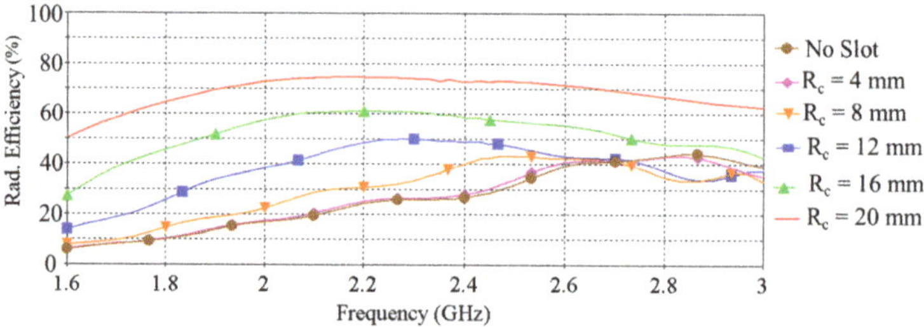

Figure 5. Radiation efficiency vs. frequency when varying the radius of the slot of the ground plane.

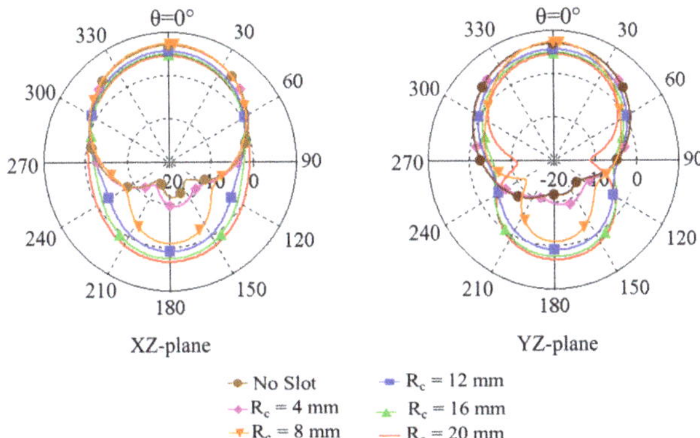

Figure 6. Radiation patterns (in dB) at the corresponding resonance frequencies when varying the radius of the slot of the ground plane.

Table 2. Summary of the antenna performance when varying the radius of the slot of the ground plane.

R_c (mm)	0	4	8	12	16	20
Res. Freq. (GHz)	2.63	2.57	2.45	2.26	2.12	2.2
Peak Gain (dBi)	1.54	1.63	2.44	2.53	2.78	3.24
Rad. Effi. (%)	40	39.7	42.6	49.8	60.1	74.5
F/B (dB): XZ-plane	16.1	14.6	8.6	5	2.8	1.6
F/B (dB): YZ-plane	16.1	14.6	8.6	5	2.7	1.5

The demonstrated antenna is conformal, which means that it can undergo bending that is crucial for wearable applications. To assess the RF performance of the antenna under bending, the antenna was simulated when bent as shown in Figure 7. "Bent Rad." denotes the bending radius of the antenna. Figures 8 and 9 show the $|S_{11}|$ and peak gain of the antenna for different bending radii ranging from 40 to 70 mm. It can be noted that the resonance frequency of the antenna is shifted slightly towards lower frequency as the bending radius decreases. Nevertheless, the target frequency of 2.4 GHz is still covered with a good matching and satisfactory gain. These numerical investigations show that the antenna is resilient against physical deformation, validating the suitability of the antenna for wearable applications.

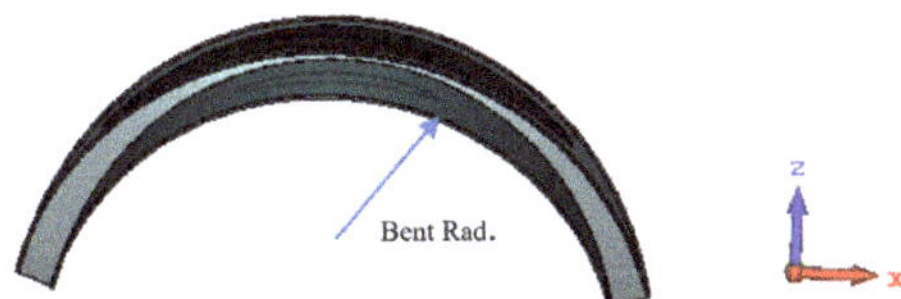

Figure 7. The antenna topology under bending.

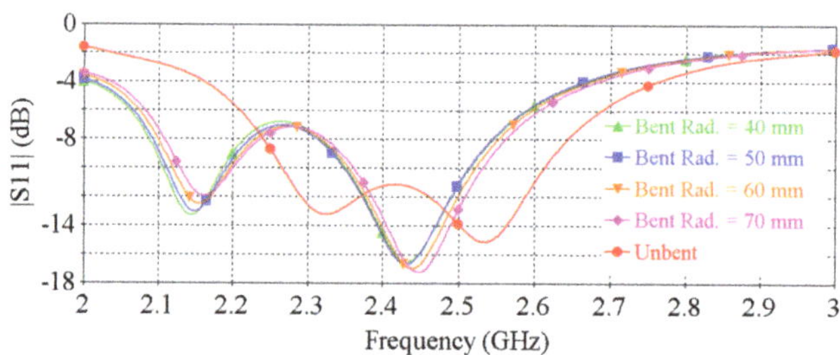

Figure 8. $|S_{11}|$ vs. frequency of the antenna for different bending radius.

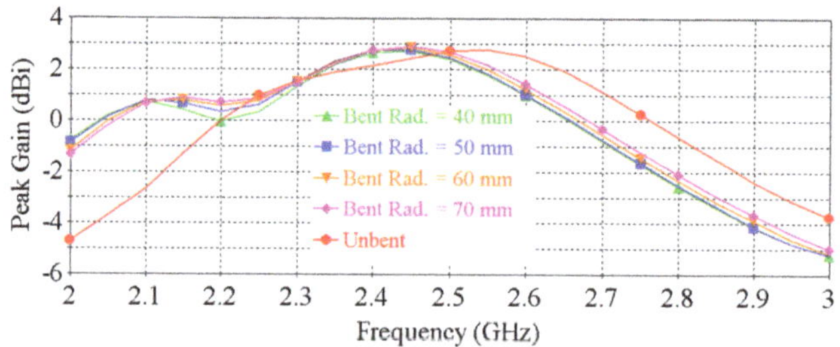

Figure 9. Peak gain vs. frequency of the antenna for different bending radius.

Lastly, Figures 10 and 11 show the impacts of PDMS encapsulation on the resonance frequency and peak gain of the antenna, respectively, demonstrating that the thin (0.2 mm) PDMS encapsulation has an insignificant effect on the antenna performance.

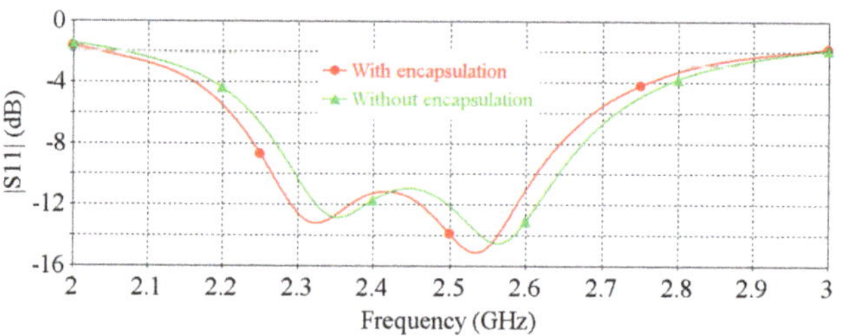

Figure 10. The effect of the PDMS encapsulation on the $|S_{11}|$ of the antenna.

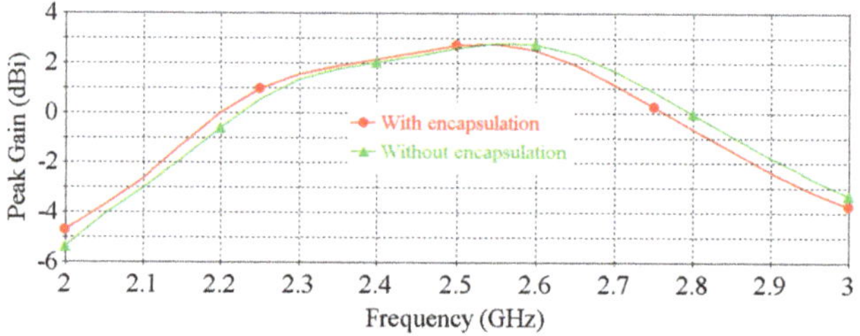

Figure 11. The effect of the PDMS encapsulation on the peak gain of the antenna.

4. Prototype Fabrication

The demonstrated transparent and non-transparent flexible antennas are fabricated by utilizing a cost-effective, straightforward and environmental-friendly method. It can be noted that the traditional physical vapor deposition [12], photo-lithography [22,41], RF sputtering [11], spray pyrolysis [13] and inkjet printing [42] methods of antenna manufacturing are often associated with fabrication complexity, high cost and involvement of hazardous chemicals and toxic by-products. In contrast to these methods, our demonstrated method is significantly cost-effective, simple and free from producing toxic chemicals. In the explored method, conductive fabric is used as the conductive parts of the antenna. Conductive fabric is certainly easy to handle, and it is easy to maintain structural regularity that is often challenging in metal deposition techniques. Moreover, the substrate is made with elastomer polymer, PDMS, which can be shaped to any pre-defined thickness by customized molds. This feature exhibits manufacturing flexibility and a simple method of maintaining accuracy. In principle, the antenna fabrication procedure is similar to our previously demonstrated works in [7,40,43]. The major difference is that, in this work, the conductive fabrics were patterned using a cutting machine, rather than through a manual cutting process with a razor blade. We found this crucial as the CP characteristic is typically very sensitive to the variation of the patch dimension. The schematic representation of the antenna fabrication process is shown in Figure 12.

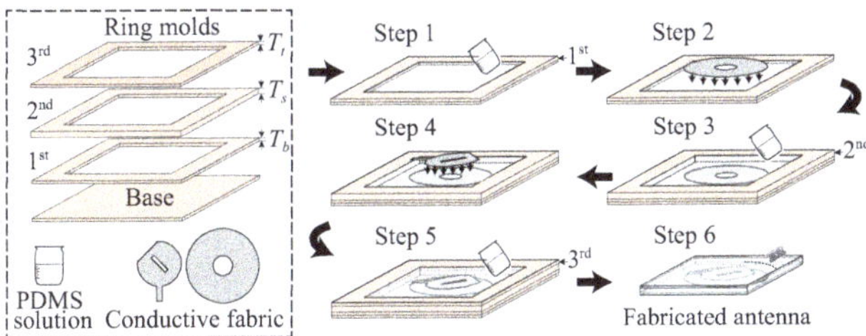

Figure 12. Schematic representation of the antenna fabrication process.

Below, the details of the manufacturing process are provided.

Detailed process: The antennas were fabricated by adopting a systemic procedure as described below:

Step-1: The first mold having the thickness 0.2 mm was attached to a base PCB board using sealant. Then, the PDMS solution was poured into the mold to create the bottom encapsulation layer. When required (e.g., if there are a lot of bubbles trapped in the PDMS solution), a vacuum desiccator can be used with approximately −80 kPa pressure to suck the air bubbles out. The mold with PDMS solution was kept in an oven at 80 °C for 30 min for curing.

Step-2: The ground plane was attached on top of the cured bottom PDMS cover, built in step 1, with small amount of uncured PDMS. Then, the molds were kept in an oven at 80 °C for 30 min for curing the attachment.

Step-3: The second mold of 3 mm thickness was attached on top of the first mold with sealant. The PDMS solution was poured into the second mold, filling up to the thickness of the mold. Then, the desiccation and curing procedures applied in step 1 were repeated to build the substrate layer of the antenna.

Step-4: The patch was attached on top of the cured substrate PDMS layer, built from step 3, with a small amount of uncured PDMS. Then, step 2 was followed to properly attach the patch on top of the substrate.

Step-5: The third mold of 0.2 mm thickness was attached on top of the second mold with sealant. The PDMS solution was poured into the third mold up to the thickness of the mold, and the desiccation and curing procedures applied in step 1 were repeated to build the top encapsulation layer.

Step-6: The prototype was carefully peeled-off from the molds and the extra PDMS were trimmed to give the exact dimensions of the antenna.

Step-7: Small section of PDMS from the top and bottom covers were etched out to connect the SMA connector to the antenna. The pin and ground of the connector were connected to the patch and ground plane of the antenna by using a CircuitWorks conductive epoxy which was cured by applying 80 °C temperature for 10 minutes in the oven. It should be noted that conductive epoxy is the most appropriate option for connecting the SMA connector to the conductive fabric because soldering has a risk of burning the fabric, and thus is not compatible. In actual wearable implementation, more compatible connection strategies can be used, such as butterfly clasps [44], contact springs [45], snap-on button [46] or hook and loop connectors [47].

In Figure 13, the photo of the fabricated transparent prototype is compared with its non-transparent counterpart fabricated through the similar process. As shown, the fabricated antenna exhibits high flexibility and transparency.

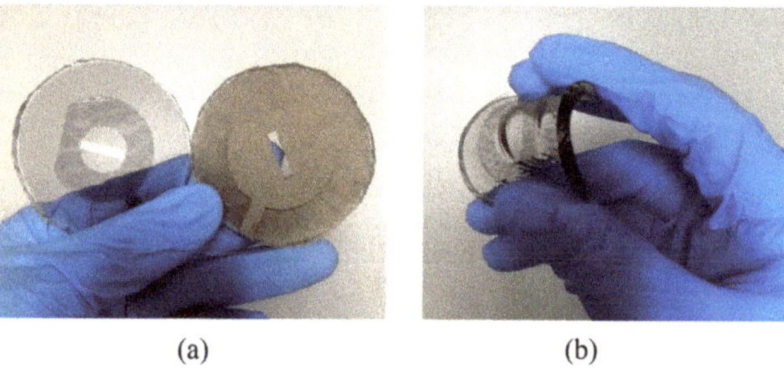

(a) (b)

Figure 13. Fabricated prototypes: (**a**) transparent and non-transparent antenna prototypes, (**b**) transparent antenna in bent state.

Three ring-shaped rectangular molds and a rectangular-shaped base were used for fabricating the antennas. Due to the availability, in this work, the molds were made from PCB boards which have the thicknesses of the target PDMS layers, cut with a slightly bigger opening than the antenna substrate aperture. The electroplated copper on the PCB was kept as it allows for an easy demolding of the built antenna at the end of the process.

PDMS solution was made with Dow Corning Sylgard 184 silicone elastomer kit. The kit comes with base silicone and curing agent. The PDMS solution was prepared by mixing the base and curing agent at the ratio of 10:1. The percentage of the curing agent in the mixture controls the flexibility level of the cured PDMS; higher percentage of the agent reduces the flexibility of the PDMS. The mixing process was undertaken in a plastic container, and the mixture was thoroughly stirred to ensure uniformity throughout the solution. The uncured PDMS solution is a clear sticky liquid, which is cured by applying heat. The curing time depends on the applied temperature.

The patch and ground plane were prepared by accurately cutting the conductive fabrics following the optimized antenna design pattern. In this work, the fabric was cut by using a cutting machine, Cricut Maker. To do this, 2D images of the patch and ground plane were exported from CST and imported into Cricut Design Space software.

5. Performance Investigation

The demonstrated transparent and non-transparent CP patch antennas were numerically investigated, and the prototypes were experimentally tested to explore the performance in different operating environments. The performance of the antennas was investigated both in free space and on-phantom. The latter is important to validate the antennas' compatibility in wearable applications. In the first scenario, the antennas were investigated in the flat condition, whereas, in the second scenario, the antennas were wrapped over the phantom. On-phantom testing was conducted with SHOGFPC-V1 forearm phantom from SPEAG [45,48]. On-phantom numerical investigation was conducted by placing the antenna on top of a double layer cylinder (Figure 14) mimicking the structure of the SPEAG phantom. The forearm bone was modelled with $\epsilon_r = 30$ and $\sigma = 2.5$ S/m, whereas the forearm tissue was modelled with $\epsilon_r = 27$ and $\sigma = 1.1$ S/m, following the averaged electrical properties of the forearm at 2.4 GHz according to the SPEAG phantom data-sheet [48]. While a more complex multilayer phantom would offer a more complete understanding on the antenna SAR performance, homogeneous phantoms or phantoms with averaged properties have also been widely employed in literature [45,49,50].

The separation between the antenna and phantom (dist) was varied and peak gain, radiation efficiency and the specific absorption rate (SAR) were calculated in simulation. The SAR was calculated by applying 0.1 W input power and averaged over 10 g of tissue. Figures 15–17 depict the antenna peak gain, radiation efficiency and SAR (at 2.4 GHz), respectively, for different separation between the antenna and phantom. From these investigations, it is revealed that, when the antenna is directly mounted on phantom (dist = 0 mm), peak gain and radiation efficiency are very low but yet the SAR values are still less than the maximum permissible SAR limit for body extremities (i.e., 4 W/kg) as specified in IEEE C95.1-2019 standard [51], demonstrating that the proposed antenna can be used safely on the human body at the target 2.4 GHz ISM band. As the separation between the antenna and phantom increases, SAR decreases further and gain and radiation efficiency also improve. It should be noted that, in a real-life scenario, there will always be a gap of approximately 2 mm to 5 mm between the antenna and body by the fabrics of the wearers' outfit and, thus, the antenna will operate well. Figure 18 shows the calculated SAR distribution in the phantom for 5 mm separation at 2.4 GHz, 2.45 GHz and 2.6 GHz. The calculated maximum 10 g-averaged SAR are 0.655 W/kg, 0.695 W/kg and 0.737 W/kg at 2.4 GHz, 2.45 GHz and 2.6 GHz, respectively. For validation, the SAR metric was further analysed by using a multilayer cylindrical phantom consisting of bone, muscle, fat and skin layers, as done in [52,53]. The configuration of the multilayer cylindrical phantom is shown in Figure 19 and the dielectric properties of each tissue layer are given in Table 3 [53]. Figure 20 illustrates the calculated SAR distribution in the multilayer cylindrical phantom for 5 mm separation at 2.4 GHz, 2.45 GHz and 2.6 GHz. The calculated maximum 10 g-averaged SAR are 0.646 W/kg, 0.715 W/kg and 0.661 W/kg at 2.4 GHz, 2.45 GHz and 2.6 GHz, respectively, which are close to the SAR values obtained with the homogeneous forearm phantom. From these investigations, it is exhibited that the ground plane provides sufficient shielding against the antenna's back radiation, demonstrating the antenna's suitability in wearable applications.

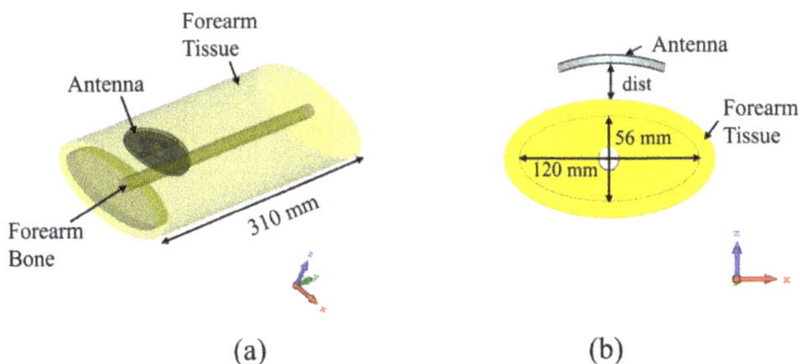

Figure 14. Antenna on top of the forearm phantom: (**a**) perspective view, (**b**) cross-sectional view.

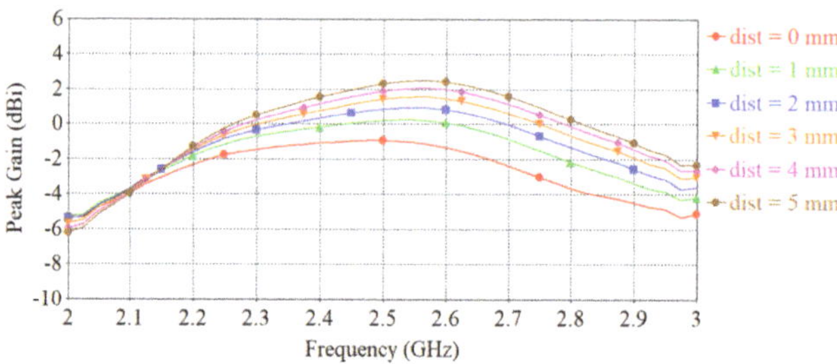

Figure 15. Peak gain vs. frequency when varying the separation between the antenna and phantom.

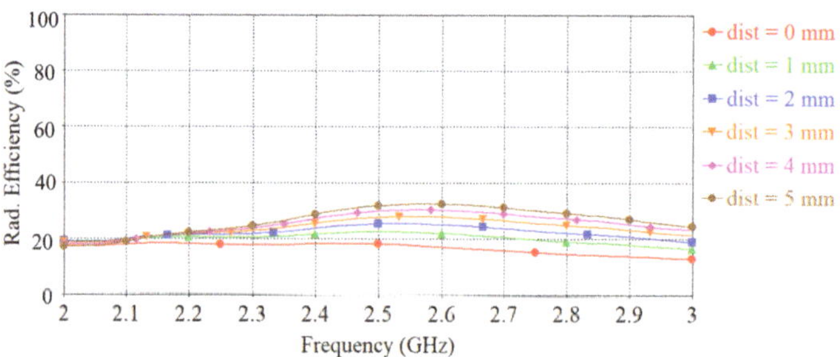

Figure 16. Radiation efficiency vs. frequency when varying the separation between the antenna and phantom.

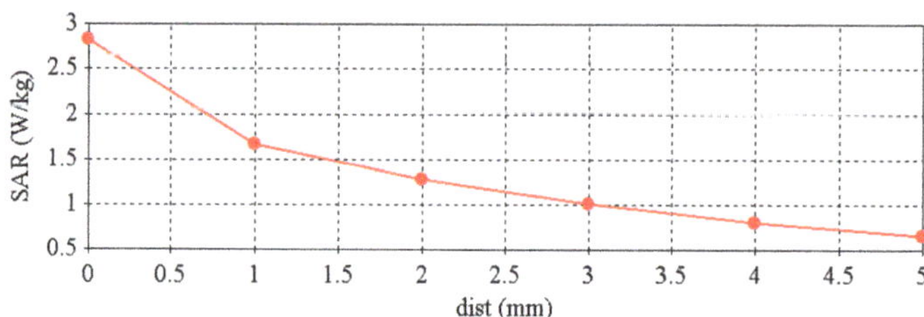

Figure 17. SAR at 2.4 GHz when varying the separation between the antenna and phantom.

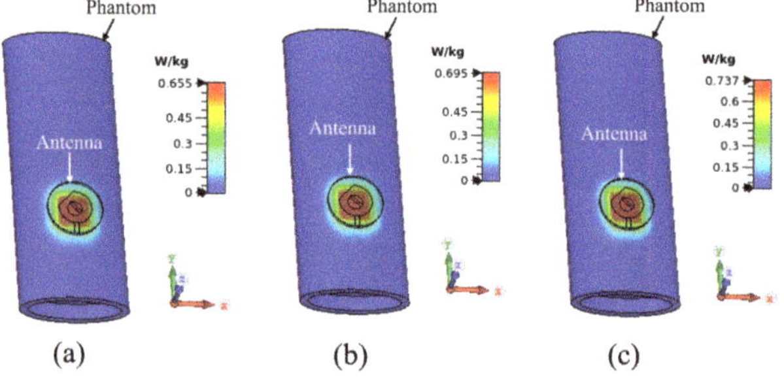

Figure 18. SAR distributions in the forearm phantom at: (**a**) 2.4 GHz, (**b**) 2.45 GHz and (**c**) 2.6 GHz.

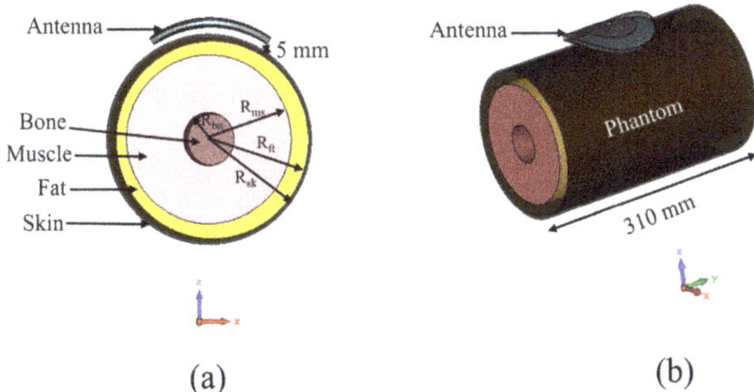

Figure 19. Antenna on top of the multilayer phantom: (**a**) cross-sectional view, (**b**) perspective view. (R_{bn} = 12.5 mm, R_{ms} = 40 mm, R_{ft} = 48.5 mm, R_{sk} = 50 mm).

Table 3. Dielectric properties of the different tissue layers at 2.4 GHz.

Tissue	Relative Permittivity	Conductivity (S/m)
Bone	11.4	0.39
Muscle	52.73	1.74
Fat	5.28	0.1
Skin	37.88	1.44

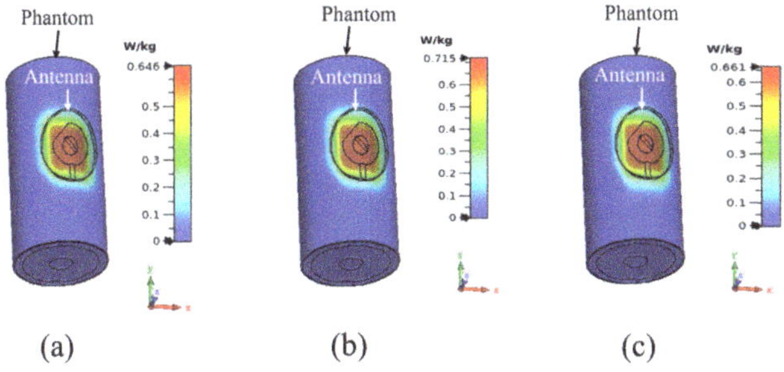

Figure 20. SAR distributions in the multilayer phantom at: (**a**) 2.4 GHz, (**b**) 2.45 GHz and (**c**) 2.6 GHz.

The performance of the antenna was experimentally tested in the worst scenario, i.e., when the antenna was directly bent around the wrist of the SPEAG forearm phantom having a circumference of approximately 15 cm. The prototype was bent in two principal planes, i.e., x-axis bent and y-axis bent. These tests were conducted for both the transparent and non-transparent antennas, and the performances were compared.

The return loss ($|S_{11}|$) measurements were accomplished by using an MS2038C Vector Network Analyzer (VNA) from Anritsu. The photos of the $|S_{11}|$ measurement set-ups of the antenna are shown in Figure 21. The antenna was fed with a 50-Ω coaxial line through a 50-Ω SMA connector. A ferrite bead was incorporated into the measurement set-up to minimize the ground current flow to the feed cable, which may affect the antenna's measured impedance and radiation characteristics. The measured and simulated $|S_{11}|$ results are displayed in Figure 22. From Figure 22, it can be seen that the measured results of the free-space case agree well with the simulated results, demonstrating a satisfactory matching ($|S_{11}| < 10$ dB) for both transparent and non-transparent antennas at the target frequency 2.4 GHz. In simulations, as the result of the differences in the characteristics of the conductive materials, it was observed that both antennas in flat free-space case exhibit slightly different input impedance, i.e., 81.6-j17.7 Ω for the transparent antenna and

84.3-j23.5 Ω for the non-transparent counterpart. The discrepancy between the simulated and measured results are most likely due to the fabrication tolerances, such as during patch and ground plane layering over PDMS layers and the antenna-SMA interconnection using conductive epoxy. The 10-dB return-loss bandwidth is around 346 MHz for both the transparent and non-transparent antenna. It can also be observed from the on-phantom $|S_{11}|$ results in Figure 22 that the resonance frequency and impedance matching of the antennas are affected by the phantom proximity and bending. Nevertheless, the antenna still operates within the desired band with satisfactory impedance matching, which demonstrate the antenna's robustness against human body loading and physical deformation.

The antenna far-field characteristics were measured in an AMS-8050 Antenna Measurement System from ETS-LINDGREN. Figure 23 displays the measurement set-up of the antenna inside the measurement chamber. The simulated and measured gains of the antennas are illustrated in Figure 24, which again shows a good agreement between the simulated and measured results of the free-space case, especially at the target operating frequency of 2.4 GHz. The discrepancy, particularly in the transparent prototype, is attributed to the fact that the transparent conductive fabric is very thin, and, hence, more vulnerable during the cutting and connecting to the SMA connector. The maximum antenna peak gain when measured flat in free space is 2.50 dBi for the transparent antenna and 4.15 dBi for the non-transparent antenna. The lower gain of the transparent antenna is expected due to the higher sheet resistance of the transparent fabric. Bending measurements over the phantom show that the gains of the antennas drop while operating on the phantom, which is the direct result of the antenna and phantom tissue coupling, particularly through the slot on the ground. Some of the radiated energy is therefore absorbed by the phantom. However, the gain performance can be improved by keeping a gap between the antenna and phantom as revealed in Figure 15.

The simulated and measured total efficiency of the antennas are depicted in Figure 25. It can be seen that, in free-space, the transparent antenna has a measured maximum total efficiency of 42.26%, and the non-transparent antenna has a measured maximum total efficiency of 54.64%. The reduced efficiency of the transparent antenna is attributed to the higher sheet resistance of the transparent fabric. When bent over the phantom, the antenna efficiency decreases due to the power loss in the phantom as described before. By keeping a distance between the antenna and phantom, the antenna efficiency can be improved as can be seen in Figure 16.

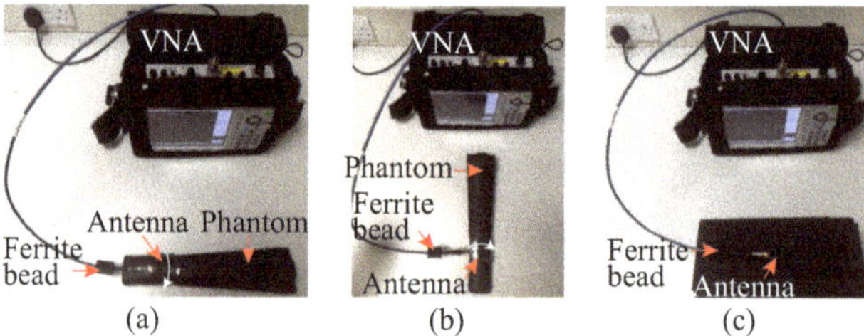

Figure 21. Antenna $|S_{11}|$ measurement set-up: (**a**) antenna bending towards the x-axis direction, (**b**) antenna bending towards the y-axis direction, and (**c**) flat antenna.

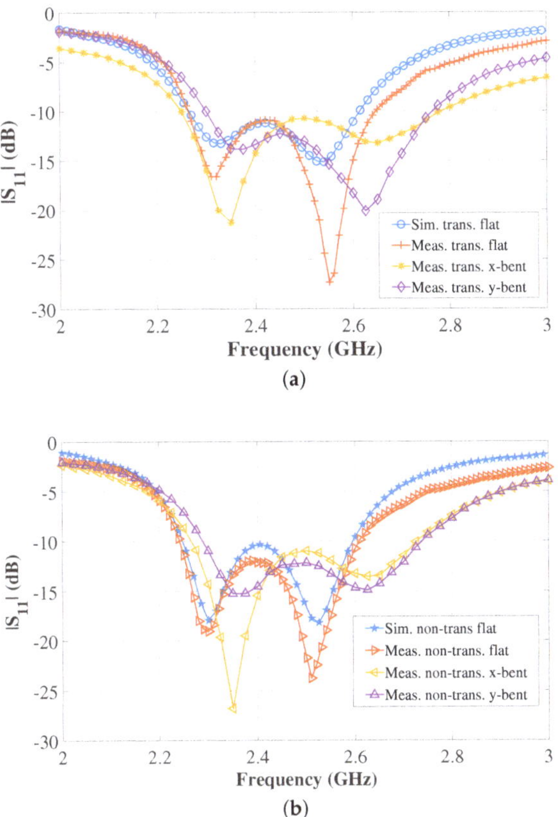

Figure 22. Simulated and measured $|S_{11}|$ vs. frequency: (**a**) transparent antenna, (**b**) non-transparent antenna.

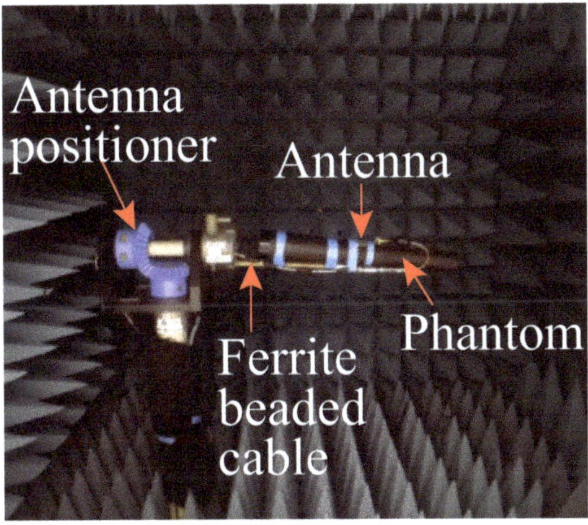

Figure 23. Antenna measurement set-up inside the AMS-8050 Antenna Measurement System.

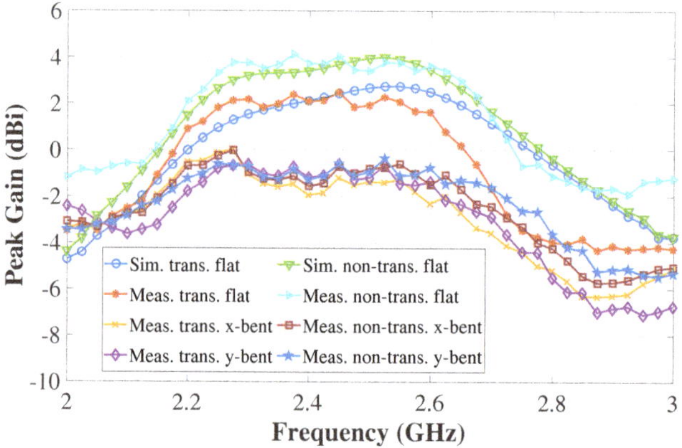

Figure 24. Simulated and measured peak gain vs. frequency of the transparent and non-transparent antennas.

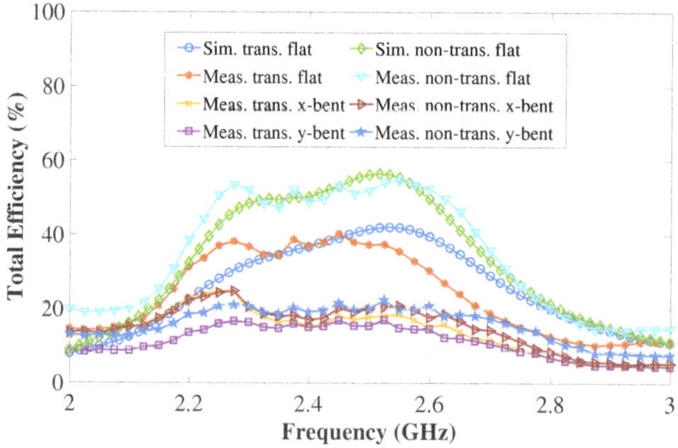

Figure 25. Simulated and measured total efficiency vs. frequency of the transparent and non-transparent antennas.

The simulated and measured axial ratio of the antennas in the broadside direction ($\theta = 0°$) are provided in Figure 26. As shown before, a good agreement is apparent in the simulated and measured results of the free-space scenario. The AR results validate that the proposed transparent antenna exhibits circular polarization at the target frequency 2.4 GHz, which is still maintained even under bending on the phantom. A shift in the frequency is expected due to the antenna physical deformation upon bending. Apparently, the change in the textile conductivity and thickness does affect the AR quality of the antenna as shown in the flat non-transparent antenna result. However, upon bending on the phantom, the AR level decreases as the result of the change in the antenna physical form. The different level of change in the AR performance incurred by the non-transparent and transparent antennas upon bending is attributed to the difference in the mechanical characteristic of the resulted PDMS-textile composites.

Figures 27 and 28 show the far-field radiation patterns of the transparent antenna and non-transparent antenna, respectively, both in free-space and when bent over the phantom. It is revealed that the measured results closely follow the simulated patterns, particularly in the free-space case. Changes in the patterns (e.g., the level of cross-polarized components and the shape of the patterns) upon bending on the phantom are expected as the results of the change in the shape of the antennas. Considering the size and shape of the phantom, there seems to be a part of the waves, possibly from the slot in the ground, crippling to the back of the forearm phantom and contributing to the increase in the antenna back radiation for the case of antenna bent on-phantom.

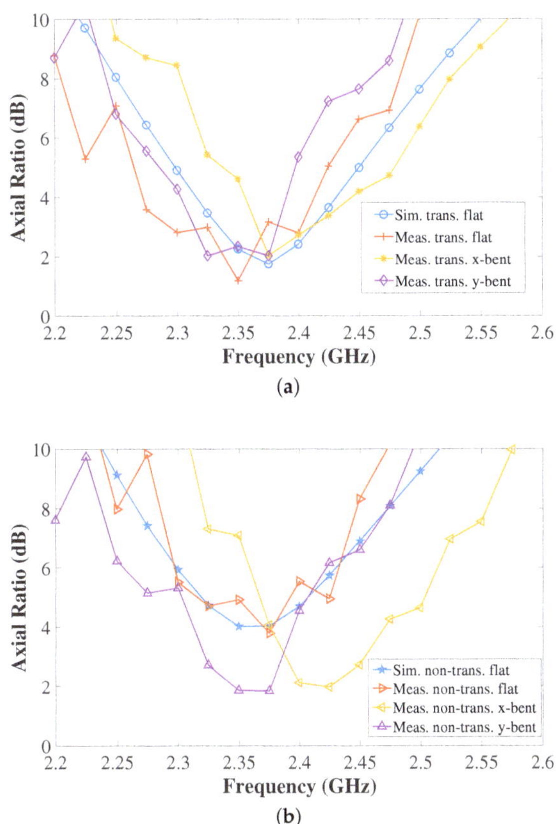

Figure 26. Simulated and measured axial ratio vs. frequency: (**a**) transparent antenna, (**b**) non-transparent antenna.

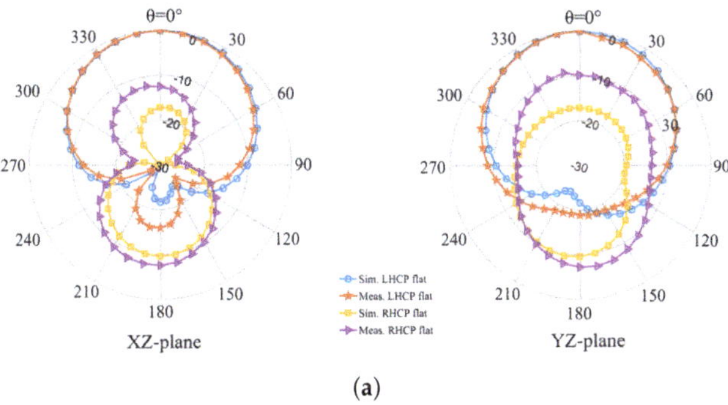

Figure 27. *Cont.*

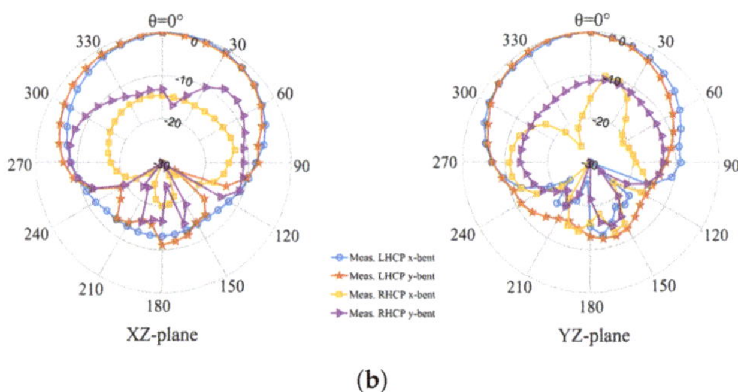

(b)

Figure 27. Simulated and measured far-field radiation patterns of the transparent antenna at 2.4 GHz: (a) unbent state, (b) bent state.

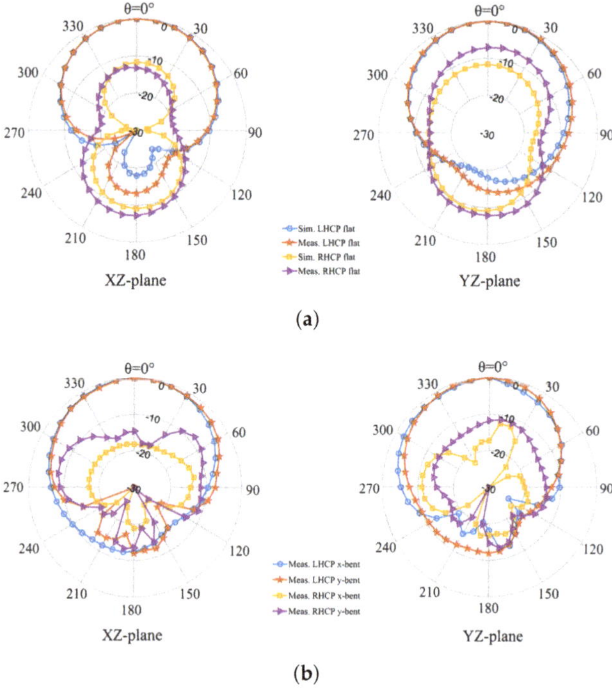

Figure 28. Simulated and measured far-field radiation patterns of the non-transparent antenna at 2.4 GHz: (a) unbent state, (b) bent state.

6. Discussion

The proposed optically transparent, flexible, and circularly polarized antenna shows promising performance. The explored antenna exhibits comparable RF performance in comparison to the non-transparent flexible antenna. The low gain and efficiency of the transparent version are expected due to to the high sheet resistance of the transparent fabric, VeilShield, compared to the non-transparent, nickel-copper coated ripstop fabric. The high sheet resistance comes as a consequence of being a highly porous fabric. Despite having slightly higher sheet resistance than non-transparent nickel-copper coated ripstop fabric, VeilShield still has significantly lower sheet resistance than the transparent thin film conductors (e.g., IZTO/Ag/IZTO has a sheet resistance of 4.99 Ω/sq [12], AgHT-8 film has a sheet resistance of 8 Ω/sq [15]). The highly porous nature of VeilShield not only allows the fabric to be see-through but also to be strongly integrated with PDMS. The latter comes as a major advantage as it is certain that metallization over PDMS is notoriously difficult. With the use of VeilShield, however, it can be done easily and the bonding between PDMS and VeilShield is very strong [19], allowing for the realization of flexible transparent antenna that is mechanically robust. Maintaining strong attachment between the flexible transparent conductive and dielectric material is one of the major challenges of flexible-transparent antenna fabrication, and the explored method works as an effective solution. Moreover, the proposed method is significantly low in cost compared to the existing techniques. Thus, the demonstrated technology appears to be a highly potential technique for fabricating flexible transparent antennas. In addition, the human body is a very dynamic lossy operating environment for an antenna, and the varying position and orientation of the antenna when worn on the body might lead to a power loss associated with the polarization mismatch. The fact that the explored antenna is circularly polarized, thus it can help to maintain a reliable communication link for on-body operations. A performance comparison of the proposed flexible and transparent CP wearable antenna with some state-of-the-art CP wearable antennas is shown in Table 4. From this comparison, it is revealed that the proposed antenna is one of the lowest in terms of profile, and it is the only antenna that is concurrently flexible and transparent. The proposed antenna also shows promising 3-dB axial ratio (AR) bandwidth. The low gain and efficiency of the demonstrated antenna are attributed to the imperfect conductivity of the transparent conductive fabric and loss of PDMS. Nevertheless, considering its low profile and size, the proposed antenna exhibits decent gain and efficiency when compared to some of the reported flexible CP wearable antennas. More importantly, the highlighted properties of excellent flexibility, transparency and RF performance have not been previously demonstrated in other reported CP wearable antennas. Thus, it is ascertained that the proposed flexible and transparent CP antenna can be a strong candidate in unobtrusive wearable applications where a communication link is more prone to rapid degradation due to polarization mismatch between the transmitter and receiver. For low power applications, the demonstrated antenna will be highly efficient. In the future, further investigations will be accomplished to improve the conductivity of the transparent conductive fabric, for instance, by applying multiple metallic coating [54] or double layering of the conductive fabric [7]. These techniques will potentially reduce the performance gap between the transparent fabric-based and non-transparent fabric-based antennas.

Table 4. Comparison of the proposed antenna with some state-of-the-art CP wearable antennas.

Ref.	Freq. (GHz)	Footprint	Profile	Gain (dBi)	Effi. (%)	3-dB AR BW (%)	Trans.	Flex.
[6]	2.45	$0.66\lambda_0{}^2$	$0.03\lambda_0$	6.03	62	2.18	No	Yes
[27]	6.15	$0.58\lambda_0{}^2$	$0.02\lambda_0$	5.7	80	30.8	No	Yes
[28]	5.8	$0.23\lambda_0{}^2$	$0.2\lambda_0$	2.1	72.6	2.93	No	No
[29]	2.45 5.8	$0.123\lambda_0{}^2$	$0.22\lambda_0$	2.2 8.6	NA	31.4	No	No
[30]	5.16	$0.05\lambda_0{}^2$	$0.25\lambda_0$	6.2	90	18.3	No	No
[31]	5.8	$0.23\lambda_0{}^2$	$0.05\lambda_0$	6	80	1.4	No	Yes
[32]	5.5	$0.1\lambda_0{}^2$	$0.22\lambda_0$	3.5	79.9	6.55	No	Yes
[33]	2.4	$0.2\lambda_0{}^2$	$0.045\lambda_0$	3.5	58	2.4	No	No
[34]	4	$0.28\lambda_0{}^2$	$0.07\lambda_0$	5.2	80	12.5	No	No
[35]	2.4	$0.17\lambda_0{}^2$	$0.045\lambda_0$	5.2	79	2.72	No	Yes
[36]	2.45	$0.24\lambda_0{}^2$	$0.025\lambda_0$	1.8	30.7	2.86	No	Yes
[37]	2.4	$0.95\lambda_0{}^2$	$0.03\lambda_0$	4.4	34.7	2.96	No	Yes
[38]	1.575 1.621	$0.27\lambda_0{}^2$	NA	6.2	70	1.23	No	Yes
This work	2.4	$0.19\lambda_0{}^2$	$0.028\lambda_0$	2.5	42.26	4.16	Yes	Yes

7. Conclusions

The explored flexible and transparent CP antenna shows excellent performance. The antenna was rigorously tested in a flat state and on a forearm phantom; the results explicitly indicate the robustness of the antenna in bending applications. The performance of the antenna has also been compared with a similar non-transparent flexible antenna and shows comparable RF performance, demonstrating the effectiveness of the proposed transparent materials in flexible-transparent antenna manufacturing. Moreover, the SAR investigation reveals its compatibility in wearable applications. In wearable applications or other similar applications where antennas are mounted on moving objects, circular polarization maintains optimum communication performance by minimizing polarization mismatch losses. Thus, it can be explicitly expressed that the demonstrated antenna can attract significant interests in unobtrusive wearable applications.

Author Contributions: Conceptualization, A.S.M.S.; Data curation, A.S.M.S. and R.B.V.B.S.; Formal analysis, A.S.M.S., R.B.V.B.S. and A.L.; Investigation, A.S.M.S., R.B.V.B.S. and K.P.E.; Methodology, A.S.M.S.; Project administration, K.P.E.; Resources, R.B.V.B.S., D.R.G., B.O. and J.L.B.; Software, A.S.M.S.; Supervision, K.P.E., B.O. and J.L.B.; Validation, A.S.M.S., R.B.V.B.S., K.P.E., A.L., D.R.G., B.O. and J.L.B.; Visualization, A.S.M.S. and A.L.; Writing—original draft preparation, A.S.M.S.; Writing—review and editing, A.S.M.S., R.B.V.B.S., K.P.E., A.L., D.R.G., B.O. and J.L.B. All authors have read and agreed to the published version of the manuscript.

Funding: This research received no external funding.

Acknowledgments: This work was supported in part by the Enterprise Ireland funded HOLISTICS DTIF project (EIDT20180291-A), as well as by Science Foundation Ireland (SFI) under the following Grant Numbers: Connect Centre for Future Networks and Communications (13/RC/2077) and the Insight Centre for Data Analytics (SFI/12/RC/2289), as well as the European Regional Development Fund.

Conflicts of Interest: The authors declare no conflict of interest.

References

1. Ding, X.; Clifton, D.; JI, N.; Lovell, N.H.; Bonato, P.; Chen, W.; Yu, X.; Xue, Z.; Xiang, T.; Long, X.; et al. Wearable Sensing and Telehealth Technology with Potential Applications in the Coronavirus Pandemic. *IEEE Rev. Biomed. Eng.* **2021**, *14*, 48–70. [CrossRef] [PubMed]
2. Movassaghi, S.; Abolhasan, M.; Lipman, J.; Smith, D.; Jamalipour, A. Wireless Body Area Networks: A Survey. *IEEE Commun. Surv. Tutorials* **2014**, *16*, 1658–1686. [CrossRef]
3. Guibert, M.; Massicart, A.; Chen, X.; He, H.; Torres, J.; Ukkonen, L.; Virkki, J. Washing reliability of painted, embroidered, and electro-textile wearable RFID tags. In Proceedings of the 2017 Progress in Electromagnetics Research Symposium-Fall (PIERS-FALL), Singapore, 19–22 November 2017; pp. 828–831. [CrossRef]
4. Toivonen, M.; Björninen, T.; Sydänheimo, L.; Ukkonen, L.; Rahmat-Samii, Y. Impact of Moisture and Washing on the Performance of Embroidered UHF RFID Tags. *IEEE Antennas Wirel. Propag. Lett.* **2013**, *12*, 1590–1593. [CrossRef]
5. Paul, D.L.; Giddens, H.; Paterson, M.G.; Hilton, G.S.; McGeehan, J.P. Impact of Body and Clothing on a Wearable Textile Dual Band Antenna at Digital Television and Wireless Communications Bands. *IEEE Trans. Antennas Propag.* **2013**, *61*, 2188–2194. [CrossRef]
6. Hertleer, C.; Rogier, H.; Vallozzi, L.; Van Langenhove, L. A Textile Antenna for Off-Body Communication Integrated Into Protective Clothing for Firefighters. *IEEE Trans. Antennas Propag.* **2009**, *57*, 919–925. [CrossRef]
7. Sayem, A.S.M.; Simorangkir, R.B.V.B.; Esselle, K.P.; Hashmi, R.M. Development of Robust Transparent Conformal Antennas Based on Conductive Mesh-Polymer Composite for Unobtrusive Wearable Applications. *IEEE Trans. Antennas Propag.* **2019**, *67*, 7216–7224. [CrossRef]
8. Lu, J.; Chang, B. Planar Compact Square-Ring Tag Antenna with Circular Polarization for UHF RFID Applications. *IEEE Trans. Antennas Propag.* **2017**, *65*, 432–441. [CrossRef]
9. Green, R.B.; Guzman, M.; Izyumskaya, N.; Ullah, B.; Hia, S.; Pitchford, J.; Timsina, R.; Avrutin, V.; Ozgur, U.; Morkoc, H.; et al. Optically Transparent Antennas and Filters: A Smart City Concept to Alleviate Infrastructure and Network Capacity Challenges. *IEEE Antennas Propag. Mag.* **2019**, *61*, 37–47. [CrossRef]
10. Sayem, A.S.M.; Simorangkir, R.B.V.B.; Esselle, K.P.; Hashmi, R.M.; Liu, H. A Method to Develop Flexible Robust Optically Transparent Unidirectional Antennas Utilizing Pure Water, PDMS, and Transparent Conductive Mesh. *IEEE Trans. Antennas Propag.* **2020**, *68*, 6943–6952. [CrossRef]

11. Colombel, F.; Castel, X.; Himdi, M.; Legeay, G.; Vigneron, S.; Cruz, E.M. Ultrathin metal layer, ITO film and ITO/Cu/ITO multilayer towards transparent antenna. *IET Sci. Measure. Tech.* **2009**, *3*, 229–234. [CrossRef]
12. Hong, S.; Kang, S.H.; Kim, Y.; Jung, C.W. Transparent and Flexible Antenna for Wearable Glasses Applications. *IEEE Trans. Antennas Propag.* **2016**, *64*, 2797–2804. [CrossRef]
13. Sheikh, S.; Shokooh-Saremi, M.; Bagheri-Mohagheghi, M. Transparent microstrip patch antenna based on fluorine-doped tin oxide deposited by spray pyrolysis technique. *IET Microwaves Antennas Propag.* **2015**, *9*, 1221–1229. [CrossRef]
14. Green, R.B.; Toporkov, M.; Ullah, M.; Avrutin, V.; Ozgur, U.; Morkoc, H.; Topsakal, E. An alternative material for transparent antennas for commercial and medical applications. *Microw. Opt. Technol. Lett.* **2017**, *59*, 773–777. [CrossRef]
15. Lee, S.Y.; Choo, M.; Jung, S.; Hong, W. Optically Transparent Nano-Patterned Antennas: A Review and Future Directions. *Appl. Sci.* **2018**, *8*, 901. [CrossRef]
16. Xi, B.; Liang, X.; Chen, Q.; Wang, K.; Geng, J.; Jin, R. Optical Transparent Antenna Array Integrated with Solar Cell. *IEEE Antennas Wirel. Propag. Lett.* **2020**, *19*, 457–461. [CrossRef]
17. Hautcoeur, J.; Talbi, L.; Hettak, K.; Nedil, M. 60 GHz optically transparent microstrip antenna made of meshed AuGL material. *IET Microwaves Antennas Propag.* **2014**, *8*, 1091–1096. [CrossRef]
18. Kim, W.K.; Lee, S.; Hee Lee, D.; Hee Park, I.; Seong Bae, J.; Woo Lee, T.; Kim, J.Y.; Hun Park, J.; Chan Cho, Y.; Ryong Cho, C.; et al. Cu Mesh for Flexible Transparent Conductive Electrodes. *Sci. Rep.* **2015**, *5*, 10715. [CrossRef]
19. Sayem, A.S.M.; Esselle, K.P.; Hashmi, R.M.; Liu, H. Experimental studies of the robustness of the conductive-mesh-polymer composite towards the development of conformal and transparent antennas. *Smart Mater. Struct.* **2020**, *29*, 085015. [CrossRef]
20. Zandvakili, M.; Honari, M.M.; Sameoto, D.; Mousavi, P. Microfluidic liquid metal based mechanically reconfigurable antenna using reversible gecko adhesive based bonding. In Proceedings of the 2016 IEEE MTT-S International Microwave Symposium (IMS), San Francisco, CA, USA, 22–27 May 2016; pp. 1–4. [CrossRef]
21. Cheng, S.; Rydberg, A.; Hjort, K.; Wu, Z. Liquid metal stretchable unbalanced loop antenna. *Appl. Phys. Lett.* **2009**, *94*, 144103. [CrossRef]
22. Jang, T.; Zhang, C.; Youn, H.; Zhou, J.; Guo, L.J. Semitransparent and Flexible Mechanically Reconfigurable Electrically Small Antennas Based on Tortuous Metallic Micromesh. *IEEE Trans. Antennas Propag.* **2017**, *65*, 150–158. [CrossRef]
23. Sayem, A.S.M.; Esselle, K.P.; Hashmi, R.M. Increasing the transparency of compact flexible antennas using defected ground structure for unobtrusive wearable technologies. *IET Microwaves Antennas Propag.* **2020**, *14*, 1869–1877. [CrossRef]
24. Sayem, A.S.M.; Simorangkir, R.B.V.B.; Esselle, K.P.; Thalakotuna, D.N.; Lalbakhsh, A. An Electronically-Tunable, Flexible, and Transparent Antenna with Unidirectional Radiation Pattern. *IEEE Access* **2021**, *9*, 147042–147053. [CrossRef]
25. Sayem, A.S.M.; Le, D.; Simorangkir, R.B.V.B.; Björninen, T.; Esselle, K.P.; Hashmi, R.M.; Zhadobov, M. Optically Transparent Flexible Robust Circularly Polarized Antenna for UHF RFID Tags. *IEEE Antennas Wirel. Propag. Lett.* **2020**, *19*, 2334–2338. [CrossRef]
26. Chen, H.; Tsai, C.; Sim, C.; Kuo, C. Circularly Polarized Loop Tag Antenna for Long Reading Range RFID Applications. *IEEE Antennas Wirel. Propag. Lett.* **2013**, *12*, 1460–1463. [CrossRef]
27. Kumar, S.; Nandan, D.; Srivastava, K.; Kumar, S.; Singh, H.; Marey, M.; Mostafa, H.; Kanaujia, B.K. Wideband Circularly Polarized Textile MIMO Antenna for Wearable Applications. *IEEE Access* **2021**, *9*, 108601–108613. [CrossRef]
28. Hu, X.; Yan, S.; Zhang, J.; Volski, V.; Vandenbosch, G.A.E. Omni-Directional Circularly Polarized Button Antenna for 5 GHz WBAN Applications. *IEEE Trans. Antennas Propag.* **2021**, *69*, 5054–5059. [CrossRef]
29. Yin, X.; Chen, S.J.; Fumeaux, C. Wearable Dual-Band Dual-Polarization Button Antenna for WBAN Applications. *IEEE Antennas Wirel. Propag. Lett.* **2020**, *19*, 2240–2244. [CrossRef]
30. Ullah, U.; Mabrouk, I.B.; Koziel, S. A Compact Circularly Polarized Antenna with Directional Pattern for Wearable Off-Body Communications. *IEEE Antennas Wirel. Propag. Lett.* **2019**, *18*, 2523–2527. [CrossRef]
31. Zu, H.R.; Wu, B.; Zhang, Y.H.; Zhao, Y.T.; Song, R.G.; He, D.P. Circularly Polarized Wearable Antenna with Low Profile and Low Specific Absorption Rate Using Highly Conductive Graphene Film. *IEEE Antennas Wirel. Propag. Lett.* **2020**, *19*, 2354–2358. [CrossRef]
32. Hu, X.; Yan, S.; Vandenbosch, G.A.E. Compact Circularly Polarized Wearable Button Antenna with Broadside Pattern for U-NII Worldwide Band Applications. *IEEE Trans. Antennas Propag.* **2019**, *67*, 1341–1345. [CrossRef]
33. Jiang, Z.H.; Gregory, M.D.; Werner, D.H. Design and Experimental Investigation of a Compact Circularly Polarized Integrated Filtering Antenna for Wearable Biotelemetric Devices. *IEEE Trans. Biomed. Circuits Syst.* **2016**, *10*, 328–338. [CrossRef]
34. Jiang, Z.H.; Werner, D.H. A Compact, Wideband Circularly Polarized Co-designed Filtering Antenna and Its Application for Wearable Devices with Low SAR. *IEEE Trans. Antennas Propag.* **2015**, *63*, 3808–3818. [CrossRef]
35. Jiang, Z.H.; Cui, Z.; Yue, T.; Zhu, Y.; Werner, D.H. Compact, Highly Efficient, and Fully Flexible Circularly Polarized Antenna Enabled by Silver Nanowires for Wireless Body-Area Networks. *IEEE Trans. Biomed. Circuits Syst.* **2017**, *11*, 920–932. [CrossRef]
36. Li, J.; Jiang, Y.; Zhao, X. Circularly Polarized Wearable Antenna Based on NinjaFlex-Embedded Conductive Fabric. *Int. J. Antennas Propag.* **2019**, *2019*, 3059480. [CrossRef]
37. Locher, I.; Klemm, M.; Kirstein, T.; Troster, G. Design and Characterization of Purely Textile Patch Antennas. *IEEE Trans. Adv. Packag.* **2006**, *29*, 777–788. [CrossRef]
38. Kaivanto, E.K.; Berg, M.; Salonen, E.; de Maagt, P. Wearable Circularly Polarized Antenna for Personal Satellite Communication and Navigation. *IEEE Trans. Antennas Propag.* **2011**, *59*, 4490–4496. [CrossRef]

39. Sun, S.; Pan, Z.; Yang, F.K.; Huang, Y.; Zhao, B. A transparent silica colloidal crystal/PDMS composite and its application for crack suppression of metallic coatings. *J. Colloid Interface Sci.* **2016**, *461*, 136–143. [CrossRef]
40. Simorangkir, R.B.V.B.; Yang, Y.; Hashmi, R.M.; Björninen, T.; Esselle, K.P.; Ukkonen, L. Polydimethylsiloxane-Embedded Conductive Fabric: Characterization and Application for Realization of Robust Passive and Active Flexible Wearable Antennas. *IEEE Access* **2018**, *6*, 48102–48112. [CrossRef]
41. Hautcoeur, J.; Colombel, F.; Castel, X.; Himdi, M.; Cruz, E.M. Radiofrequency performances of transparent ultra-wideband antennas. *Prog. Electromag. Res. C* **2011**, *22*, 259–271. [CrossRef]
42. Yasin, T.; Baktur, R. Inkjet printed patch antennas on transparent substrates. In Proceedings of the 2010 IEEE Antennas and Propagation Society International Symposium, Toronto, ON, Canada, 11–17 July 2010; pp. 1–4.
43. Mohamadzade, B.; Simorangkir, R.B.V.B.; Hashmi, R.M.; Gharaei, R.; Lalbakhsh, A.; Shrestha, S.; Zhadobov, M.; Sauleau, R. A Conformal, Dynamic Pattern-Reconfigurable Antenna Using Conductive Textile-Polymer Composite. *IEEE Trans. Antennas Propag.* **2021**, *69*, 6175–6184. [CrossRef]
44. Pinapati, S.P.; Kaufmann, T.; Linke, I.; Ranasinghe, D.; Fumeaux, C. Connection strategies for wearable microwave transmission lines and antennas. In Proceedings of the 2015 International Symposium on Antennas and Propagation (ISAP), Hobart, Australia, 9–12 November 2015; pp. 1–4.
45. Kumar, S.; Buckley, J.L.; Barton, J.; Pigeon, M.; Newberry, R.; Rodencal, M.; Hajzeraj, A.; Hannon, T.; Rogers, K.; Casey, D.; et al. A Wristwatch-Based Wireless Sensor Platform for IoT Health Monitoring Applications. *Sensors* **2020**, *20*, 1675. [CrossRef]
46. Chen, S.J.; Fumeaux, C.; Ranasinghe, D.C.; Kaufmann, T. Paired Snap-On Buttons Connections for Balanced Antennas in Wearable Systems. *IEEE Antennas Wirel. Propag. Lett.* **2015**, *14*, 1498–1501. [CrossRef]
47. Seager, R.; Chauraya, A.; Zhang, S.; Whittow, W.; Vardaxoglou, Y. Flexible radio frequency connectors for textile electronics. *Electron. Lett.* **2013**, *49*, 1371–1373. [CrossRef]
48. Speag Phantom Arm, SHO-GFPC-V1. Available online: https://speag.swiss/products/em-phantoms/ctia-sub-10/sho-gfpc-v1/ (accessed on 12 December 2021).
49. Lee, H.; Tak, J.; Choi, J. Wearable Antenna Integrated into Military Berets for Indoor/Outdoor Positioning System. *IEEE Antennas Wirel. Propag. Lett.* **2017**, *16*, 1919–1922. [CrossRef]
50. Soh, P.J.; Vandenbosch, G.; Wee, F.H.; van den Bosch, A.; Martinez-Vazquez, M.; Schreurs, D. Specific Absorption Rate (SAR) Evaluation of Textile Antennas. *IEEE Antennas Propag. Mag.* **2015**, *57*, 229–240. [CrossRef]
51. *IEEE Std C95.1-2019 (Revision of IEEE Std C95.1-2005/ Incorporates IEEE Std C95.1-2019/Cor 1-2019)*; IEEE Standard for Safety Levels with Respect to Human Exposure to Electric, Magnetic, and Electromagnetic Fields, 0 Hz to 300 GHz. IEEE: Piscataway, NJ, USA, 2019; pp. 1–312. [CrossRef]
52. Alemaryeen, A.; Noghanian, S. On-Body Low-Profile Textile Antenna with Artificial Magnetic Conductor. *IEEE Trans. Antennas Propag.* **2019**, *67*, 3649–3656. [CrossRef]
53. Kaim, V.; Kanaujia, B.K.; Rambabu, K. Quadrilateral Spatial Diversity Circularly Polarized MIMO Cubic Implantable Antenna System for Biotelemetry. *IEEE Trans. Antennas Propag.* **2021**, *69*, 1260–1272. [CrossRef]
54. Bayram, Y.; Zhou, Y.; Shim, B.S.; Xu, S.; Zhu, J.; Kotov, N.A.; Volakis, J.L. E-textile conductors and polymer composites for conformal lightweight antennas. *IEEE Trans. Antennas Propag.* **2010**, *58*, 2732–2736. [CrossRef]

Communication

Implementation of a Miniaturized Planar Tri-Band Microstrip Patch Antenna for Wireless Sensors in Mobile Applications

Ahmed Saad Elkorany [1], Alyaa Nehru Mousa [1], Sarosh Ahmad [2,3,*], Demyana Adel Saleeb [4], Adnan Ghaffar [5], Mohammad Soruri [6], Mariana Dalarsson [7,*], Mohammad Alibakhshikenari [3,*] and Ernesto Limiti [8]

1 Department of Electronics and Electrical Communication Engineering, Faculty of Electronic Engineering, Menoufia University, Menouf 32952, Egypt; elkoranyahmed@el-eng.menofia.edu.eg (A.S.E.); alyaanehru27@gmail.com (A.N.M.)
2 Department of Electrical Engineering and Technology, Government College University Faisalabad (GCUF), Faisalabad 38000, Pakistan
3 Department of Signal Theory and Communications, Universidad Carlos III de Madrid, Leganés, 28911 Madrid, Spain
4 Faculty of Engineering, Kafrelsheikh University, Kafrelsheikh 33516, Egypt; demyanasaleeb@eng.kfs.edu.eg
5 Department of Electrical and Electronic Engineering, Auckland University of Technology, Auckland 1010, New Zealand; aghaffar@aut.ac.nz
6 Technical Faculty of Ferdows, University of Birjand, Birjand 9717434765, Iran; mohamad.soruri@birjand.ac.ir
7 School of Electrical Engineering and Computer Science, KTH Royal Institute of Technology, SE 100-44 Stockholm, Sweden
8 Electronic Engineering Department, University of Rome "Tor Vergata", Vial Del Politecnico 1, 00133 Rome, Italy; limiti@ing.uniroma2.it
* Correspondence: saroshahmad@ieee.org (S.A.); mardal@kth.se (M.D.); mohammad.alibakhshikenari@uc3m.es (M.A.)

Citation: Elkorany, A.S.; Mousa, A.N.; Ahmad, S.; Saleeb, D.A.; Ghaffar, A.; Soruri, M.; Dalarsson, M.; Alibakhshikenari, M.; Limiti, E. Implementation of a Miniaturized Planar Tri-Band Microstrip Patch Antenna for Wireless Sensors in Mobile Applications. *Sensors* **2022**, *22*, 667. https://doi.org/10.3390/s22020667

Academic Editors: Pedro Pinho and Zvonimir Šipuš

Received: 9 December 2021
Accepted: 13 January 2022
Published: 16 January 2022

Abstract: Antennas in wireless sensor networks (WSNs) are characterized by the enhanced capacity of the network, longer range of transmission, better spatial reuse, and lower interference. In this paper, we propose a planar patch antenna for mobile communication applications operating at 1.8, 3.5, and 5.4 GHz. A planar microstrip patch antenna (MPA) consists of two F-shaped resonators that enable operations at 1.8 and 3.5 GHz while operation at 5.4 GHz is achieved when the patch is truncated from the middle. The proposed planar patch is printed on a low-cost FR-4 substrate that is 1.6 mm in thickness. The equivalent circuit model is also designed to validate the reflection coefficient of the proposed antenna with the S_{11} obtained from the circuit model. It contains three RLC (resistor–inductor–capacitor) circuits for generating three frequency bands for the proposed antenna. Thereby, we obtained a good agreement between simulation and measurement results. The proposed antenna has an elliptically shaped radiation pattern at 1.8 and 3.5 GHz, while the broadside directional pattern is obtained at the 5.4 GHz frequency band. At 1.8, 3.5, and 5.4 GHz, the simulated peak realized gains of 2.34, 5.2, and 1.42 dB are obtained and compared to the experimental peak realized gains of 2.22, 5.18, and 1.38 dB at same frequencies. The results indicate that the proposed planar patch antenna can be utilized for mobile applications such as digital communication systems (DCS), worldwide interoperability for microwave access (WiMAX), and wireless local area networks (WLAN).

Keywords: triband antenna; wireless sensors; planar patch antenna; mobile applications; DCS; WLAN; WiMAX

1. Introduction

The need for mobile communication systems has risen dramatically in the last decade, and it continues to rise. The important standards in mobile communication are GPS,

Wi-MAX, and WLAN. These wireless applications require efficient small size antennas. Portable antenna technology has grown, along with cellular and mobile technologies [1–4].

The antenna is a crucial component of a communication system, and it is the device that transfers electromagnetic wave into free space in transmitting modes and vice versa [5–7]. Different antennas are required to support multiband systems. Multiband antennas play an important role in mobile communications, because they can be used in various frequency bands such as DCS, Wi-Fi, WLAN bands (802.11 b/n/g), and WiMAX (IEEE 802.16) [8–10]. There are different types of antennas such as PIFA, dipole, monopole, etc., that can be used in these applications. In general, the microstrip patch antenna (MPA) is an essential part of the communication system since it possesses several distinct and appealing characteristics. Compact size, low-cost, simple structure, minimal weight, ease of manufacture, and a wide bandwidth are some of these characteristics. MPA is a common choice for systems with a variety of features and the capacity to support many frequency bands at the same time [11]. The size of the patch antenna is determined by the dielectric constant of the substrate [12,13]. MPAs are chosen as the best antenna design because they are easy to implement with integrated circuits [14–16]. Such an antenna should provide high gain, wide impedance bandwidth, suitable return loss, and improved efficiency [17,18]. The MPA's biggest drawback is its limited impedance bandwidth. There are numerous methods for resolving this issue [19,20]. The U-slots technique is a popular patch-etching technique for obtaining multiband operation, and it was primarily utilized for increasing bandwidth [21]. Several articles have discussed various design and analytic techniques for improving performance [22–24]. The E-shaped, H-shaped, and U-slotted patch MPAs are very common with interesting characteristics [25–28]. Many researchers have recently reported on the design of MPAs that include slots or many layers. Li et al. [25] reported double and triple resonant frequencies for single and double patch antennas with an air substrate. The performance of equal-sized rectangular patch antennas with and without two included L-shaped strips was compared by Khunead et al. [29]. Prasad et al. [30] reported a triband heart-shaped MPA for wireless communications that can cover the 2.4, 5.4, and 7.6 GHz frequency bands. Darimireddy et al. [21] reported that wide bandwidth and triple bands at 1.6 GHz, 1.9 GHz, and 3.8 GHz are achieved by using a combination of dual U-slot and multiple layers. Ghalibafan et al. [31] described a multiband microstrip patch antenna for WLAN, Wi-MAX, and X-band use. Chitra et al. [32] introduced a new design technique for a microstrip patch antenna with an E-slot and an a-slot at the radiator's edge for Worldwide Interoperability for Microwave Access (WiMAX) application by using the Rogers Duroid 5880 substrate.

Osama et al. [26] reported a Double U-slot rectangular patch antenna for multiband applications to obtain thee resonant frequencies. Asif et al. [27] designed a printed microstrip patch antenna with two rectangular U-shaped parasitic elements to operate three resonant frequencies at 6.2, 4.52, 6.9 GHz. Gupta et al. [33] reported a multiband frequency with dimensions $60 \times 55 \times 1.59$ mm^3. The developed antenna efficiently operates at 4.3 GHz, 5.0 GHz, 6.1 GHz, 7.4 GHz, 8.9 GHz, and 9.2 GHz. Roopa et al. [34] proposed a square fractal antenna, which can operate in multiband frequency in the range of 2 GHz to 8.2 GHz with dimension $70 \times 70 \times 1.58$ mm^3. Dabas et al. [35] presented a microstrip patch antenna for wireless application, which can operate at 2.313, 2.396, and 2.478 GHz with dimension $70 \times 70 \times 1.6$ mm^3. Mazen et al. [36] reported a microstrip antenna to operate at multiple frequencies with dimensions $94 \times 76 \times 3.18$ mm^3.

Wireless sensor networks (WSNs) allow innovative applications and involve non-conventional design models due to some limitations. Antennas in wireless sensor networks (WSNs) have several advantages, such as enhanced capacity of the network, longer range of transmission, better spatial reuse, and lower interference. The reliability requirements and energy concerns make that antenna technology more advantageous. The proposed antenna is tri-band and works at 1.8 GHz, 3.5 GHz, and 5.4 GHz frequency bands. At the frequency of 1.8 GHz, it is applicable for digital control systems (DCS); at 3.5 GHz, it is applicable for the worldwide interoperability for microwave access (WiMAX); and at 5.4 GHz, it is applicable for wireless local area network (WLAN). In this article, a planar MPA fed by a coaxial probe is proposed for the applications in mobile communications. In the suggested design, we used horizontal slots, a single patch, and a single layer to make fabrication simple and easy. The radiating planar patch is located at the center of the substrate backed by a partial ground plane. The planar patch consists of two F-shaped resonators with a truncated patch from its middle point to obtain multiband operation. The proposed antenna is implemented on an FR-4 substrate with the following characteristics: height = 1.6 mm, dielectric constant ε_r = 4.3, and loss tangent of 0.025. In this study, a new approach for the planar patch antenna is utilized in order to obtain triband characteristics covering DCS (1.4–2 GHz), WiMAX (3.4–3.8 GHz), and WLAN (5.2–5.6 GHz) frequency bands. The size of the antenna is calculated as 60 mm × 50 mm. The proposed antenna is simulated using CST Microwave Studio (CST MWS) software. The outline of the paper is as follows: The design principles and recommended antenna geometry are described in Section 2. Section 3 explains the proposed measurement findings, while Section 4 presents conclusions.

2. Antenna Design Methodology

2.1. Proposed Single Antenna Design

Figure 1 illustrates the geometric configuration of the proposed triband flexible single element antenna. The planar patch is simple with two F-shaped resonators relative to a resonator at the lower and middle frequency bands while the middle-truncated patch helps operations at 5.4 GHz. Ground plane dimensions are reported as Lg × Wg = 50 mm × 40.8 mm; the dielectric material used above the rectangular ground plane is FR-4 possessing a height of Hs = 1.6 mm and a relative permittivity of ε_r = 4.3. Generally, the overall dimensions of the designed antenna are 60 × 50 × 1.6 mm^3. The antenna is fed by using a 50-ohm coaxial probe. The feed point is 2 mm above from the center of the feed patch. Figure 2 shows the simulated S_{11} of the triband antenna operating at 1.8 GHz, 3.5 GHz, and 5.4 GHz. The dimensions of the antenna are listed in Table 1. CST Studio Suite was used to simulate and analyze the performance of the antenna under bending conditions. CST Studio Suite is a powerful multilayer 3D full-wave electromagnetic solver that uses method of moments (MoM) technique to accurately solve Maxwell's equations. The simulations include the thickness of the conductor.

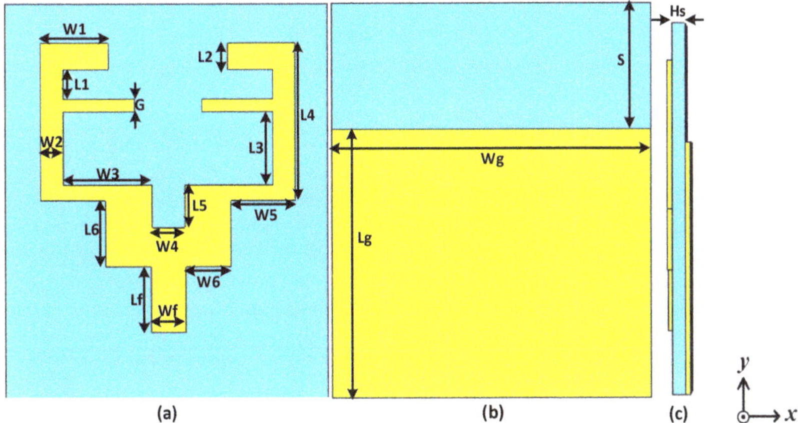

Figure 1. Structure of proposed triband antenna: (**a**) front view, (**b**) back view, and (**c**) side view.

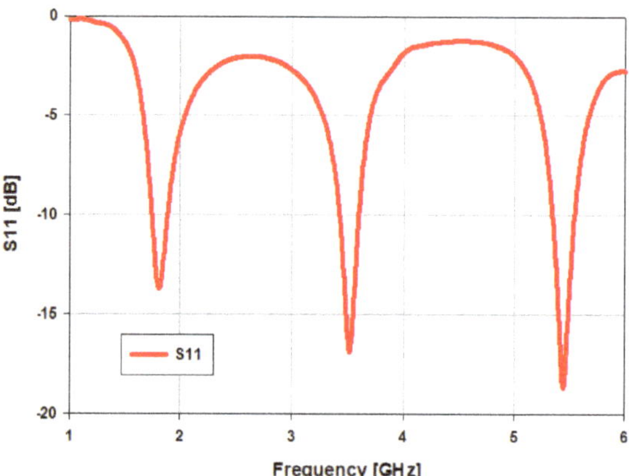

Figure 2. Reflection coefficient S_{11} of the proposed triband antenna.

Table 1. Parameters of the proposed triband antenna.

Parameters	Value (mm)	Parameters	Value (mm)
Lg	40.8	Wg	50
L1	4.45	W1	10.5
L2	4.0	W2	3.5
L3	11.05	W3	13.70
L4	21.50	W4	5.10
L5	6.50	W5	10.0
L6	10.0	W6	7.0
Lf	10.0	Wf	5.0
G	2.0	S	19.2

2.2. Design Methodology

As illustrated in Figure 3, the proposed planar patch antenna for mobile applications is designed in four steps. First, in order to create a resonance band, a rectangular planar patch is designed to possess a width of Wp = 39.5 mm and a length of Lp = 41.5 mm centered at the middle of the FR-4 substrate. After that, a feed extension is designed to possess a length of Lf = 10 mm and a width of Wf = 5 mm starting from the center of lower edge of the rectangular planar patch in ANT I, as shown in Figure 3a. In this case, the MPA operated at 3.3 and 5 GHz. Then, in the next step, both bottom sides of the patch are truncated, and the patch width is kept unchanged in the case of ANT II, as presented in Figure 3b. A partial ground plane is created in each step of the design procedure, posessing a length of lg = 40.8 mm and width of wg = 50 mm. In the third step, a U-shaped patch is introduced, which is obtained from ANT II. Then, a small size rectangular slot is created in the middle portion of the planar patch, as shown in Figure 3c. In the final step, two F-shaped resonators are utilized, and the proposed planar patches are introduced (ANT IV), and each resonator has a length of L4 = 21.5 mm, as depicted in Figure 3d. The comparisons of S11 obtained from ANT I, II, III, and IV are illustrated in Figure 4.

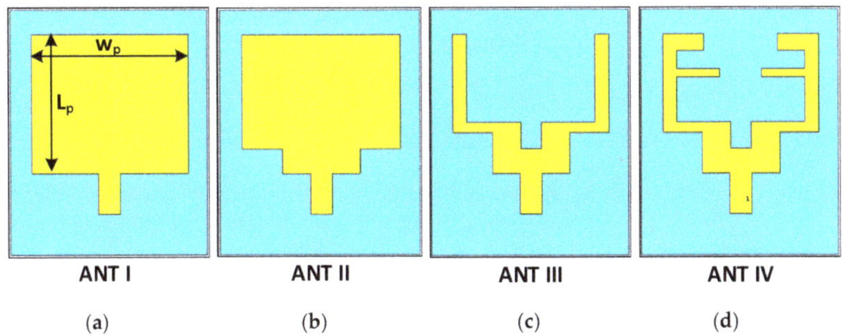

Figure 3. Design steps of the proposed triband antenna: (**a**) ANT I (patch only), (**b**) ANT II (Truncated patch), (**c**) ANT III (U-shaped patch), and (**d**) ANT IV (proposed antenna).

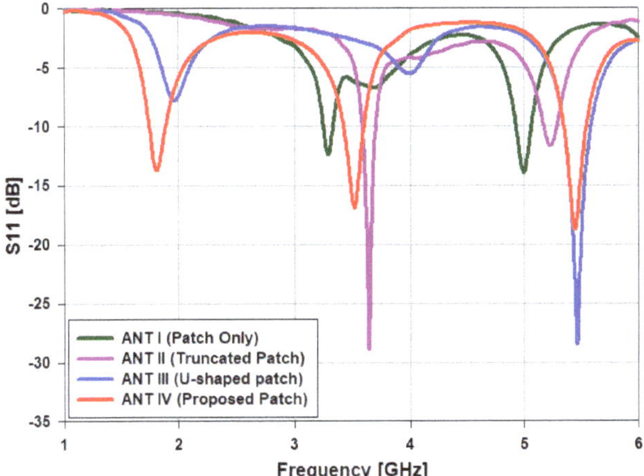

Figure 4. Simulated S_{11} for the corresponding design steps.

The antenna design consists of a 50-ohm uniform coaxial probe, a ground plane, and a radiating patch. Based on the transmission line model, the dimensions of a rectangular patch antenna can be calculated with Equations (1)–(4) [26]. The width of the patch (Wp) can be calculated by using the following expression:

$$Wp = \frac{\lambda_o}{2\sqrt{0.5(\varepsilon_r + 1)}} \tag{1}$$

where λ_o is the free space wavelength, and ε_r is the substrate's relative permittivity. The patch length Lp can be found by utilizing the following:

$$Lp = \frac{c_o}{2f_o\sqrt{\varepsilon_{eff}}} - 2\Delta L_p \tag{2}$$

where the light speed is c_o, the extra length due to fringing effect is ΔL_p, and ε_{eff} is the effective dielectric constant. The ε_{eff} can be found by using the following:

$$\varepsilon_{eff} = \frac{\varepsilon_r + 1}{2} + \frac{\varepsilon_r - 1}{2}\left(1/\sqrt{1 + 12\frac{Hs}{W_p}}\right) \tag{3}$$

where h is the thickness of the substrate. The change in length resulting from fringing fields can be calculated by using the following.

$$\Delta Lp = 0.421h\left(\frac{\varepsilon_{eff}+0.3}{\varepsilon_{eff}-0.3}\right)\left(\frac{W_p/Hs+0.264}{W_p/Hs+0.813}\right) \quad (4)$$

The initial dimensions of the patch design at 1.8 GHz, 3.5 GHz, and 5.4 GHz using the above expressions for implementation on an FR-4 substrate of εr = 4.3 and Hs = 1.6 mm are Lp = 156 41.5 mm and Wp = 39.5 mm.

The current distributions at different frequencies can be observed in Figure 5. In order to better comprehend current density paths, the antenna has been separated into several regions. In Figure 5, the current has very low flow and magnitude in many regions, which means that the patch radiator is not excited at this frequency as expected. However, when the antenna operates at the desired frequency, the current flows have the largest magnitude. The current density of the triband antenna is shown in Figure 5. At the lower band (1.8 GHz), the current mostly flows around the T-shaped patch of the antenna, while surface current circulates in F-shaped resonators at 3.5 GHz, and some amount of current flows through the partial ground plane at 5.4 GHz.

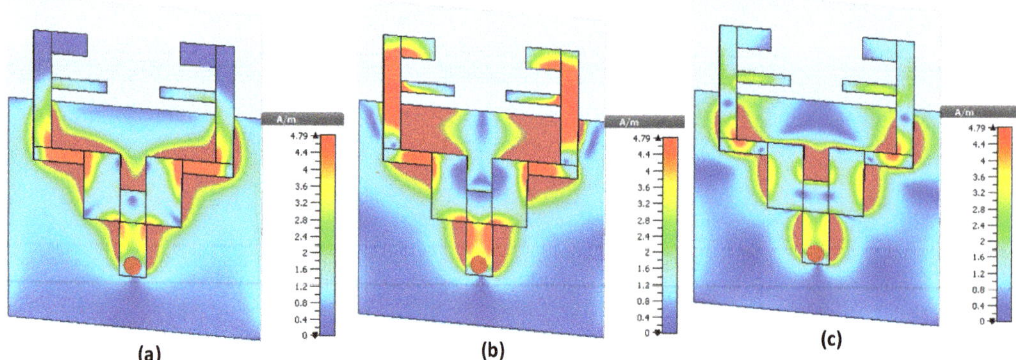

Figure 5. Surface current distribution (**a**) at 1.8 GHz, (**b**) at 3.5 GHz, and (**c**) at 5.4 GHz.

2.3. Equivalent Circuit Model

A circuit model for the triband MPA fed by a coaxial probe technique is designed using advanced design system (ADS) software. The main purpose of the equivalent circuit model is to validate the scattering parameters of the antenna as well as to prove that our proposed design is theoretically sound. The circuit model consists of five inductors, five capacitors, four resistors, and three resistor–inductor–capacitor (RLC) circuits connected in series with each other, as shown in Figure 6a. By varying the values of the resistors, the return loss of the circuit model can be varied, while the S_{11} of the MPA can be tuned by changing the values of the capacitors and inductors. The return loss of the circuit model is illustrated in Figure 6b. It covers bandwidths from 1.79 GHz to 2.01 GHz (220 MHz bandwidth) at 1.8 GHz; 3.36 GHz to 3.72 GHz (360 MHz) at 3.5 GHz; and 5.28 GHz to 5.52 GHz (240 MHz) at 5.4 GHz. The lumped element component values of the equivalent circuit model have been tabulated in Table 2.

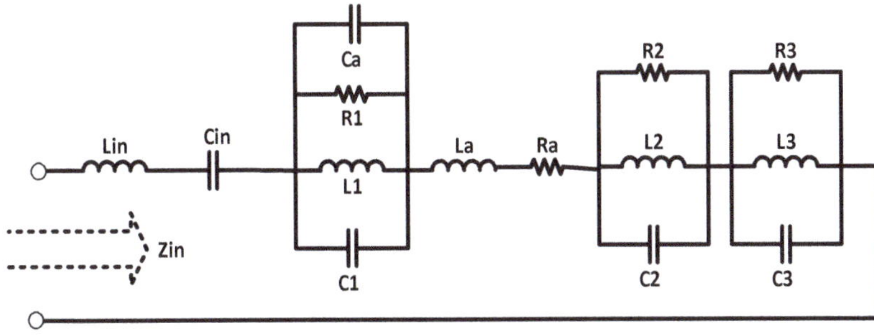

(a)

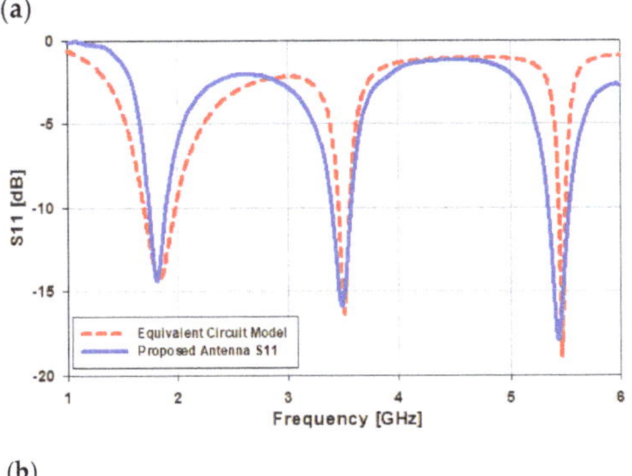

(b)

Figure 6. (**a**) Proposed equivalent circuit model for the triband antenna; (**b**) S_{11} obtained from the circuit model.

Table 2. Lumped element component values of the equivalent circuit model.

Capacitor	Value (pF)	Inductor	Value (nH)	Resistor	Value (Ohm)
Cin	2.3	Lin	1	R1	50
C1	3.0	L1	1.3	R2	47
C2	20	L2	100	R3	45
C3	35	L3	24	Ra	2
Ca	2	La	1	Zin	50

3. Results and Discussion

Fabrication and Measurements

The proposed triband MPA was fabricated by using a computer numerical control (CNC) machine that specialized in PCB board manufacturing utilizing FR-4 substrate material (t = 1.6 mm). Figure 7 shows a fabricated antenna with a 50-ohm feed probe. The S_{11} performance of the triband MPA shows resonance frequencies at 1.8, 3.5, and 5.4 GHz with very good agreement with simulated ones, as in Figure 8. In the case of the simulation results, the S11 of the designed antenna covers the bandwidth from 1.74 GHz to 1.88 GHz (140 MHz or 7.7%); 3.42 GHz to 3.6 GHz (180 MHz or 5.14%); and 5.34 GHz to 5.54 GHz (200 MHz or 3.7%) at 1.8 GHz, 3.5 GHz, and 5.4 GHz, respectively, while in the case of measurement results, the antenna covers bandwidths 1.73–1.86 GHz (130 MHz or 7.22%) at 1.8 GHz; 3.4–3.54 GHz (140 MHz or 4%) at 3.5 GHz; and 5.2–5.45 GHz (250 MHz or 4.6%).

Figure 7. Front and back view of the fabricated prototype of the antenna.

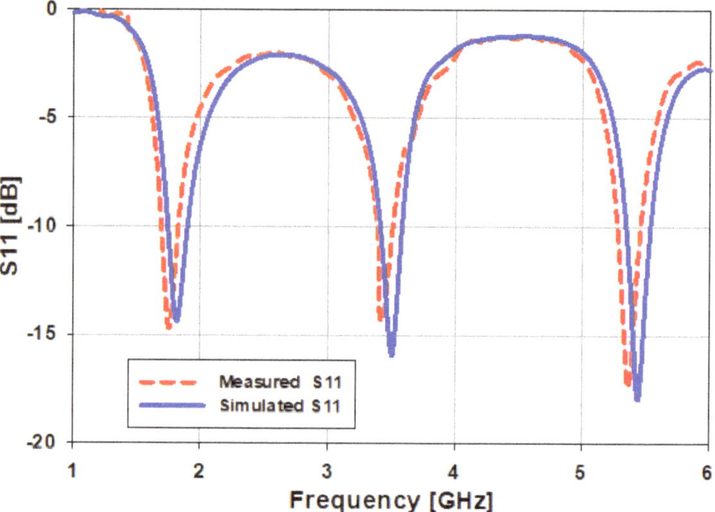

Figure 8. Simulated and measured return losses of the proposed antenna at three operating frequencies.

The power transmitted in the direction of peak radiation relative to that of an isotropic source is referred to as antenna gain. The two-dimensional radiation pattern of the triband MPA is shown in Figure 9. The antenna has an elliptical shaped radiation pattern at 1.8 and 3.5 GHz, while a broadside directional pattern is obtained at a frequency band of 5.4 GHz. This antenna has been measured in an anechoic chamber by the Agilent N5227A vector network analyzer. The anechoic chamber is a reflection-free room and prevents surrounding waves from affecting ongoing measurements. This combination means that a person or detector perceives only direct sounds (no reverberant sounds), imitating the experience of being within an indefinitely large room. Anechoic chambers are structures that simulate testing in free space and are used to simulate and measure results including gains, S-parameters, and normalized antenna patterns.

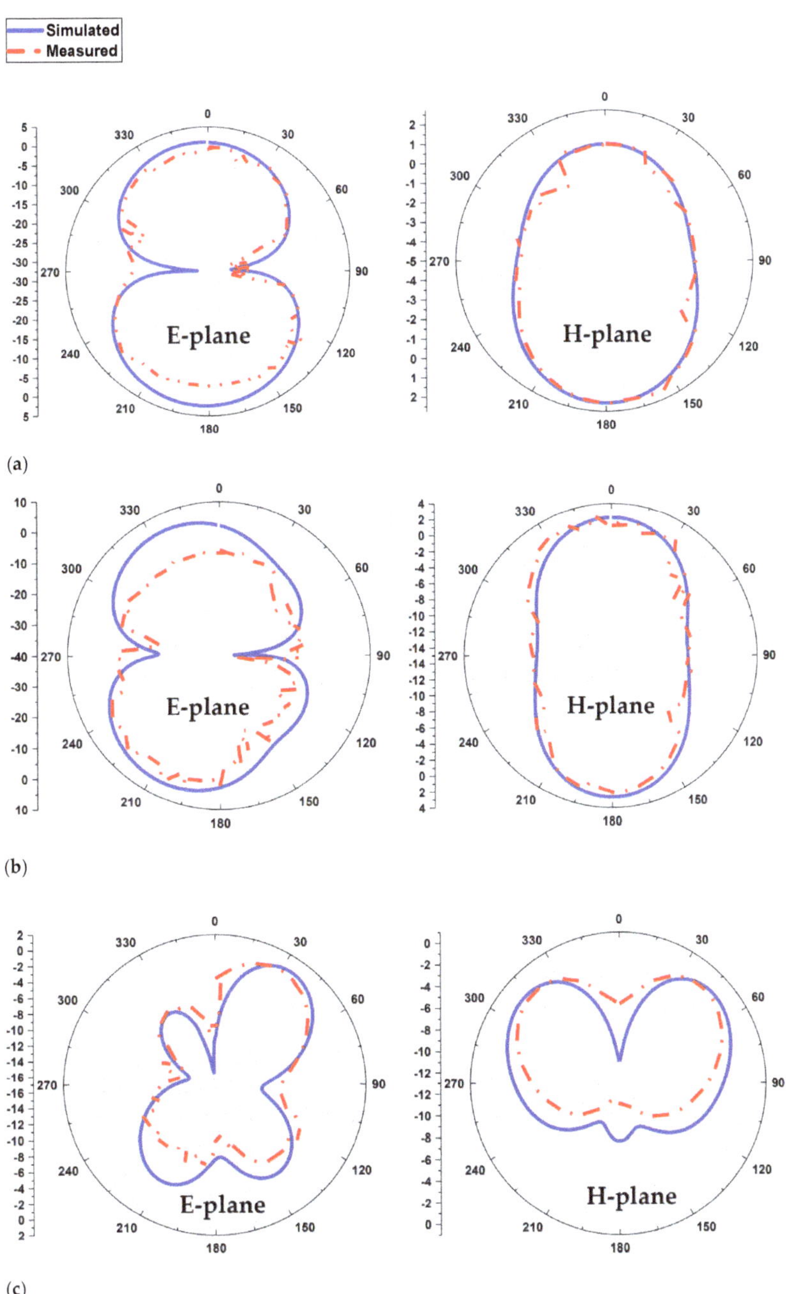

Figure 9. 2D radiation pattern (**a**) at 1.8 GHz, (**b**) at 3.5 GHz, and (**c**) at 5.4 GHz.

Radiated efficiency compares the power delivered to the antenna terminals to the power radiated through the antenna as an electro-magnetic wave. If an antenna could be designed to be a perfect electrical component, it could convert all of the power provided to its terminals into radiating electromagnetic energy that could travel into space. This is only theoretically conceivable; thus, so some of the power sent to the antenna terminals is always lost in practice. For example, power losses are caused by a mismatch between the antenna element and the feeding network. Furthermore, the antenna material itself loses energy and generates undesired heat by its very nature. All of these losses add up to a situation where the antenna's actual radiated efficiency is always less than 100% (equals 0 dB). By providing some power to the antenna feed pads and measuring the strength of the radiated electromagnetic field in the surrounding environment, antenna efficiency is measured in an anechoic chamber. In general, a good antenna transmits 50–60% of the energy provided to it (−3 to −2.2 dB).

The simulated peak gain reported as 2.34, 5.2, and 1.42 dB at 1.8, 3.5, and 5.4 GHz frequency bands, respectively, while the measured realized antenna gain is calculated as 2.22 dB at 1.8 GHz; 5.18 dB at 3.5 GHz; and 1.38 dB at 5.4 GHz. Realized gains and efficiency vs. frequency are presented in Figure 10. In an anechoic chamber, antenna efficiency is determined by applying power to the antenna feed pads and measuring the intensity of the emitted electromagnetic field in the surrounding region. An efficiency of 73% at 1.8 GHz; 68% at 3.5 GHz; and 59% at 5.4 GHz is obtained experimentally.

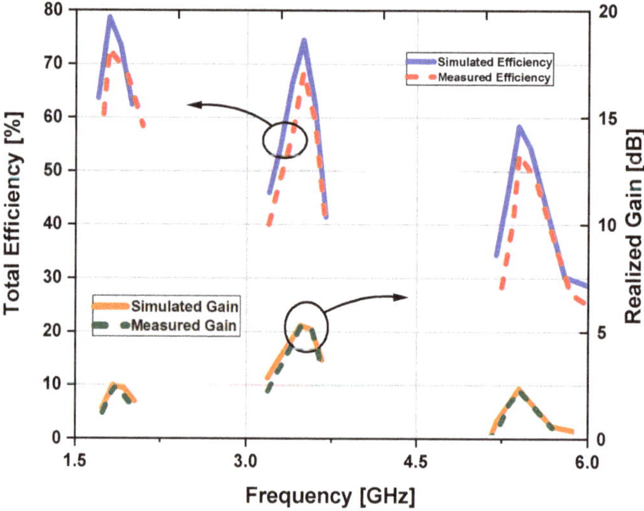

Figure 10. Simulated and measured gain and efficiency graph.

4. Comparison with State-of-the-Art Antennas

The proposed antenna is compared with other reported multiband antennas in Table 3. The proposed antenna has a relatively small fractional bandwidth compared to the other antennas cited in Table 3. However, it offers higher gain and radiation efficiency performance. The main advantage of the proposed design is that it has higher gain, large bandwidth, compact size, simple design, and ease of implementation in practice. In addition to that, the low-cost material called FR-4 is used as a substrate, which means that the proposed antenna will be less costly but result in non-flexibility, as FR-4 is a duroid substrate with non-flexible materials.

Table 3. Comparison of the proposed antenna to other multiband antennas.

Ref. No.	Size (mm^3)	Operating Frequency (GHz)	Bandwidth (MHz)	Peak Gain (dB)	Substrate Material	Proposed Technique
[27]	40 × 40 × 1.52	2.6, 6, 8.5	50, 22.8, 30	6.2, 4.52, 6.9	FR-4	Microstrip Patch
[31]	80 × 78.93 × 1.7	1.429, 1.839	NA	2.9, 4.3	FR-4	U-shaped patch
[33]	60 × 55 × 1.59	4.3, 5.0, 6.1, 7.4, 8.9, 9.2	68.6, 126.7, 132, 124.3, 191.2, 530.6	1.08, 3.23, 3.36, 2.77, 3.07, 4.87	FR-4	Square shaped microstrip patch
[34]	70 × 70 × 1.58	1.75, 3.65, 5.55, 6.6	170, 60, 140, 120	7.2, 11.2, 11.3, 7	FR-4	Sierpeinsiki-shaped patch
[35]	70 × 60 × 1.6	1, 1.2, 0.7	50, 60, 60	2.313, 2.396, 2.478	FR-4	Microstrip patch
[36]	94 × 78 × 3.18	2.53, 3.86, 6.45, 6.93	50.70, 410, 1250	8.18, 7.97, 10.56, 22, 5	Rogors RT5880	Microstrip Patch
[37]	50 × 50 × 1.5	1.2, 2.4, 5.6	12.76, 52.979, 52.979	NA	FR-4	Patch with defected ground
[This work]	60 × 50 × 1.6	Sim: 1.8, 3.5, 5.4 Meas: 1.7, 3.39, 5.38	Sim: 140, 180, 200 Meas: 140.2, 180.1, 200.2	Sim: 2.34, 5.2, 1.42 Meas: 2.22, 5.18, 1.38	FR-4	F-shaped Planar patch

5. Conclusions

A triband planar patch antenna with two F-shaped patches is designed, fabricated, and measured to operate at three bands, i.e., 1.8 GHz, 3.5 GHz, and 5.4 GHz. The proposed MPA is printed on a low-cost FR-4 substrate with a standard thickness of 1.6 mm. An equivalent circuit model is also designed to validate the reflection coefficient of the proposed antenna with the S_{11} obtained from the circuit model. A prototype of the presented triband MPA is built, and its characteristics are measured. Excellent agreements between the simulated and measured return loss results have been obtained. The antenna has an elliptically shaped radiation pattern at 1.8 GHz and 3.5 GHz, while a broadside directional pattern is obtained at the 5.4 GHz frequency band. Simulated peak-realized gains of 2.34, 5.2, and 1.42 dB are obtained at 1.8, 3.5, and 5.4 GHz respectively, while experimental peak realized gains of 2.22, 5.18, and 1.38 dB are obtained at the same frequencies. An efficiency of 73% at 1.8 GHz; 68% at 3.5 GHz; and 59% at 5.4 GHz is obtained experimentally. The results indicate that the proposed planar patch antenna can be utilized for wireless sensors in mobile applications. The antenna resonant frequency is suitable for distributed control system (DCS) applications at 1800 MHz, WiMAX applications at 3.5 GHz, and Wireless LAN applications at 5.4 GHz. The proposed patch antenna is a perfect candidate for wireless sensors in the applications of mobile phones.

Author Contributions: Conceptualization, A.S.E., A.N.M., S.A., D.A.S., A.G., M.S., M.D., M.A. and E.L.; data curation, A.S.E., A.N.M., S.A., D.A.S., A.G., M.D. and M.A.; formal analysis, A.S.E., A.N.M., S.A., D.A.S., A.G., M.S., M.D. and M.A.; funding acquisition, M.D., M.A. and E.L.; investigation, A.S.E., A.N.M., S.A., D.A.S., A.G., M.S., M.D., M.A. and E.L.; methodology, A.S.E., A.N.M., S.A., D.A.S., A.G., M.S., M.D., M.A. and E.L.; project administration, A.S.E., A.N.M., A.G., M.A. and E.L.; resources, A.S.E., A.N.M., A.G., M.S., M.D., M.A. and E.L.; software, A.S.E., A.N.M., S.A., D.A.S. and A.G.; supervision, A.G., M.A. and E.L.; validation, A.S.E., A.N.M., S.A., D.A.S., A.G., M.S., M.D., M.A. and E.L.; visualization, A.S.E., A.N.M., S.A., D.A.S., A.G., M.S., M.D., M.A. and E.L.; writing—original draft, A.S.E., A.N.M., S.A. and D.A.S.; writing—review and editing, A.S.E., A.N.M., S.A., D.A.S., A.G., M.S., M.D., M.A. and E.L. All authors have read and agreed to the published version of the manuscript.

Funding: This project has received funding from Universidad Carlos III de Madrid and the European Union's Horizon 2020 research and innovation program under the Marie Sklodowska-Curie Grant 801538.

Institutional Review Board Statement: Not applicable.

Informed Consent Statement: Not applicable.

Data Availability Statement: All data are included within the manuscript.

Acknowledgments: The authors appreciate the financial support from Universidad Carlos III de Madridand and the European Union's Horizon 2020 research and innovation program under Marie Sklodowska-Curie Grant 801538.

Conflicts of Interest: The authors declare no conflict of interest.

References

1. Balanis, C.A. *Antenna Theory: Analysis and Design*; John Wiley & Sons: New York, NY, USA, 2005.
2. Nella, A.; Gandhi, A.S. A survey on microstrip antennas for portable wireless communication system applications. In Proceedings of the 2017 International Conference on Advances in Computing, Communications and Informatics (ICACCI), Udupi, India, 13–16 September 2017; pp. 2156–2165. [CrossRef]
3. Kadir, E.A.; Shamsuddin, S.M.; Rahman, E.S.T.A.; Rahim, S.K.A.; Rosa, S.L. Multi Bands Antenna for Wireless Communication and Mobile System. *Int. J. Circuits Syst. Signal Process.* **2014**, *8*, 563–568.
4. Laheurte, J.-M. *Compact Antennas for Wireless Communications and Terminals: Theory and Design*; Wiley-ISTE: Hoboken, NJ, USA, 2012; 272p. [CrossRef]
5. Dakulagi, V.; Bakhar, M. Advances in Smart Antenna Systems for Wireless Communication. *Wireless Pers. Commun.* **2020**, *110*, 931–957. [CrossRef]
6. Sharma, S.K.; Chieh, J.S. *Multifunctional Antennas and Arrays for Wireless Communication Systems*; Wiley-IEEE Press: Hoboken, NJ, USA, 2021; 464p.
7. Goudarzi, A.; Honari, M.M.; Mirzavand, R. Resonant Cavity Antennas for 5G Communication Systems: A Review. *Electronics* **2020**, *9*, 1080. [CrossRef]
8. Ullah, S.; Ahmad, S.; Khan, B.; Flint, J. A multi-band switchable antenna for Wi-Fi, 3G Advanced, WiMAX, and WLAN wireless applications. *Int. J. Microw. Wirel. Technol.* **2018**, *10*, 991–997. [CrossRef]
9. Chetal, S.; Nayak, A.K.; Panigrahi, R.K. Multiband antenna for WLAN, WiMAX and future wireless applications. In Proceedings of the 2019 URSI Asia-Pacific Radio Science Conference (AP-RASC), New Delhi, India, 9–15 March 2019; pp. 1–4. [CrossRef]
10. Afzal, W.; Rafique, U.; Ahmed, M.M.; Khan, M.A.; Mughal, F.A. A tri-band H-shaped microstrip patch antenna for DCS and WLAN applications. In Proceedings of the 2012 19th International Conference on Microwaves, Radar & Wireless Communications, Warsaw, Poland, 21–23 May 2012; pp. 256–258. [CrossRef]
11. Lee, K.-F.; Tong, K.-F. Microstrip Patch Antennas—Basic Characteristics and Some Recent Advances. *Proc. IEEE* **2012**, *100*, 2169–2180. [CrossRef]
12. Waterhouse, R. *Microstrip Patch Antennas: A Designer's Guide*; Springer: Boston, MA, USA, 2003; 421p. [CrossRef]
13. Malik, P.K.; Padmanaban, S.; Holm-Nielsen, J.B. *Microstrip Antenna Design for Wireless Applications*; CRC Press: Boca Raton, FL, USA, 2021; 352p, ISBN 9780367554385.
14. Liu, Y.; Si, L.; Wei, M.; Yan, P.; Yang, P.; Lu, H.; Zheng, C.; Yuan, Y.; Mou, J.; Lv, X.; et al. Some Recent Developments of Microstrip Antenna. *Int. J. Antennas Propag.* **2012**, *2012*, 428284. [CrossRef]
15. Cui, Y.; Wang, X.; Shen, G.; Li, R. A Triband SIW Cavity-Backed Differentially Fed Dual-Polarized Slot Antenna for WiFi/5G Applications. *IEEE Trans. Antennas Propag.* **2020**, *68*, 8209–8214. [CrossRef]
16. Wahab, W.M.A.; Safavi-Naeini, S.; Busuioc, D. Low cost microstrip patch antenna array using planar waveguide technology for emerging millimeter-wave wireless communication. In Proceedings of the 2010 14th International Symposium on Antenna Technology and Applied Electromagnetics & the American Electromagnetics Conference, Ottawa, ON, Canada, 5–8 July 2010; pp. 1–4. [CrossRef]
17. Davoudabadifarahani, H.; Ghalamkari, B. High efficiency miniaturized microstrip patch antenna for wideband terahertz communications applications. *Optik* **2019**, *194*, 163118. [CrossRef]
18. Belen, M.A. Performance enhancement of a microstrip patch antenna using dual-layer frequency-selective surface for ISM band applications. *Microw. Opt. Technol. Lett.* **2018**, *60*, 2730–2734. [CrossRef]
19. Chen, D.; Che, W.; Yang, W. High-efficiency microstrip patch antennas using non-periodic artificial magnetic conductor structure. In Proceedings of the 2015 Asia-Pacific Microwave Conference (APMC), Nanjing, China, 6–9 December 2015; pp. 1–3. [CrossRef]
20. Alibakhshikenari, M.; Virdee, B.S.; Azpilicueta, L.; Naser-Moghadasi, M.; Akinsolu, M.O.; See, C.H.; Liu, B.; Abd-Alhameed, R.A.; Falcone, F.; Huynen, I.; et al. A Comprehensive Survey of "Metamaterial Transmission-Line Based Antennas: Design, Challenges, and Applications". *IEEE Access* **2020**, *8*, 144778–144808. [CrossRef]
21. Darimireddy, N.; Mallikarjuna, A. Design of triple-layer double U-slot patch antenna for wireless applications. *J. Appl. Res. Technol.* **2015**, *13*, 526–534. [CrossRef]
22. Tan, Q.; Chen, F.-C. Triband Circularly Polarized Antenna Using a Single Patch. *IEEE Antennas Wirel. Propag. Lett.* **2020**, *19*, 2013–2017. [CrossRef]

23. Alibakhshikenari, M.; Limiti, E.; Naser-Moghadasi, M.; Virdee, B.S.; Sadeghzadeh, R.A. A New Wideband Planar Antenna with Band-Notch Functionality at GPS, Bluetooth and WiFi Bands for Integration in Portable Wireless Systems. *AEU—Int. J. Electron. Commun.* **2017**, *72*, 79–85. [CrossRef]
24. Naser-Moghadasi, M.; Alibakhshi-Kenari, M.; Sadeghzadeh, R.A.; Virdee, B.S.; Limiti, E. New CRLH-Based Planar Slotted Antennas with Helical Inductors for Wireless Communication Systems, RF-Circuits and Microwave Devices at UHF-SHF Bands. *Wirel. Pers. Commun.* **2017**, *92*, 1029–1038.
25. Li, E.; Li, X.J.; Zhao, Q. A Design of Ink-Printable Triband Slot Microstrip Patch Antenna for 5G Applications. In Proceedings of the 4th Australian Microwave Symposium (AMS), Sydney, Australia, 13–14 February 2020; pp. 1–2. [CrossRef]
26. Osama, W.; Khaleel, A. Double U-slot rectangular patch antenna for multiband applications. *Comput. Electr. Eng.* **2020**, *84*, 106608.
27. Asif, S.; Rafiq, M. A compact multiband microstrip patch antenna with U-shaped parasitic elements. In Proceedings of the IEEE International Symposium on Antennas and Propagation & USNC/URSI National Radio Science Meeting, Vancouver, BC, Canada, 19–24 July 2015; pp. 617–618.
28. Alibakhshi-Kenari, M.; Naser-Moghadasi, M.; Sadeghzadah, R. The Resonating MTM Based Miniaturized Antennas for Wide-band RF-Microwave Systems. *Microw. Opt. Technol. Lett.* **2015**, *57*, 2339–2344. [CrossRef]
29. Khunead, G.; Nakasuwan, J.; Songthanapitak, N.; Anantrasirichai, N. Investigate Rectangular Slot Antenna with L-shape Strip. *Piers Online* **2007**, *3*, 1076–1079. [CrossRef]
30. Prasad, M.; Khasim, S. A Triband Heart Shaped Microstrip Patch antenna. *Int. J. Recent Innov. Trends Comput. Commun.* **2015**, *3*, 1070–1073. [CrossRef]
31. Ghalibafan, J.; Farrokh, H. A new dual-band microstrip antenna with U-shaped slot. *Prog. Electromagn. Res. C* **2010**, *12*, 215–223. [CrossRef]
32. Chitra, R.; Nagarajan, V. Design of E slot rectangular microstrip slot antenna for WiMAX application. In Proceedings of the IEEE International Conference on Communication and Signal Processing, Melmaruvathur, India, 3–5 April 2013; pp. 1048–1052.
33. Gupta, M.; Vinita, M. Koch boundary on the square patch microstrip antenna for ultra-wideband applications. *Alex. Eng. J.* **2018**, *57*, 2113–2122. [CrossRef]
34. Roopa, R.; Kumarswamy, Y. Enhancement of performance parameters of sierpeinsiki antenna using computational technique. In Proceedings of the IEEE International Conference on Wireless Communications, Signal Processing and Networking, Chennai, India, 23–25 March 2016; pp. 7–10.
35. Dabas, T.; Kanaujia, B. Design of multiband multipolarised single feed patch antenna. *IET Microw. Antennas Propag.* **2018**, *12*, 2372–2378. [CrossRef]
36. Mazen, K.; Emran, A. Design of Multi-band Microstrip Patch Antennas for Mid-band 5G Wireless Communication. *Int. J. Adv. Comput. Sci. Appl.* **2021**, *12*, 458–469. [CrossRef]
37. Mabaso, M.; Pradeep, K. A Microstrip Patch Antenna with Defected Ground Structure for Triple Band Wireless Communications. *J. Commun.* **2019**, *14*, 684–688. [CrossRef]

Article

Low Discrepancy Sparse Phased Array Antennas

Travis Torres [1], Nicola Anselmi [2], Payam Nayeri [1,*], Paolo Rocca [2] and Randy Haupt [3]

1 Electrical Engineering Department, Colorado School of Mines, Golden, CO 80401, USA; travistorres@mines.edu
2 Consorzio Nazionale Interuniversitario per le Telecomunicazioni (CNIT), "University of Trento" ELEDIA Research Unit, via Sommarive 9, 38123 Trento, Italy; nicola.anselmi.1@unitn.it (N.A.); paolo.rocca@unitn.it (P.R.)
3 Haupt Associates, Boulder, CO 80303, USA; randyhaupt@gmail.com
* Correspondence: pnayeri@mines.edu

Abstract: Sparse arrays have grating lobes in the far field pattern due to the large spacing of elements residing in a rectangular or triangular grid. Random element spacing removes the grating lobes but produces large variations in element density across the aperture. In fact, some areas are so dense that the elements overlap. This paper introduces a low discrepancy sequence (LDS) for generating the element locations in sparse planar arrays without grating lobes. This nonrandom alternative finds an element layout that reduces the grating lobes while keeping the elements far enough apart for practical construction. Our studies consider uniform sparse LDS arrays with 86% less elements than a fully populated array, and numerical results are presented that show these sampling techniques are capable of completely removing the grating lobes of sparse arrays. We present the mathematical formulation for implementing an LDS generated element lattice for sparse planar arrays, and present numerical results on their performance. Multiple array configurations are studied, and we show that these LDS techniques are not impacted by the type/shape of the planar array. Moreover, in comparison between the LDS techniques, we show that the Poisson disk sampling technique outperforms all other approaches and is the recommended LDS technique for sparse arrays.

Keywords: phased array; sensor array; sparse array; nonuniform array; planar array; random array; low discrepancy sequence

Citation: Torres, T.; Anselmi, N.; Nayeri, P.; Rocca, P.; Haupt, R. Low Discrepancy Sparse Phased Array Antennas. *Sensors* **2021**, *21*, 7816. https://doi.org/10.3390/s21237816

Academic Editors: Pedro Pinho and Andrea Randazzo

Received: 28 October 2021
Accepted: 18 November 2021
Published: 24 November 2021

1. Introduction

Sensing applications, such as radio telescopes, satellite communications, sonars, and defense radars, require large antenna arrays. Physically large arrays offer high resolution as well as high directivity (as long as the element spacing, or sampling remains small). Since the array cost is proportional to the number of elements in the array, designers try to minimize the number of elements in the aperture. However, if the array has a uniform grid of elements that undersamples the aperture (large element spacing), then grating lobes (extra main beams) result due to aliasing and are predictable from theory [1]. Most arrays are designed with an element spacing of $\lambda/(1+\sin\theta_{\max})$ or less, where λ is the wavelength and $\theta_{\max}$ is the maximum scan angle from broadside.

Thinned and aperiodic arrays have fewer elements than dense periodic arrays [2–4]. For a dense array, thinning (removing elements from a regular grid [5]) and aperiodic spacing (spacing between elements are not constant [6]) mimic low sidelobe amplitude distributions through an amplitude density across the aperture. These arrays have far field patterns with low sidelobes near the main beam and increased sidelobe levels farther from the main beam.

Sparse arrays fill an antenna aperture with elements that are widely separated from each other in order to reduce the cost but maintain a narrow beamwidth. The definition of a sparse array in both the antenna and signal processing literature is vague. For instance, the IEEE standard defines a sparse antenna array as [7]: "An array antenna that contains substantially fewer driven radiating elements than a conventional uniformly spaced array with the same beamwidth having identical elements. Interelement spacings in the sparse array can be chosen such that no large grating lobes are formed and sidelobes are reduced." On the other hand, mathematics has a different definition: a sparse matrix has most elements equal to zero [8]. Sparsity of a matrix equals the number of zero valued elements divided by the total number of elements. The words "fewer" and "most" do not specify a sharp dividing line between dense and sparsity. It is important to note here that while these two definitions of sparsity are different, both have been used in designing sparse arrays. For the latter, compressive sensing (CS) approaches aim at solving a system of linear equations, forcing the solution to be maximally sparse, namely, to have the minimum number of nonzero coefficients, with respect to an expansion basis [9]. Accordingly, CS-based methods have been applied to the synthesis of sparse arrays by properly reformulating the design problem as a pattern matching one [10]. The problem unknowns are the set of complex (amplitude and phase) excitation coefficients of the "candidate" array elements, the positions of these latter obtained through a dense sampling of the array aperture. The CS solution is the sparse complex-valued vector of the excitations, and the positions of the array elements are obtained as a byproduct and correspond to the candidate locations having non-null coefficient [10]. CS has been applied to the design of both linear [11,12] and planar [13,14] sparse arrays, considering symmetric [11,13] as well as asymmetric [12,14] pattern shapes. The obtained results have shown achieving up to 40% elements reduction with respect to regular/uniform array arrangements. It is important to note that CS approaches do not change the definition of sparse arrays; but deal with minimizing the number of nonzero coefficients in the system of equations in designing the array.

The authors of this paper propose the following definition of sparsity, based on the IEEE definition [7], but we add that: a sparse array has an average element spacing greater than λ. If the array has a uniform square grid of isotropic elements, then grating lobes exist when the main beam points at broadside. Periodic sparse arrays have grating lobes with the same gain as the main beam. A random distribution of elements in the sparse array lowers the grating lobes to a level of the surrounding sidelobes. Low sidelobes are not an option for sparse arrays.

Figure 1 distinguishes between dense and sparse arrays. Given this definition, our literature review of sparse arrays emphasizes papers that present arrays with an average element spacing of at least one wavelength. Sparse arrays (by our definition) in the literature generally have random element spacing [15–17].

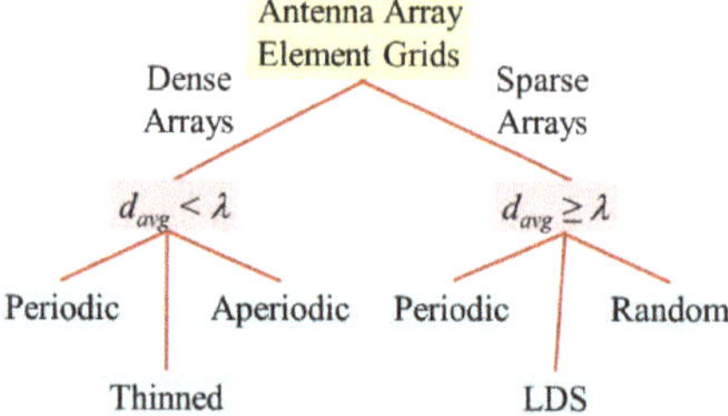

Figure 1. Categories of antenna array grids.

Sparse arrays are widely used in radio telescopes and MIMO systems, as well as other sensing systems. The Long Wavelength Array Station 1 (LWA1) is an aperture synthesis imaging array (20–80 MHz) for radio astronomy [18]. It contains 512 antenna elements inside an ellipse that has a 110 m major axis and a 100 m minor axis. To reduce grating lobes in this sparse array at 80 MHz, the elements are pseudorandomly distributed with a 5 m minimum spacing constraint that results in an average element spacing of about 5.4 m or 1.44λ at 80 MHz. The LOw Frequency ARray (LOFAR) Low Band Antenna (LBA) in the Square Kilometer Array (SKA) consists of 96 dual polarized crossed dipole active antennas operating from 30 to 80 MHz [19]. An inner array of 46 randomly spaced dual-polarized elements have average element spacings between 0.4λ and 0.8λ within a 30 m radius, while the outer array of 48 randomly spaced dual-polarized elements have average element spacings between 0.8λ and 1.7λ within an annulus having an inner diameter of 30 m and outer diameter of 85 m [20].

Some examples of very sparse radio telescopes include the Very Long Baseline Array (VLBA) that has 10 parabolic reflectors that are 25 m in diameter forming a total collecting area of 19,635 m^2 [21], The Atacama Large Millimeter/submillimeter Array (ALMA) that has 10 parabolic reflectors that are 12 m in diameter forming a total collecting area of 6600 m^2 [22], and the Very Large Array (VLA) that has 27 parabolic reflectors that are 25 m that extend in a "Y" shape with each arm over 15 km long [23].

Microwave imaging experiments were performed using an array aperture of 320 $\times$ 320 mm operating over 17–20 GHz [24]. The first model was a fully populated planar array of 64 $\times$ 64 elements on a square grid. The second model was a sparse array of randomly placed elements. An x-y positioner moved a single antenna to the designated element positions in the aperture to form the dense and sparse arrays. A compressive sensing algorithm (average sampling rate that is less than the Nyquist rate) outperformed other reconstruction algorithms for sparse arrays having 1024, 400, and 160 antennas which correspond to 25%, 10%, and 4% of the elements in the fully populated array.

Sparse antenna arrays with random element spacing significantly improve the sum rate capacity (maximum aggregation of all the users' data rates) of a MIMO base station antenna system [25]. The sparse array aperiodicity spreads the grating lobe (GL) energy over all the lower sidelobes [26].

The difference between the uniform element grid and the randomized element grid is mathematically known as the discrepancy. Using a low discrepancy sequence (LDS) to place elements in an array aperture ensures that elements do not overlap while keeping a uniform sampling of the aperture. An LDS produces a random-like equidistribution of elements using a deterministic generating formula. Equidistributed means that if the aperture is divided into equal subareas, then the number of elements in all subareas is the same. The aperture discrepancy approaches zero as the number of elements approaches infinity. Random element spacing has the highest discrepancy, because large areas of empty space as well as high densities of elements exist within the aperture. In contrast, elements on a regular grid have the lowest possible discrepancy. A low discrepancy element distribution appears random, but the elements also appear to be evenly distributed across the aperture.

Discrepancy theory has its origin in a paper by H. Weyl on the uniform distribution of sequences [27]. Different LDSs have been introduced from the early 1960s, including the Hammersley point set [28], the Sobol sequence [29], the Faure sequence [30], and the Niederreiter sequence [31]. LDSs found their first applications in the 1990s for numerical analysis and integration for numerical simulation, in the fields of computer graphics [25], computational physics [32], and finance engineering [33]. The application of LDS to the generation of sample points for Monte Carlo sampling (i.e., the quasi-Monte Carlo approach) is theoretically superior to a standard Monte Carlo technique.

To our knowledge, the first paper to apply an LDS (Hammersley sequence) to element spacing explored sparse aperiodic spacing on a spherical array [34]. The advantages of LDS for sparse phased array design demonstrated that the Hammersley sequence maintains the large separation between the elements, while reducing the grating lobes compared to element spacings derived from random, pseudo-random, and uniform plus jitter sequences [35].

In this paper we use low discrepancy sequences to distribute elements in a sparse planar array aperture while maintaining the following properties:

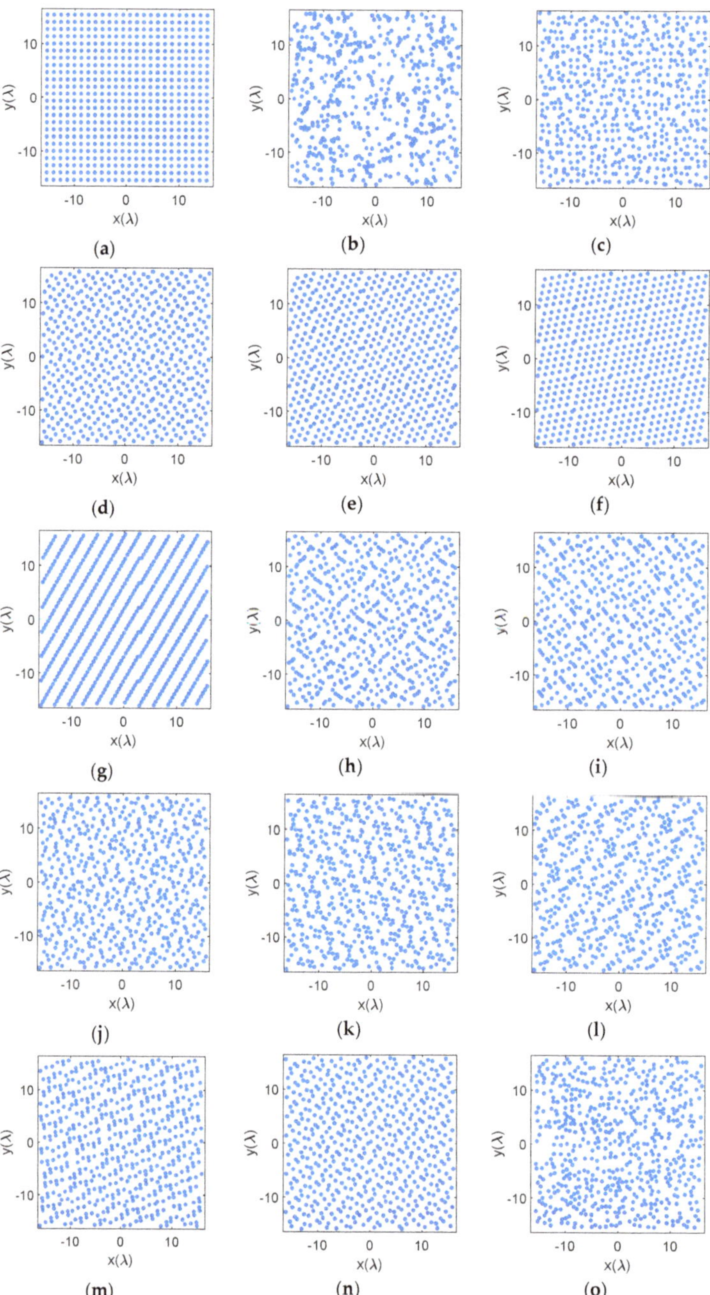

Figure 3. Position of elements on the aperture with different distributions: (**a**) uniform, (**b**) random, (**c**) random with jitter, (**d**) Hammersley (base 2), (**e**) Hammersley (base 3), (**f**) Hammersley (base 5), (**g**) Hammersley (base 7), (**h**) Halton (bases 2, 3), (**i**) Halton (bases 2, 5), (**j**) Halton (bases 2, 7), (**k**) Halton (bases 3, 5), (**l**) Halton (bases 3, 7), (**m**) Halton (bases 5, 7), (**n**) Sobol, (**o**) Poisson disk.

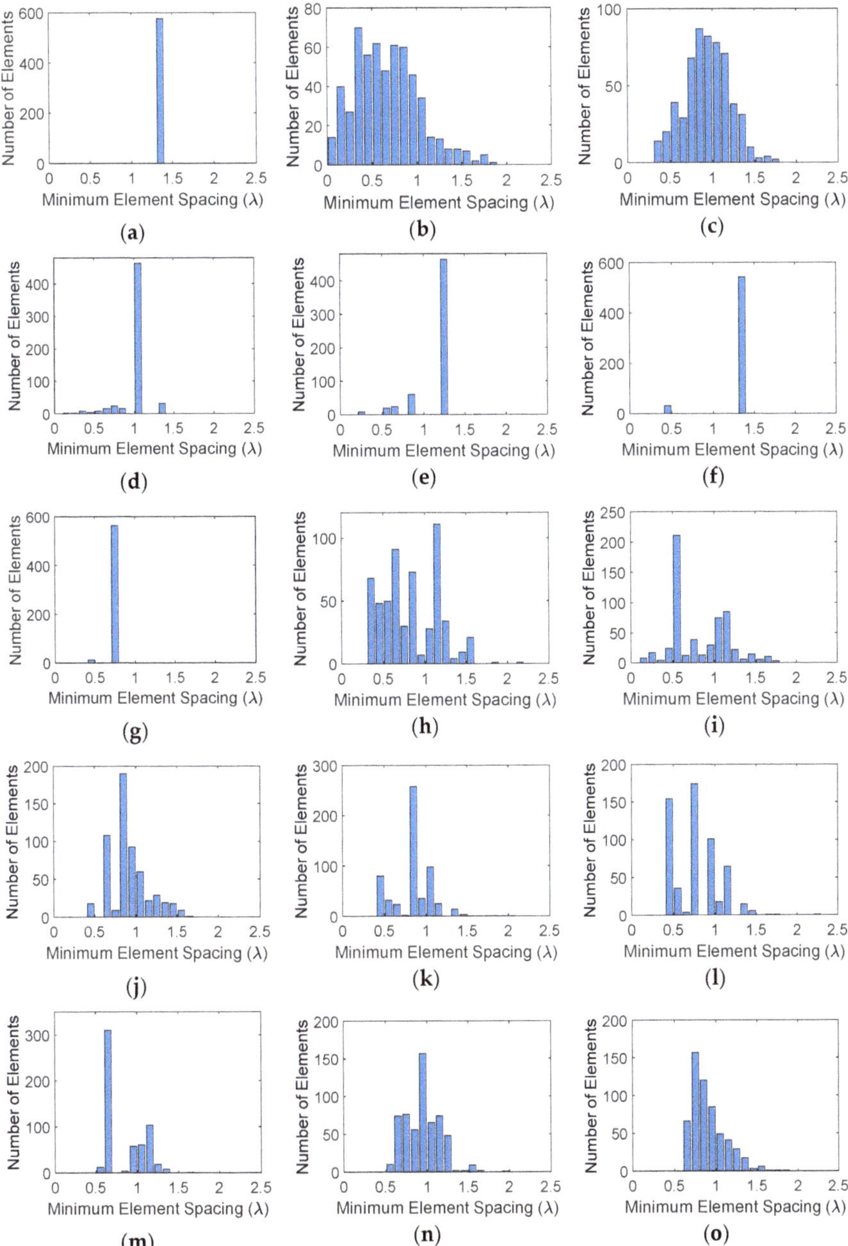

Figure 4. Bar plot of minimum element spacings with different distributions: (**a**) uniform, (**b**) random, (**c**) random with jitter, (**d**) Hammersley (base 2), (**e**) Hammersley (base 3), (**f**) Hammersley (base 5), (**g**) Hammersley (base 7), (**h**) Halton (bases 2, 3), (**i**) Halton (bases 2, 5), (**j**) Halton (bases 2, 7), (**k**) Halton (bases 3, 5), (**l**) Halton (bases 3, 7), (**m**) Halton (bases 5, 7), (**n**) Sobol, (**o**) Poisson disk.

3.2. Radiation Patterns of the Sparse Phased Array Antennas

The complete far-field radiation pattern of an antenna array with identical elements is given by

$$F(u,v) = E(u,v) \sum_{n=1}^{N} a_n e^{-jk[x_n u + y_n v]} \tag{15}$$

where $E(u, v)$ is the element pattern in u-v space, the summation represents the array factor, and $F(u, v)$ is the far-field radiation pattern of the antenna array. In this expression, a_n is the element weight, and the other terms are as defined in (1). Here we consider elements that have unity amplitude and zero phase, so $a_n = 1$. The elements are also isotropic, so $E(u, v) = 1$.

The radiation patterns of the arrays in the u-v space are computed using (15) and are given in Figure 5. We note that in these graphs the visible region is a circle with $\sqrt{u^2+v^2} \leq 1$. It can be seen that the uniform case has the worst performance and four grating lobes with sidelobe level of 0 dB appear in the visible space. Random sampling completely removes the grating lobes, however as we saw in Section 2, it places many elements too close to each other. Random sampling with jitter reduces the grating lobes but cannot break the grating lobes completely because its element spacing is too regular. On the other hand, all LDS methods are effective in reducing the SLLs while maintaining low discrepancy. In comparison between these methods, Hammersley sampling has the poorest performance due to its hierarchical problem. For Hammersley sampling, performance degrades as the prime base number is increased. Halton sampling on the other hand shows a much better performance, and all 6 cases studied here show that they can break the grating lobes. Due to its binary implementation, Sobol sampling shows a similar performance to Hammersley sampling with a prime base of 2. Poisson sampling appears to outperform all other LDS methods as well as the random distribution. The 2D u-v graphs in Figure 5, allow one to visually compare the radiation performance of all the sampling techniques, however, for better comparison, a quantitative analysis is also provided in the next section.

3.3. Quantitative Analysis of Sparse Array Performances

As discussed earlier in this work, for sparse arrays we are interested in removing the grating lobes, while at the same time having sampling that avoids too sparse or too dense element distributions. In order to quantitatively analyze these arrays, here we look at some metrics for element distributions and radiation pattern performances.

For element distributions, we look at an important statistical parameter, i.e., average minimum element spacing. This average value is obtained by computing the minimum element spacing for each element of the array, and then averaging the sum of those numbers over the total number of elements. These results are given in Table 1 for all 15 cases studied here. All LDS techniques provide a larger value of average minimum element spacing compared to the random technique, with Hammersley technique yielding the largest in comparison. From an element distribution perspective, it can be seen that LDS methods are very effective in providing a physically realizable distribution and outperform the random approach. For a quantitative study of radiation performance, we look at peak sidelobe level (SLL), as well as directivity and aperture efficiency. Peak SLL is the ratio of the pattern of the sidelobe peak (F_{SLL}), to the pattern value of the main lobe (F_{max}). We note that for a boresight beam F_{max} is the value of the far-field radiation pattern at $F(0, 0)$. The directivity is defined as

$$D = \frac{4\pi F_{max}}{\int_{0}^{\sqrt{u^2+v^2}\leq 1} \int_{0}^{\sqrt{u^2+v^2}\leq 1} F(u,v)\, du\, dv} \tag{16}$$

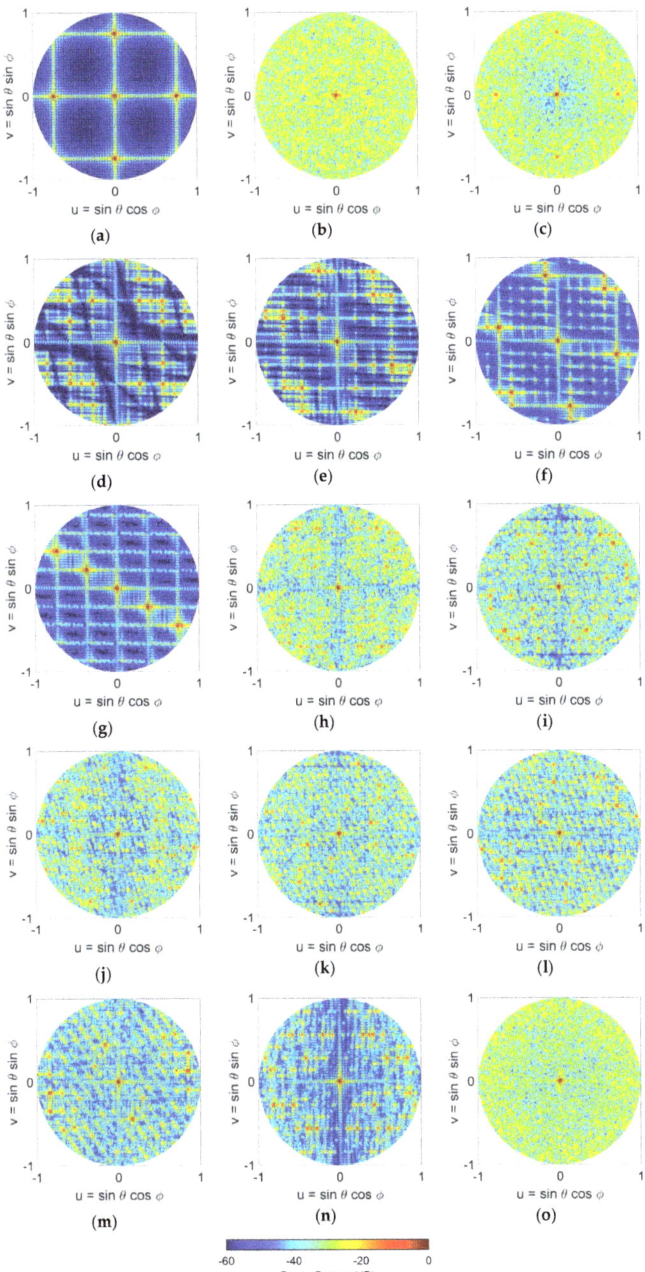

Figure 5. Normalized power patterns of the antenna arrays with different element distributions in the u-v space: (**a**) uniform, (**b**) random, (**c**) random with jitter, (**d**) Hammersley (base 2), (**e**) Hammersley (base 3), (**f**) Hammersley (base 5), (**g**) Hammersley (base 7), (**h**) Halton (bases 2, 3), (**i**) Halton (bases 2, 5), (**j**) Halton (bases 2, 7), (**k**) Halton (bases 3, 5), (**l**) Halton (bases 3, 7), (**m**) Halton (bases 5, 7), (**n**) Sobol, (**o**) Poisson disk.

Table 1. Performance metrics for sparse arrays with different element distributions.

Method	Average Minimum Element Spacing (λ)	Peak SLL (dB)	Directivity (dB)	Aperture Efficiency (%)
Uniform	1.3333	0	31.5478	11.168
Random	0.6667	−10.90	31.9619	12.209
Random with Jitter	0.9325	−9.36	32.5775	14.068
Hammersley (base 2)	1.0037	−7.0	32.7075	14.166
Hammersley (base 3)	1.1688	−2.69	32.6226	14.215
Hammersley (base 5)	1.2624	−0.55	31.6892	11.466
Hammersley (base 7)	0.7538	−0.25	33.8949	19.054
Halton (bases 2, 3)	0.8436	−10.10	32.4609	13.696
Halton (bases 2, 5)	0.8115	−4.33	32.3534	13.361
Halton (bases 2, 7)	0.9172	−12.35	32.8188	14.872
Halton (bases 3, 5)	0.8430	−8.10	32.8611	15.018
Halton (bases 3, 7)	0.7663	−5.90	32.2561	13.065
Halton (bases 5, 7)	0.8633	−5.60	32.9511	15.332
Sobol	0.9307	−6.58	32.7127	14.513
Poisson Disk	0.9031	−12.28	32.8212	14.880

Aperture efficiency is defined as directivity divided by maximum aperture directivity, where the maximum aperture directivity is given by

$$D_{aperture} = \frac{4\pi A}{\lambda^2} \tag{17}$$

Here, A is the size of the array aperture, which in our study in this section is $32\lambda \times 32\lambda$, and λ is the wavelength.

These results are also given in Table 1 for all 15 cases. It can be seen that the uniform case has the poorest performance in terms of SLL as well as directivity and efficiency. Random distribution can notably improve these, but as discussed earlier, the element distribution is undesirable. Random distribution with jitter improves directivity and efficiency but degrades SLL. In comparison between the LDS methods, Hammersley sampling has the poorest SLL performance, which degrades as the prime base number is increased. Halton sampling shows a better performance in comparison. The prime bases of 2 and 7 yield the best performance for this technique. Sobol sampling shows a similar performance to Hammersley with a prime base of 2. Poisson disk sampling shows the best performance of all LDS techniques, with a SLL of −12.28 dB and close to 15% aperture efficiency. While all these LDS techniques outperform the uniform case, the best performances come from Halton and Poisson disk methods that outperform the random technique.

4. Aperture Shape Effects on the Performance of Sparse Phased Array Antennas

To see the impact of aperture shape on the performance of these LDS arrays, in this section we study three different types of arrays with rectangular, circular, and elliptical apertures. All apertures are designed with the same surface area of $1024\lambda^2$ as the square aperture studied in Section 3. The rectangular aperture has an aspect ratio of 9/4, i.e., the ratio of its longer side to its shorter side, corresponding to longer and shorter side lengths of 48λ and $64\lambda/3$, respectively. The circular aperture has a radius of 18λ. To maintain the same aspect ratio as the rectangle and the same aperture size as the other arrays, the elliptical array major and minor axis are 27λ and 12λ, respectively. Here we compare the performance of two LDS sparse arrays that showed the best performance, namely Halton (with bases of 2 and 7) and Poisson distributions, with uniform distribution. The element distributions for these arrays are given in Figure 6.

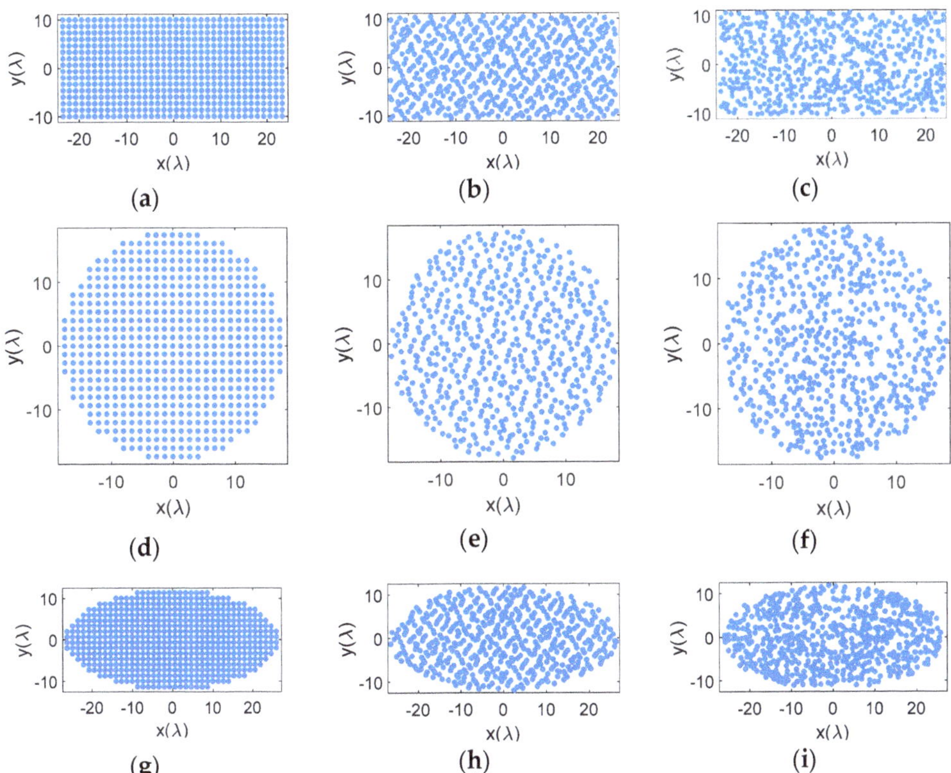

Figure 6. Position of elements on rectangular, circular, and elliptical apertures with different distributions: (**a**) uniform (rectangular), (**b**) Halton (rectangular), (**c**) Poisson disk (rectangular), (**d**) uniform (circular), (**e**) Halton (circular), (**f**) Poisson disk (circular), (**g**) uniform (elliptical), (**h**) Halton (elliptical), (**i**) Poisson disk (elliptical).

Bar plots in Figure 7 show the number of elements that fall within a minimum range of element spacing for the different distributions. Similar to Figure 4, for the uniform case, all elements have the same minimum separation between them. Halton sampling avoids too close placement of elements and distributes them in the range around $\lambda/2$ to 2λ, while Poisson sampling provides a Gaussian distribution of elements with and minimum element spacing of $4\lambda/6$. We note that these observations are similar to the square aperture studies given in Section 3. The radiation patterns of these arrays in the u-v space are given in Figure 8, where it can be seen that similar to the square aperture array, the uniform case has the worst performance and four grating lobes with sidelobe level of 0 dB appear in the visible space. Both LDS techniques remove the grating lobes; however, it can be seen that the Poisson technique outperforms Hammersley. Nonetheless, these studies show that the performance of LDS sampling techniques are not impacted by the shape of the array aperture, and in general these sampling approaches can be used for arbitrary shaped arrays.

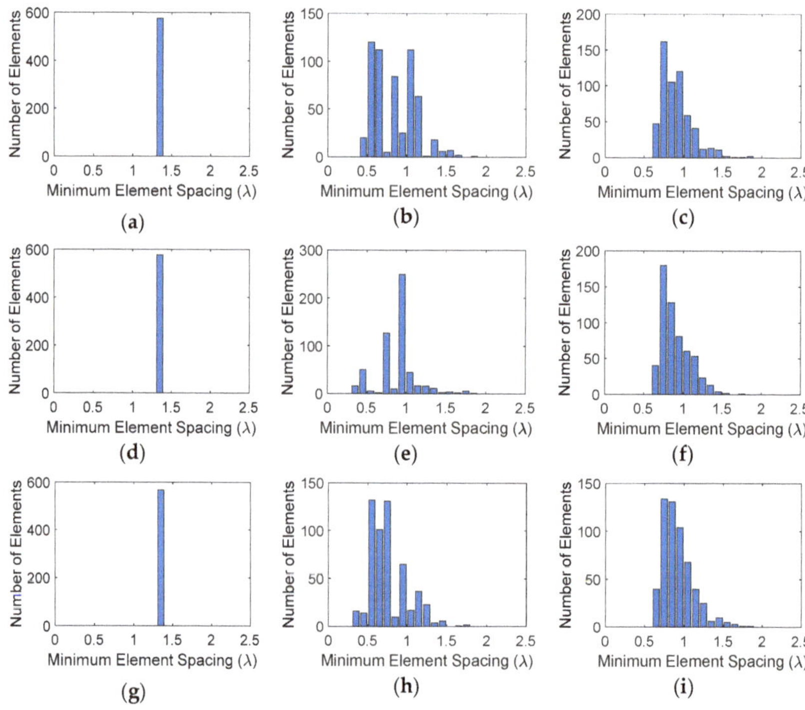

Figure 7. Bar plot of minimum element spacings on rectangular, circular, and elliptical apertures with different distributions: (**a**) uniform (rectangular), (**b**) Halton (rectangular), (**c**) Poisson disk (rectangular), (**d**) uniform (circular), (**e**) Halton (circular), (**f**) Poisson disk (circular), (**g**) uniform (elliptical), (**h**) Halton (elliptical), (**i**) Poisson disk (elliptical).

Table 2 summarizes the performance factors for these three types of arrays. As expected, the uniform case has the poorest performance in terms of SLL as well as directivity and efficiency. Both LDS techniques are effective in removing the grating lobes, but it can be seen that the Poisson technique provides the best results. It should be noted that in comparison between array aperture types, the circular aperture provides the best performance in terms of peak SLL, directivity, and efficiency.

Table 2. Performance metrics for sparse antenna arrays with rectangular, circular, and elliptical apertures and different element distributions.

Method/Aperture Type	Average Minimum Element Spacing (λ)	Peak SLL (dB)	Directivity (dB)	Aperture Efficiency (%)
Uniform/Rectangular	1.3333	0	31.5778	11.175
Uniform/Circular	1.3333	0	31.5848	11.261
Uniform/Elliptical	1.3333	0	31.5181	11.228
Halton/Rectangular	0.8481	−6.74	32.4857	13.774
Halton/Circular	0.8967	−9.27	32.6474	14.383
Halton/Elliptical	0.7724	−8.94	32.5883	14.365
Poisson Disc/Rectangular	0.9094	−12.18	32.8062	14.829
Poisson Disc/Circular	0.9016	−15.30	32.8643	15.119
Poisson Disc/Elliptical	0.9245	−14.34	32.6634	14.616

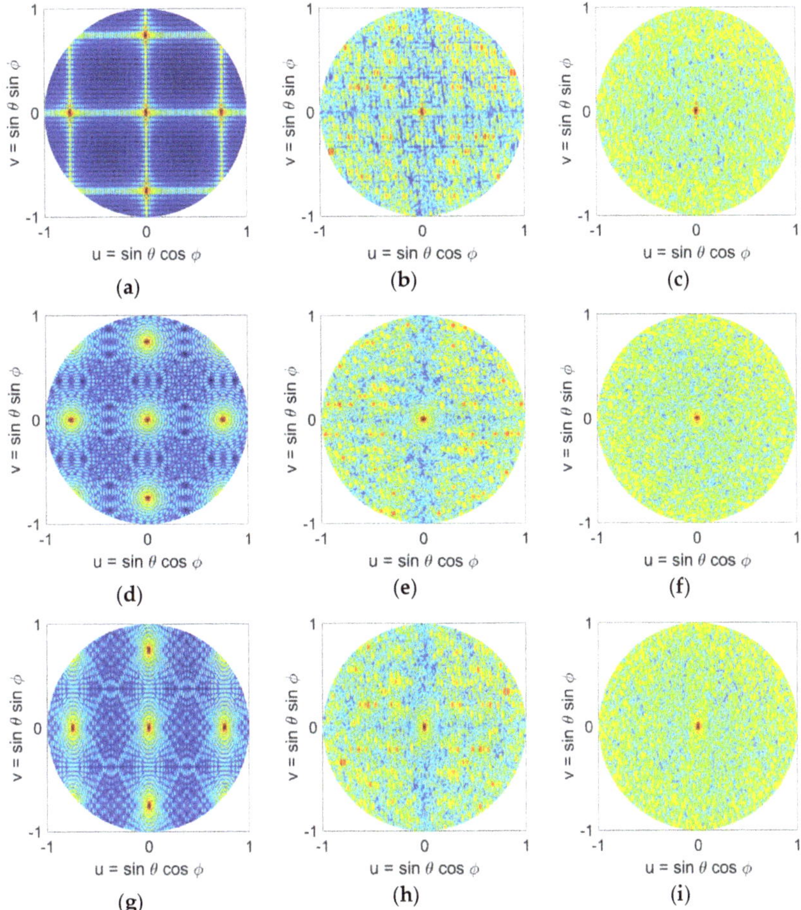

Figure 8. Normalized power patterns of sparse antenna arrays with rectangular, circular, and elliptical apertures and different element distributions in the u-v space: (**a**) uniform (rectangular), (**b**) Halton (rectangular), (**c**) Poisson disk (rectangular), (**d**) uniform (circular), (**e**) Halton (circular), (**f**) Poisson disk (circular), (**g**) uniform (elliptical), (**h**) Halton (elliptical), (**i**) Poisson disk (elliptical).

5. Beam-Scanning Performance of Sparse Phased Array Antennas

In this section, we study the beam-scanning performance of sparse phased array antennas. The studies conducted in Sections 3 and 4 showed that the performance of LDS distributions are not impacted by the type/shape of the array aperture, so for brevity we only consider the square aperture configuration in Section 3. By adding a progressive phase shift to the elements on the aperture, the sparse arrays can provide 2D beam scanning. Mathematically, the phase shift needed to scan the beam is given by

$$\varphi(x_n, y_n) = k(x_n \sin\theta_s \cos\phi_s + y_n \sin\theta_s \sin\phi_s) \quad (18)$$

where, $k = 2\pi/\lambda$, λ is the wavelength, (x_n, y_n) is the location of element n, θ_s and ϕ_s are the elevation and azimuth angle of the desired scanned beam direction, respectively.

Here we investigate 1-D and 2-D scanning performance of the arrays by studying scanning in the elevation plane at $\phi = 0°$ and $\phi = 45°$ directions, but similar results are observed for other scan directions. We note that scanning is essentially shifting the observation window. With sparse arrays this means that if grating lobes are created, scanning may result in more grating lobes appearing in the visible space.

Here we compare the scan performance of two LDS sparse arrays that showed the best performance, namely Halton (with bases of 2 and 7), and Poisson distributions, along with uniform and random distributions. Scanned patterns of each of these arrays are given in Figure 9 (ϕ = 0° direction) and 10 (ϕ = 45° direction), for 20°, 40°, and 60° elevation scans. We note that the array elements are isotropic point sources. For the uniform array, four grating lobes are observed when the array beam is at boresight, Figure 5a. When the array is scanned in 1-D, Figure 9a–c, the number of grating lobes increases to five at 20° and to seven when pointing at 40° and 60°. When the array is scanned in 2-D, Figure 10a–c, the number of grating lobes first reduces to three at 20° and 40°, and then increases to five when pointing at 60°. This change in the number of grating lobes degrades directivity, and in general the uniform array is not suitable for beam-scanning. The other three arrays however, i.e., random, Halton, and Poisson, show a good beam-scanning performance. These arrays do not have grating lobes, and scanning does not change that. In particular, we note that Poisson distribution, Figures 9 and 10j–l, shows a performance better than random distribution, i.e., Figures 9 and 10d–f, with the added advantage of having a physically realizable element distribution. It is important to note that while Halton sampling does have a higher SLL in the visible space, it does not have grating lobe issues, and again provides a notably better element distribution compared to the random case. Nonetheless, these studies also confirm that the Poisson disk sampling provides the best performance.

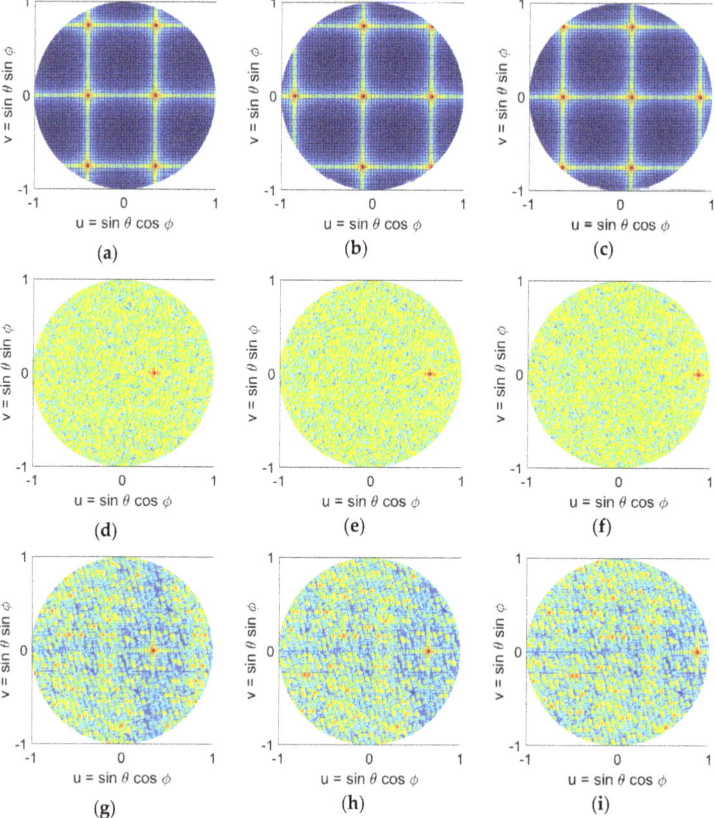

Figure 9. *Cont.*

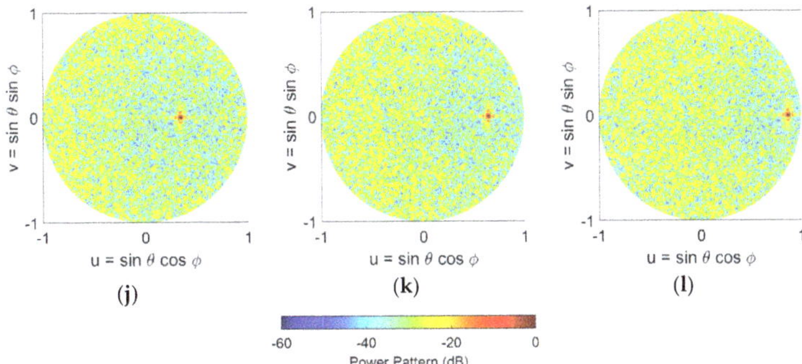

Figure 9. Normalized power patterns of the antenna arrays in the u-v space scanned along the elevation plane in $\phi = 0°$ direction with different element distributions: (**a**) uniform (20° scan), (**b**) uniform (40° scan), (**c**) uniform (60° scan), (**d**) random (20° scan), (**e**) random (40° scan), (**f**) random (60° scan), (**g**) Halton (20° scan), (**h**) Halton (40° scan), (**i**) Halton (60° scan), (**j**) Poisson disk (20° scan), (**k**) Poisson disk (40° scan), (**l**) Poisson disk (60° scan). The Halton sampling uses prime bases of 2 and 7.

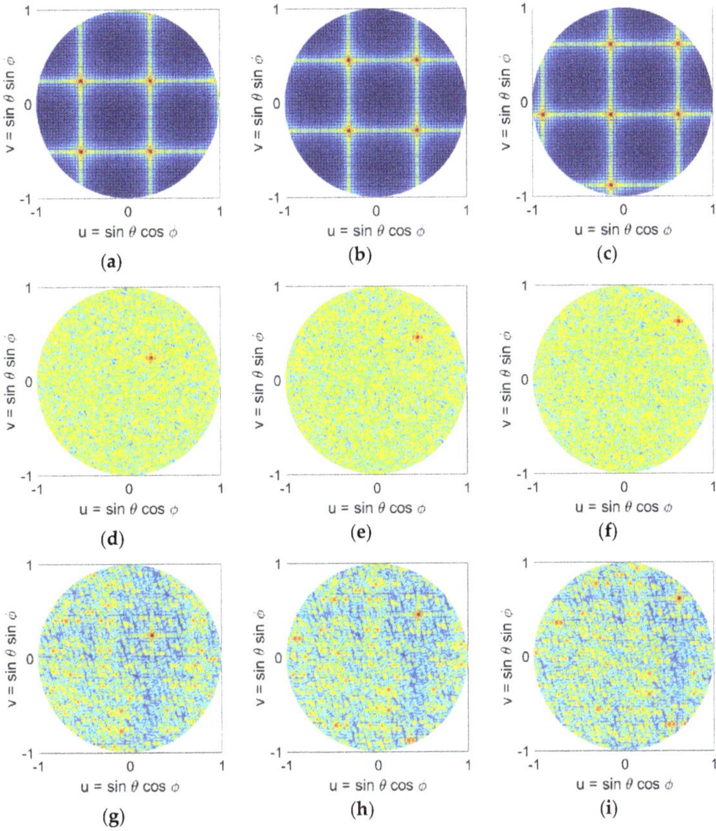

Figure 10. *Cont.*

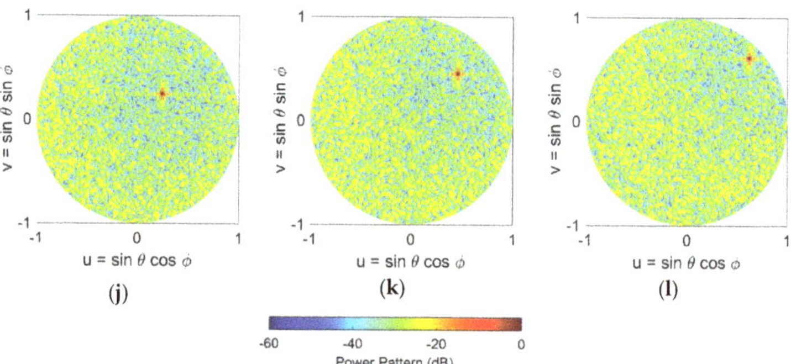

Figure 10. Normalized power patterns of the antenna arrays in the u-v space scanned along the elevation plane in ϕ = 45° direction with different element distributions: (**a**) uniform (20° scan), (**b**) uniform (40° scan), (**c**) uniform (60° scan), (**d**) random (20° scan), (**e**) random (40° scan), (**f**) random (60° scan), (**g**) Halton (20° scan), (**h**) Halton (40° scan), (**i**) Halton (60° scan), (**j**) Poisson disk (20° scan), (**k**) Poisson disk (40° scan), (**l**) Poisson disk (60° scan). The Halton sampling uses prime bases of 2 and 7.

6. Conclusions

Sparse arrays have a significantly smaller number of elements compared to traditional dense arrays (with $\lambda/2$ element spacing) which minimizes their cost and complexity. A comprehensive study of sparse phased arrays using low-discrepancy sequence (LDS) element distribution is presented. We show that these LDS element distributions remove the grating lobes associated with large element spacing in sparse arrays, while at the same time avoid undesirable and impractical element distributions in random arrays. The mathematical formulation for implementing LDS for sparse planar arrays is presented, along with numerical studies on their performance. Our studies considered sparse arrays with 86% less elements than a fully populated array. The performance factors compared equidistribution on the aperture of the array by looking at average minimum element spacing, as well as array pattern peak SLL, directivity, and aperture efficiency. Different array aperture configurations were also studied, and we show that the performance of LDS distributions is not impacted by the type/shape of the array. Our studies concluded that in comparison between the LDS techniques, the Poisson disk sampling technique outperforms all other approaches and is the recommended LDS technique for sparse arrays.

Author Contributions: Conceptualization, P.N., P.R., R.H.; writing—original draft preparation, T.T., P.N.; writing—review and editing, T.T., N.A., P.N., P.R., R.H.; supervision, P.N., P.R., R.H.; project administration, R.H.; funding acquisition, P.N. All authors have read and agreed to the published version of the manuscript.

Funding: This research was funded in part by the National Institute for Occupational Safety and Health (NIOSH) grant 75D30119C05413.

Institutional Review Board Statement: Not applicable.

Informed Consent Statement: Not applicable.

Acknowledgments: The authors wish to thank the reviewers whose thoughtful comments helped improve the quality and efficacy of this paper.

Conflicts of Interest: The authors declare no conflict of interest.

References

1. Haupt, R.L. *Timed Arrays Wideband and Time Varying Antenna Arrays*; Wiley: Hoboken, NJ, USA, 2015.
2. Haupt, R.L. *Antenna Arrays: A Computational Approach*; Wiley: Hoboken, NJ, USA, 2010.
3. Willey, R.E. Space tapering of linear and planar arrays. *IRE Trans. Antenna Propag.* **1962**, *10*, 369–377. [CrossRef]

4. Skolnik, M.; Sherman, J., III; Ogg, F., Jr. Statistically designed density-tapered arrays. *IEEE AP-S Trans.* **1964**, *12*, 408–417. [CrossRef]
5. Haupt, R.L. Thinned arrays using genetic algorithms. *IEEE Trans. Antennas Propag.* **1997**, *42*, 993–999. [CrossRef]
6. Kurup, D.G.; Himdi, M.; Rydberg, A. Synthesis of uniform amplitude unequally spaced antenna arrays using the differential evolution algorithm. *IEEE Trans. Antennas Propag.* **2003**, *51*, 2210–2217. [CrossRef]
7. *IEEE Standard for Definitions of Terms for Antennas*; IEEE Std 145-2013 (Revision of IEEE Std 145-1993); IEEE: New York, NY, USA, 6 March 2014. [CrossRef]
8. Available online: https://xlinux.nist.gov/dads/HTML/sparsematrix.html (accessed on 5 November 2020).
9. Candès, E.J.; Wakin, M.B. An introduction to compressive sampling. *IEEE Signal Process. Mag.* **2008**, *25*, 21–30. [CrossRef]
10. Rocca, P.; Oliveri, G.; Mailloux, R.J.; Massa, A. Unconventional Phased Array Architectures and Design Methodologies—A Review. *Proc. IEEE* **2016**, *104*, 544–560. [CrossRef]
11. Oliveri, G.; Massa, A. Bayesian compressive sampling for pattern synthesis with maximaly sparse non-uniform linear arrays. *IEEE Trans. Antennas Propag.* **2011**, *59*, 467–481. [CrossRef]
12. Oliveri, G.; Carlin, M.; Massa, A. Complex-weight sparse linear array synthesis by Bayesian compressive sampling. *IEEE Trans. Antennas Propag.* **2012**, *60*, 2309–2326. [CrossRef]
13. Zhang, W.; Li, L.; Li, F. Reducing the number of elements in linear and planar antenna arrays with sparseness constrained optimization. *IEEE Trans. Antennas Propag.* **2011**, *59*, 3106–3111. [CrossRef]
14. Viani, F.; Oliveri, G.; Massa, A. Compressive sensing pattern matching techniques for synthesizing planar sparse arrays. *IEEE Trans. Antennas Propag.* **2013**, *61*, 4577–4587. [CrossRef]
15. Buchanan, K.; Rockway, J.; Sternberg, O.; Mai, N.N. Sum-difference beamforming for radar applications using circularly tapered random arrays. In Proceedings of the 2016 IEEE Radar Conference (RadarConf), Philadelphia, PA, USA, 2–6 May 2016; pp. 1–5.
16. Leeper, D.G. Isophoric arrays-massively thinned phased arrays with well-controlled sidelobes. *IEEE Trans. Antennas Propag.* **1999**, *47*, 1825–1835. [CrossRef]
17. Christodoulou, C.G.; Ciaurriz, M.; Tawk, Y.; Costantine, J.; Barbin, S.E. Recent advances in randomly spaced antenna arrays. In Proceedings of the 8th European Conference on Antennas and Propagation (EuCAP 2014), The Hague, The Netherlands, 6–11 April 2014; pp. 732–736.
18. Ellingson, S.W.; Taylor, G.B.; Craig, J.; Hartman, J.; Dowell, J.; Wolfe, C.N.; Clarke, T.E.; Hicks, B.C.; Kassim, N.E.; Ray, P.S.; et al. The LWA1 Radio Telescope. *IEEE Trans. Antennas Propag.* **2013**, *61*, 2540–2549. [CrossRef]
19. van Haarlem, M.P.; Wise, M.W.; Gunst, A.W.; Heald, G.; McKean, J.P.; Hessels, J.W.; de Bruyn, A.G.; Nijboer, R.; Swinbank, J.; Fallows, R.; et al. LOFAR: The Low-Frequency Array. *Astron. Astrophys.* **2013**, *556*, 1–54. [CrossRef]
20. di Ninni, P.; Bolli, P.; Paonessa, F.; Pupillo, G.; Virone, G.; Wijnholds, S.J. Electromagnetic Analysis and Experimental Validation of the LOFAR Radiation Patterns. *Int. J. Antennas Propag.* **2019**, *2019*, 1–12. [CrossRef]
21. Available online: https://public.nrao.edu/telescopes/vlba/ (accessed on 14 November 2020).
22. Available online: https://public.nrao.edu/telescopes/alma/ (accessed on 14 November 2020).
23. Available online: https://public.nrao.edu/telescopes/vla/ (accessed on 14 November 2020).
24. Scott, S.; Wawrzynek, J. Compressive sensing and sparse antenna arrays for indoor 3-D microwave imaging. In Proceedings of the 2017 25th European Signal Processing Conference (EUSIPCO), Kos, Greece, 28 August–2 September 2017; pp. 1314–1318.
25. Amani, N.; Maaskant, R.; van Cappellen, W.A. On the Sparsity and Aperiodicity of a Base Station Antenna Array in a Downlink MU-MIMO Scenario. In Proceedings of the 2018 International Symposium on Antennas and Propagation (ISAP), Busan, Korea, 23–26 October 2018; pp. 1–2.
26. Cappellen, W.A.V.; Wijnholds, S.J.; Bregman, J.D. Sparse antenna array configurations in large aperture synthesis radio telescopes. In Proceedings of the 2006 European Radar Conference, Manchester, UK, 13–15 September 2006; pp. 76–79.
27. Hermann, W. Über die Gleichverteilung von Zahlen mod. Eins [About the equal distribution of numbers]. *Math. Ann.* **1916**, *77*, 313–352.
28. Hammersley, J.M. Monte Carlo methods for solving multivariable problems. *Ann. N. Y. Acad. Sci.* **1960**, *86*, 844–874. [CrossRef]
29. Sobol, I.M. The distribution of points in a cube and the approximate evaluation of integrals. *USSR Comput. Math. Math. Phys.* **1967**, *7*, 86–112. [CrossRef]
30. Faure, H. Discrépance de suites associées à un système de numération (en dimensions). *Acta Arith.* **1982**, *41*, 337–351. [CrossRef]
31. Niederreiter, H. *Random Number Generation and Quasi-Monte Carlo Methods, CBMS-NSF*; SIAM: Philadelphia, PA, USA, 1992; Volume 63.
32. Morokoff, W.J.; Caflisch, R.E. Quasi-Monte Carlo integration. *J. Comput. Phys.* **1995**, *122*, 218–230. [CrossRef]
33. Paskov, S.H.; Traub, J.F. Faster valuation of financial derivatives. *J. Portf. Manag.* **1995**, 113–120. [CrossRef]
34. Haupt, R.L. A Sparse Hammersley Element Distribution on a Spherical Antenna Array for Hemispherical Radar Coverage. In Proceedings of the IEEE Radar Conference, Seattle, WA, USA, 8–12 May 2017.
35. Bui, P.L.-T.; Rocca, P.; Haupt, R.L. Aperiodic planar array synthesis using pseudo-random sequences. In Proceedings of the 2018 International Applied Computational Electromagnetics Society Symposium, ACES 2018, Beijing, China, 29 July–1 August 2018; pp. 1–2.
36. Torquato, S. Hyperuniform states of matter. *Phys. Rep.* **2018**, *745*, 1–95. [CrossRef]
37. Torquato, S. Disordered hyperuniform heterogenous materials. *J. Phys. Condens Matter* **2016**, *28*, 414012. [CrossRef]

38. Ding, Z.; Zheng, Y.; Xu, Y.; Jiao, Y.; Li, W. Hyperuniform flow fields resulting from hyperuniform configurations of circular disks. *Phys. Rev. E* **2018**, *98*, 063101. [CrossRef]
39. Di Battista, D.; Ancora, D.; Zacharakis, G.; Ruocco, G.; Leonetti, M. Hyperuniformity in amorphous speckle patterns. *Opt. Express* **2018**, *26*, 15594–15608. [CrossRef] [PubMed]
40. DiStasio, R.A., Jr.; Zhang, G.; Stillinger, F.H.; Torquato, S. Rational design of stealthy hyperuniform two-phase media with tunable order. *Phys. Rev. E* **2018**, *97*, 023311. [CrossRef]
41. Kocis, L.; Whiten, W.J. Computational Investigations of Low-Discrepancy Sequences. *ACM Trans. Math. Softw.* **1997**, *23*, 266–294. [CrossRef]
42. van der Corput, J.G. Verteilungsfunktionen I. *Akad. Van Wet.* **1935**, *38*, 813–821.
43. Wong, T.T.; Luk, W.S.; Heng, P.A. Sampling with Hammersley and Halton points. *J. Graph. Tools* **1997**, *2*, 9–24. [CrossRef]
44. Sobol, I.M.; Levitan, Y.L. A pseudo-random number generator for personal computers. *Comput. Math. Appl.* **1999**, *37*, 33–40. [CrossRef]
45. Bratley, P.; Fox, B.L. Algorithm 659 Implementing Sobol's Quasirandom Sequence Generator. *ACM Trans. Math. Softw.* **1988**, *14*, 88–100. [CrossRef]
46. Hong, H.S.; Hickernell, F.J. Algorithm 823: Implementing Scrambled Digital Sequences. *ACM Trans. Math. Softw.* **2003**, *29*, 95–109. [CrossRef]
47. Joe, S.; Kuo, F.Y. Remark on Algorithm 659: Implementing Sobol's Quasirandom Sequence Generator. *ACM Trans. Math. Softw.* **2003**, *29*, 49–57. [CrossRef]
48. Available online: https://www.mathworks.com/help/stats/sobolset.html (accessed on 12 October 2021).
49. Gamito, M.N.; Maddock, S. Accurate multidimensional poisson disk sampling. *ACM Trans. Graph.* **2009**, *29*, 1–19. [CrossRef]
50. Lagae, A.; Dutre, P. A comparison of methods for generating poisson disk distributions. *Comput. Graph. Forum* **2008**, *27*, 114–129. [CrossRef]

Communication

A Novel Method of Transmission Enhancement and Misalignment Mitigation between Implant and External Antennas for Efficient Biopotential Sensing

Md Shifatul Islam [1], Asimina Kiourti [2] and Md Asiful Islam [1,*]

[1] Department of Electrical and Electronic Engineering, Bangladesh University of Engineering and Technology, Dhaka 1205, Bangladesh; shifatbuet@gmail.com

[2] Electroscience Laboratory, Department of Electrical and Computer Engineering, The Ohio State University, Columbus, OH 43212, USA; kiourti.1@osu.edu

* Correspondence: maislam@eee.buet.ac.bd

Citation: Islam, M.S.; Kiourti, A.; Islam, M.A. A Novel Method of Transmission Enhancement and Misalignment Mitigation between Implant and External Antennas for Efficient Biopotential Sensing. *Sensors* **2021**, *21*, 6730. https://doi.org/10.3390/s21206730

Academic Editor: Pedro Pinho

Received: 5 September 2021
Accepted: 9 October 2021
Published: 11 October 2021

Abstract: The idea of passive biosensing through inductive coupling between antennas has been of recent interest. Passive sensing systems have the advantages of flexibility, wearability, and unobtrusiveness. However, it is difficult to build such systems having good transmission performance. Moreover, their near-field coupling makes them sensitive to misalignment and movements. In this work, to enhance transmission between two antennas, we investigate the effect of superstrates and metamaterials and propose the idea of dielectric fill in between the antenna and the superstrate. Preliminary studies show that the proposed method can increase transmission between a pair of antennas significantly. Specifically, transmission increase of $\approx$5 dB in free space and $\approx$8 dB in lossy media have been observed. Next, an analysis on a representative passive neurosensing system with realistic biological tissues shows very low transmission loss, as well as considerably better performance than the state-of-the-art systems. Apart from transmission enhancement, the proposed technique can significantly mitigate performance degradation due to misalignment of the external antenna, which is confirmed through suitable sensitivity analysis. Overall, the proposed idea can have fascinating prospects in the field of biopotential sensing for different biomedical applications.

Keywords: biopotential sensing; Fabry-Perot resonator; antenna; superstrate; metamaterials; passive sensing

1. Introduction

Biopotentials are electric signals which are generated inside a biological substance through the electro-chemical activities of a cluster of cells. Through the sensing of biopotentials, it is often possible to describe the physiological state of a group of cells which are of interest. These biopotentials are extremely weak, with magnitudes in the order of μV [1]. The biosignals are often modulated by a higher frequency for efficient transmission, and this modulation frequency ranges from MHz to GHz [2]. In general, almost all biosensing tools involve measuring electrodes, whose task is to exploit the electrochemical activities of the cell and transform to equivalent voltage signals for detection.

Depending on the extraction mechanism of the signal, sensing can be either "active" or "passive". In active sensing, biosignals are processed and received directly by external devices, which can be different integrated circuits (ICs), harvesters, and sensors. In the more recently explored passive sensing, there is an implanted/interrogator antenna pair that communicates signals between the two. The implanted antenna is connected to the electrode and is buried inside the tissue, while the interrogator antenna is an external antenna to which the biosignals are transmitted. The passive sensing mechanism has been tried on different biopotential sensing systems [3–6]. The main advantage of the passive sensing systems compared with the active sensing systems is that passive systems do not require obtrusive circuitry, and such wireless links are easily portable and wearable. In passive sensing, typically, the two antennas are placed close to each other [3–7] so that they can inductively couple with each other. This requires the two antennas to be separated by a small distance, leaving almost no room for engineering improvements that could boost transmission. By contrast, since the antennas are tightly coupled, change in the dimension or parameters of one antenna directly influences the behavior of the second antenna. This makes the passive sensing systems both difficult to engineer for transmission enhancement and susceptible to misalignment, which is a very likely event if the system is used as a wearable device [3]. Most of the antennas in the literature for passive sensing rely on near-field coupling, and the associated issue of misalignment remains largely unaddressed to date.

On the other hand, not related to such passive sensing application, a significant amount of work has been done to enhance antenna gain for efficient transmission in the far field region. Theoretical and experimental methods include the use of reflecting planes [8,9], use of superstrate dielectric [10,11], and metamaterials with or without superstrates [12–18]. At first, 'metamaterial' was defined as periodic structures capable of showing negative electromagnetic properties [19–22], but, today, any external tampering with materials which results in the change of effective electromagnetic properties is termed as metamaterial [23,24]. The enhancement of directivity in the aforementioned works is mainly based on one of the following three principles: (a) the model of Fabry-Perot resonator cavities, where the superstrate or the reflecting planes induce multiple reflections inside the antenna ground-superstrate gap [8–11,15,16,25], (b) the idea of negative refraction and inverse focusing, which is induced by left-handed materials (LHM) [12–14], and (c) the ability of different periodic structures to induce a small frequency band just above the resonant frequency, where the refractive index is less than that of air [18,20–22]. Published results in these works have shown considerable enhancement in the directive gain of the antenna.

While none of these works directly provide any insight on transmission between antenna pairs for passive sensing, it is, of course, encouraging to try similar concepts and see how they contribute to the problem of interest. However, attempting these concepts for passive biosensing is not straightforward. For example, both the principles (b) and (c) mentioned above are highly dependent on operating frequency and work in a very small frequency band. Since the passive sensing antennas are closely coupled with each other and the resonant frequency changes with small misalignment and fabrication imperfections, it is indeed challenging to design structures that operate at a particular frequency (or in a small band). Furthermore, the direction of periodicity and the period along the propagation direction requires large spatial requirement, and complex fabrication, which is not desirable in passive sensing. These leave idea (a) (Fabry-Perot resonators) to have some probable positive effect on transmission enhancement. The merit of this idea is that the governing principle only depends on the reflective nature of the designs, not on frequencies, and is, therefore, more convenient to implement. While we focus on this approach to enhance transmission, there is also a major hurdle to overcome. To incorporate the superstrate and the metamaterials in between the antenna pairs, the separation between them needs to be increased. However, increasing this separation reduces the transmission between the antenna pair. So, the gain in performance has to surpass the loss of transmission due to increasing the antenna pair separation, ultimately achieve an overall increase in transmission, and, of course, fill up the transmission requirement of the desired application.

In this work, we propose a new technique of increasing transmission between the implanted/interrogator antenna pair for biopotential sensing, adopting the concept of Fabry-Perot resonators and metamaterials. The obtained transmission results are compared with the existing literature, to demonstrate the method's efficacy. In addition, a sensitivity analysis is carried out, and it is shown that the transmission performance remains nearly unaltered for reasonable misalignment between the two antennas.

The rest of the paper is divided into four sections. In Section 2, we present the concept of the Fabry-Perot resonator and the motivation of our proposed approach. In

Section 3, we describe two simulation environments where we test our ideas and observe the transmission characteristics. In Section 4, we perform a simulation and subsequent analysis with practical materials and biological media models to demonstrate a passive biosensing application. We summarize the work with final remarks in Section 5.

2. Prior Art

The method of improving the directive gain of an antenna with an ideal (infinitely large) ground was first mathematically explored in Reference [8]. There, a partially reflecting sheet was placed above the infinite antenna ground plane at a distance $\frac{\lambda}{2}$ to form a resonant cavity. The sheet has a planewave reflection coefficient of magnitude R. The cavity enhances the gain of the antenna by a maximum factor of:

$$gain = \frac{1+R}{1-R}. \quad (1)$$

This formula suggests that the more reflective the sheet is, the higher the gain enhancement becomes. A study, where multiple superstrate materials were used instead of the sheet [10], also agrees with the idea. However, in this approach, there are two issues:

- A physical approximation of the infinite ground requires a large antenna plane, which is not feasible for small domain applications, such as biosensing. Therefore, we need to investigate what the superstrate can offer in transmission for finite size grounds.
- The replacement of the reflecting sheet with the superstrate adds another unknown, which is the thickness of the superstrate, which is needed to be optimized for practical and finite sized patches.

In a subsequent work [15], it has been shown that the inclusion of metamaterials on the superstrate surface can enhance the reflectivity; therefore, the superstrate will demonstrate higher reflection than the material itself can offer. Later, in Reference [25], the effect of metallic imprints on gain enhancement was examined in further detail for circular patch antennas, and it has been shown that the gain enhancement is not indefinitely proportional with the surface reflection as Equation (1) would suggest, and the idea of "optimum reflection coefficient" was introduced. The claim was that, up to a certain value of superstrate reflection, gain will increase, and then the gain will decrease very rapidly. So, the task of the metamaterial designer is to make the meta imprinted superstrate acquire that optimum reflection, for which maximum gain will be achieved.

Since, in this work, we investigate the same concept on the transmission characteristics between two antennas, the above discussions boil down to the following expectations:

- With the increase of superstrate permittivity, transmission will enhance but only up to a given maximum superstrate permittivity.
- Insertion of metamaterial imprints on low permittivity superstrates can produce the optimum performance at a higher permittivity. This tool can be specially useful since practical high permittivity materials are not always commercially available, and they provide relatively large dispersive loss.

Along with exploring these ideas, we also propose and investigate the effect of filling the patch-superstrate gap with dielectric materials. The idea came up with two motivations:

- The insertion of the dielectric material turns the gap into a dielectric waveguide channel, where some of the leaky waves can reflect into the channel and increase the number of reflections (see Figure 1b), and possibly overall transmission.
- The effective wavelength distance will be $\frac{\lambda}{2\sqrt{\epsilon_{gap}}}$, ϵ_{gap} being the permittivity of the gap filled material. Hence, the system can be miniaturized, which is always desirable in implementation of biosensing systems.

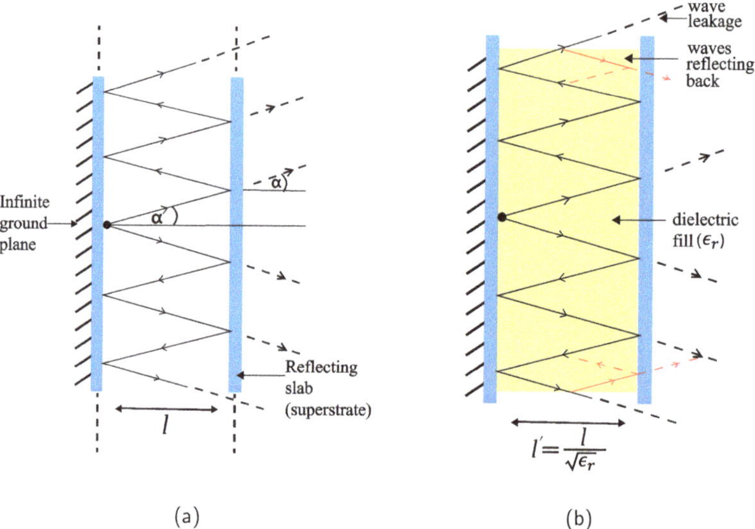

Figure 1. Geometric interpretation of the Fabry-Perot resonator antenna concept. (**a**) Waves leak from the top and the bottom of the cavity. (**b**) Leaked waves are reflected back to induce further reflection.

3. Examining Transmission Enhancement

In this section, we present simulation analysis on the performance enhancement in transmission due to variation in each variable, namely the dielectric properties of the superstrate and the gap insertion material, and the effect of metamaterials. First, we describe the simulation environment in which we will explore the effects, and then provide the analysis. Throughout the work, the frequency of interest is 4.8 GHz, similar to the implant operating frequency in Reference [3].

3.1. Simulation Systems

For our analysis, we will explore two different antenna pair systems and examine transmission characteristics on each of them. The first system is the simpler one, where the two antennas are placed in free space, whereas, for the second system, one antenna is placed in air, and the second antenna is placed in a medium other than air. This new medium has an assumed dielectric constant of $\epsilon_{tissue} = 40$ and a conductivity of $\sigma_{tissue} = 3\ \text{Sm}^{-1}$ at 4.8 GHz. We choose these values for simulation because most biological tissues have dielectric properties of similar range [26]. For clarity's sake, we denote these two systems as "system 1" and "system 2", respectively (see Figure 2).

For each of the two systems, we analyze the transmission characteristics in the following steps:

- *Step 1:* First, we take note of the transmission loss for the two systems, without any engineering in between. For system 1, the two antennas are separated by a wavelength distance in air $\lambda_{air} \approx 62.5$ mm at 4.8 GHz For system 2, the medium boundary is placed at a distance $\frac{3\lambda_{air}}{4} \approx 46.875$ mm from the external antenna. The internal antenna is buried further $\frac{\lambda_{tissue}}{2} = \frac{\lambda_{air}}{2\sqrt{\epsilon_{tissue}}} \approx 4.94$ mm inside (see Figure 2a).
- *Step 2:* Since the air-superstrate cavity needs to have a half-wavelength distance, we place a superstrate material at a distance $\frac{\lambda_{air}}{2} \approx 31.25$ mm above antenna 1, for both systems. The location of the antennas will remain unchanged (see Figure 2b). From there on, we vary the superstrate dielectric constant (ϵ_{sup}) and take note of the transmission characteristics.

- *Step 3:* Next, we fill the gap between the patch and the superstrate with a dielectric having $\epsilon_{gap} = 2$. To maintain the same half wavelength cavity, the new air-superstrate gap is now $\frac{\lambda_{gap}}{2} = \frac{\lambda_{air}}{2\sqrt{\epsilon_{gap}}} \approx 22.09$ mm (see Figure 2c). In this way, although the distance between the antenna and the superstrate is reduced, the overall distance is the same if measured in wavelengths. We again vary the dielectric constant of the superstrate to observe the transmission performance.
- *Step 4:* Finally, we insert circular ring metamaterial imprints on both sides of the superstrate for a relatively low $\epsilon_{sup} = 3$ to induce high permittivity substrate transmission from the previous step.

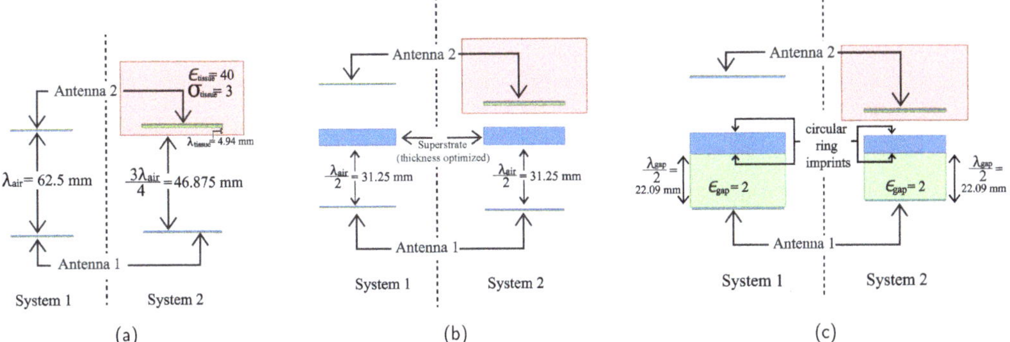

Figure 2. Step-by-step schematic diagram for the analysis of the two defined different antenna systems. (**a**) The antenna pair is placed without any engineering in between. (**b**) Superstrate material is inserted at a distance $\frac{\lambda_{air}}{2}$ in front of antenna 1. (**c**) The gap between antenna 1 and the superstrate is filled with a dielectric of $\epsilon_{gap} = 2$.

3.2. Antenna Pairs

Throughout the analysis, we have noted that the antenna frequency change in each of the four steps mentioned above, and, for proper comparison at the desired frequency, we tune the antenna pair regularly. Therefore, we have used simplified patch antennas, which could be tuned to the desired frequency of operation rather easily using parametric analysis.

Both of the two antenna substrates are assumed to be made of Rogers RO4003 material ($\epsilon_r = 3.55, tan\delta = 0.0027$), with thickness of 0.762 mm. The width of the feedline is 0.5 mm, and the length of the feedline is 5.24 mm (see Figure 3a). For both the antennas, the substrate has the dimension $2W \times 2L$. The parameters of the two antennas, in each of the steps to tune at the desired frequency, are given in Table 1. $L1, W1$ and $L2, W2$ are the L, W values for antenna 1 and 2, respectively.

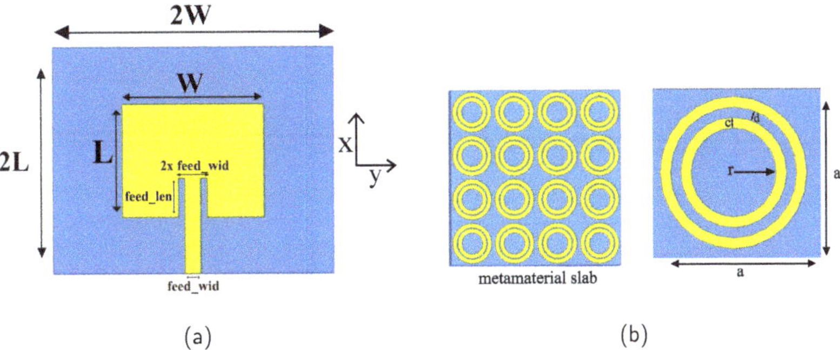

Figure 3. (**a**) The common geometry of the two antenna configurations for the analysis. (**b**) The view of circular ring imprints on both sides of the superstrate with tunable parameters.

Table 1. Dimensions of the two antennas for each system, and for each step of the analysis. All units are in mm.

	System 1	System 2
Step 1	$L1 = 15.97$, $W1 = 20.21$ $L2 = 15.97$, $W2 = 20.21$	$L1 = 16.03$, $W1 = 20.29$, $L2 = 15.87$, $W2 = 20.29$
Step 2	$L1 = 15.97$, $W1 = 20.21$ $L2 = 16.03$, $W2 = 20.29$	$L1 = 16.03$, $W1 = 20.29$, $L2 = 15.87$, $W2 = 20.29$
Step 3	$L1 = 15.4$, $W1 = 19.5$ $L2 = 16.03$, $W2 = 20.29$	$L1 = 15.71$, $W1 = 19.88$, $L2 = 15.87$, $W2 = 20.29$
Step 4	$L1 = 15.45$, $W1 = 19.5$ $L2 = 16.03$, $W2 = 20.29$	$L1 = 15.71$, $W1 = 19.88$, $L2 = 15.87$, $W2 = 20.29$

In addition, for system 2, we have also covered the implant patch (antenna 2) with a 1-mm thick coverage of lossless dielectric material ($\epsilon_r = 2$). It has been shown that such coating not only ensures biocompatibility but also improves the transmission significantly by reducing dispersion loss inside the lossy medium [3,27]. To illustrate this, the amount of transmission with and without the presence of the coating is shown in Figure 4, from which transmission enhancement of $\approx$14 dB is observed when insulation is added.

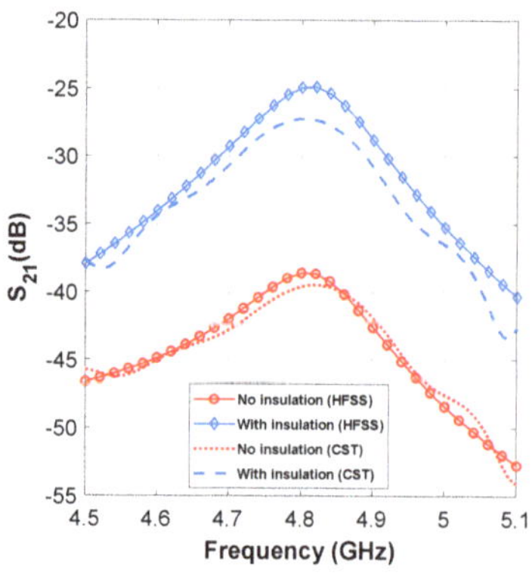

Figure 4. Effect of insulation to enhance transmission gain.

The metamaterial imprints are consisting of simple two concentric rings, the reflectivity is modified through parametric analysis by increasing/decreasing the ring width c (see Figure 3b).

3.3. Transmission Analysis

In this section, we present the simulation results for each of the two systems mentioned before. For the analyses, we use two widely used EM simulation tools: ANSYS HFSS and CST microwave studio. In the following discussions, we present the simulation results from HFSS, whilst mentioning the CST verified results in parenthesis.

For system 1, the initial transmission loss is found to be $-9.33(-9.30)$ dB. After adding the superstrate, at first the performance improves with the increase of superstrate permittivity. The best transmission of $-5.54(-5.67)$ dB is found for $\epsilon_{sup} = 11$, which is an improvement of $\approx$3.8 dB (see Figure 5a). After $\epsilon_{sup} > 11$, we see a fast reduction in transmission. Therefore, we identify the maximum limit of superstrate reflection (permittivity) for which the transmission is maximum. Next, after filling up the patch-superstrate gap with the dielectric, we see that the transmission improves further, for all values of superstrate dielectric constant. The best transmission is found to be $-4.08(-3.78)$ dB for $\epsilon_{sup} = 5$. Later, after adding circular rings on both sides of the superstrate, we obtain a similar transmission performance of $-3.96(-3.63)$ dB for $\epsilon_{sup} = 3$ (see Figure 5b). From the final transmission curves, we find that an improvement of more than 5 dB has been achieved after all these steps.

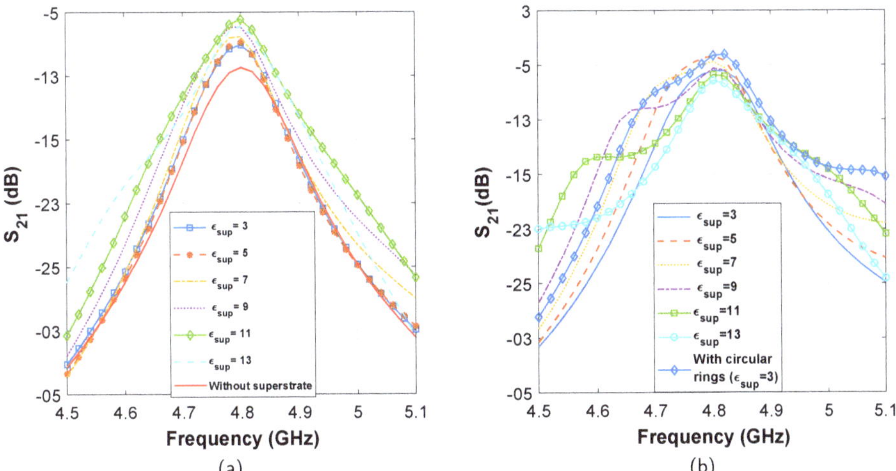

Figure 5. Effect of transmission enhancement with varying superstrate dielectric constant in system 1. (**a**) Gain enhancement in steps 1 and 2. (**b**) Gain enhancement in steps 3 and 4.

For system 2, the initial transmission loss after the inclusion of insulation layer was found to be $-24.91(-27.2)$ dB (see Figure 3). From there on, the performance improved with the inclusion of superstrates, and reached a maximum of $-21.7(-23.9)$ dB for $\epsilon_{sup} = 5$ (see Figure 6a). Further increase of ϵ_{sup} degraded transmission. Later, after filling the patch superstrate gap with dielectric material, the transmission improved yet again. The maximum transmission was noted to be $-17.06(-19.36)$ dB for $\epsilon_{sup} = 5$. Finally, after including circular rings on the superstrate, we obtain a transmission of $-17.11(-18.53)$ dB for $\epsilon_{sup} = 3$ (see Figure 6b). Therefore, a total transmission enhancement of $\approx$8 dB has been achieved for system 2.

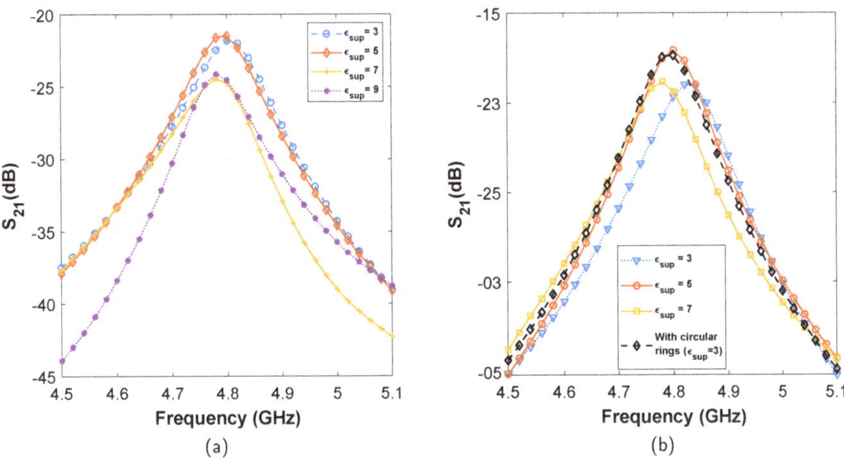

Figure 6. Effect of transmission enhancement with varying superstrate dielectric constant in system 2. (**a**) Gain enhancement in steps 1 and 2. (**b**) Gain enhancement in steps 3 and 4.

Simulations using both HFSS and CST agree with almost identical amount of gain enhancement through all the steps. However, the simulated transmission losses differ slightly for the two softwares: by ≈0.3 dB in system 1, and ≈2 dB in system 2. This can be attributed to the difference between the difference in the solver systems and excitations.

After exploring the effects of superstrate, gap filler, and metamaterials for different systems, we can draw the following conclusions:

- With the inclusion of the superstrate material, the transmission improves, but only up to a certain limit.
- Filling up the patch-superstrate gap with dielectric material further enhances the transmission.

4. Implementation: A Passive Neurosensing System

In this section, we design a passive neuropotential recording system [3,4,7] as a potential application of biosensing with the developed method. In the design, the biological media consist of a layered structure of skin, bone, brain grey matter, and brain white matter [4]. Each of these tissues are frequency dispersive, whose dielectric properties are presented in Figure 7. In the arrangement, the implant antenna is placed 1 mm inside the skin tissue. The external antenna is placed 3 mm outside the skin to form the passive sensing system.

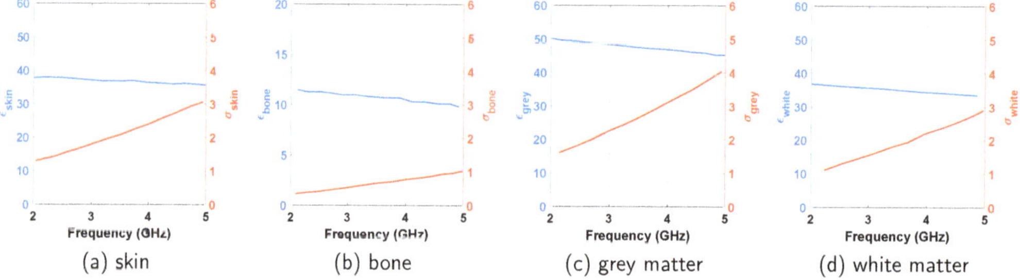

Figure 7. Frequency dependent dielectric constant (ϵ) and conductivity (σ) of the tissues to design the neurosensing system.

4.1. Antenna Design

As in the previous section, the two antennas here are simple patch antennas, and tuned to operate at 4.8 GHz (see Figure 3a). As antenna substrate, Rogers RO4003 dielectric has been used, with thickness of 0.762 mm Each of the conducting surfaces (patch, ground, circular rings) of the arrangement is coated with PDMS ($\epsilon_r = 2.8, tan\delta = 0.002$) with a chosen thickness of 1.5 mm.

For the two antennas, $L1 = L2 = 15.24$ mm, $W1 = W2 = 19.31$ mm. The length of the feed ($feed\_len$) is 5.24 mm, and the width ($feed\_wid$) is 0.5 mm. The dimension of the external and implant antenna substrates are $2W1 \times 2L1$ and $1.2W2 \times 1.2L2$, respectively.

Rogers RO4003 has been used as the superstrate material with a thickness of 8.5 mm. Metamaterials are printed on both sides of the superstrate. The gap between the external patch and the superstrate is filled with PDMS material which is 18.675-mm thick. The metamaterial imprint is similar to the circular rings in Figure 3b, where we choose $r = 2.1$ mm, $c = 0.8$ mm, and $d = 0.2$ mm. A sectional view of the whole arrangement is presented in Figure 8a.

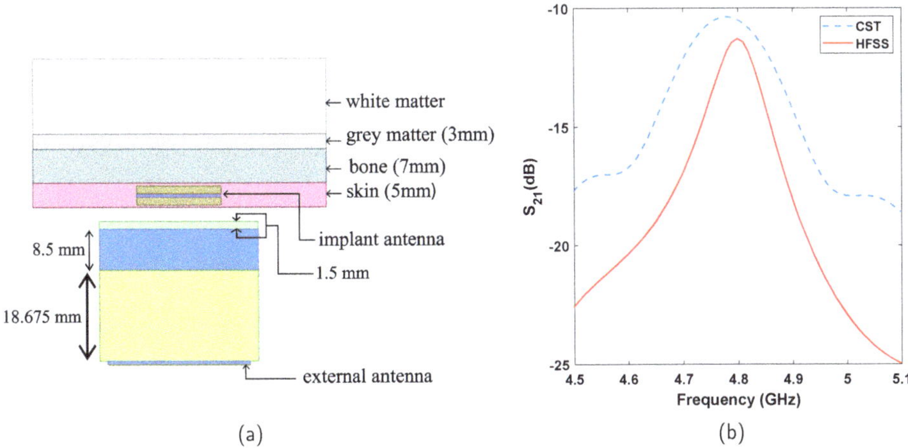

Figure 8. (**a**) Sectional view of the proposed neuropotential recording system. (**b**) Transmission performance of the design.

4.2. Performance Analysis

After designing a biopotential recording system with the proposed concept, we simulate the transmission characteristics between the two antennas. The simulated transmission plots are shown in Figure 8b. At the resonant frequency of 4.8 GHz, the system shows a transmission loss of −11.27 dB in the simulation in HFSS (−10.50 dB in the simulation in CST). This transmission performance compares very well with some of the existing works as presented in Table 2. The dimension of the is also comparable with what has been reported in those works. However, the inclusion of superstrate and filling material makes the proposed design slightly more bulky and complex compared to those.

Table 2. Comparison of the performance of the proof of concept design with some of the existing biosensing works.

Reference	Operating Frequency	Maximum S_{21}	External Antenna Surface	Implant Antenna Surface
[3]	≈4.8 GHz	−19 dB (−17 dB in simulation)	145 mm diameter	15 × 16 mm^2
[28]	2.4 GHz	−22.5 dB	12 × 12 mm^2 9 × 10 mm^2, etc.	same as the external antenna
[29]	2.4 GHz	−21.4 dB	24.9 × 24.9 mm^2	N/A
[30]	≈4.8 GHz	−18.2 dB	40 × 40 mm^2	19.94 × 29.17 mm^2
[31]	400 MHz (2.4) GHz	−33 dB	22 × 23 mm^2	N/A
[32]	400 MHz	−24 dB	26.8 × 28 mm^2	N/A
proposed design	4.8 GHz	−11.27 dB (in HFSS) −10.50 dB (in CST)	30.48 × 38.62 mm^2	18.29 × 23.172 mm^2

An important analysis for such neurosensing applications is the performance degradation of the system under possible misalignment of the external antenna. For the proposed antenna system, we also explore this performance degradation misalignment from the center position. The proposed system appears to be very stable with the misalignment along both the x (along the feedline) and y axis (perpendicular to the feedline) (see Table 3), with maximum performance degradation of only 2.41 dB, for 1 cm offset.

Another key consideration is the implementation of the proposed concept requires a layered geometry with fine tuned parameters, and a slight change in any of the thicknesses (superstrate, dielectric fill, air gaps, etc.) should also influence the performance of the design. Therefore, we present a short sensitivity analysis with the change of these parameters. For the dielectric gap between the source and the superstrate, a 1 mm increase (decrease) of the dielectric thickness degrades the performance by 0.1(0.48) dB. For the superstrate, similar increase (decrease) of the thickness decreases transmission by 0.15(0.15) dB. Such low performance degradation is encouraging as in practice, it may not always be possible to assemble the system with strict dimension requirements. However, maintaining a 3 mm air gap may also be challenging, and, for the gap thickness, a 1 cm increase results in a more prominent degradation of around 5.9 dB. This is comparatively large, but the systems still maintains a very high transmission performance.

Table 3. Performance degradation of the proof of concept design with misalignment of the external antenna.

Misalignment along x Axis		**Misalignment along y Axis**	
Distance (mm)	S_{21} degradation (dB)	Distance (mm)	S_{21} degradation (dB)
−10	2.41	−10	1.11
−5	0.67	−5	0.14
5	0.62	5	0.14
10	2.33	10	1.11

5. Conclusions

A new method of transmission enhancement and misalignment mitigation between internal and external antenna pair for biopotential sensing is proposed in this paper. A full-wave simulation of a typical neuropotential system is carried out, taking into account the realistic permittivity and conductivity of biological tissues, along with proper modeling of the antenna pair. Exploiting the theory of Fabry-Perot resonators and metamaterials, the proposed technique achieves a very high transmission gain of around −11 dB. The design also appear to be very stable under possible misalignment of the antenna pairs, and small dimension mismatch from the optimum design. The proposed technique is anticipated to find its application in the area of biosensing for the diagnosis, monitoring, and treatment of several critical health conditions.

Author Contributions: Conceptualization, M.S.I. and M.A.I.; methodology, M.S.I. and M.A.I.; software, M.S.I., A.K. and M.A.I.; validation, A.K. and M.A.I.; formal analysis, M.S.I.; investigation, M.S.I. and M.A.I.; resources, M.S.I. and M.A.I.; data curation, M.S.I., A.K. and M.A.I.; writing—original draft preparation, M.S.I.; writing—review and editing, A.K. and M.A.I.; visualization, M.S.I.; supervision, M.A.I.; project administration, M.A.I. All authors have read and agreed to the published version of the manuscript.

Funding: This research received no external funding.

Institutional Review Board Statement: Not applicable.

Informed Consent Statement: Not applicable.

Data Availability Statement: Not applicable.

Conflicts of Interest: The authors declare no conflict of interest.

References

1. De Melo, J.L.; Querido, F.; Paulino, N.; Goes, J. A 0.4-V 410-nW opamp-less continuous-time Σ Δ modulator for biomedical applications. In Proceedings of the 2014 IEEE International Symposium on Circuits and Systems (ISCAS), Melbourne, VIC, Australia, 1–5 June 2014; IEEE: Piscataway, NJ, USA, 2014; pp. 1340–1343.
2. Poon, A.S.; O'Driscoll, S.; Meng, T.H. Optimal frequency for wireless power transmission into dispersive tissue. *IEEE Trans. Antennas Propag.* **2010**, *58*, 1739–1750. [CrossRef]
3. Lee, C.W.; Kiourti, A.; Chae, J.; Volakis, J.L. A high-sensitivity fully passive neurosensing system for wireless brain signal monitoring. *IEEE Trans. Microw. Theory Tech.* **2015**, *63*, 2060–2068. [CrossRef]

4. Kiourti, A.; Lee, C.W.; Chae, J.; Volakis, J.L. A wireless fully passive neural recording device for unobtrusive neuropotential monitoring. *IEEE Trans. Biomed. Eng.* **2015**, *63*, 131–137. [CrossRef] [PubMed]
5. Schwerdt, H.N.; Xu, W.; Shekhar, S.; Abbaspour-Tamijani, A.; Towe, B.C.; Miranda, F.A.; Chae, J. A fully passive wireless microsystem for recording of neuropotentials using RF backscattering methods. *J. Microelectromech. Syst.* **2011**, *20*, 1119–1130. [CrossRef] [PubMed]
6. Schwerdt, H.N.; Miranda, F.A.; Chae, J. A fully passive wireless backscattering neurorecording microsystem embedded in dispersive human-head phantom medium. *IEEE Electron Device Lett.* **2012**, *33*, 908–910. [CrossRef]
7. Lee, C.W.; Kiourti, A.; Volakis, J.L. Miniaturized fully passive brain implant for wireless neuropotential acquisition. *IEEE Antennas Wirel. Propag. Lett.* **2016**, *16*, 645–648. [CrossRef]
8. Trentini, G.V. Partially reflecting sheet arrays. *IRE Trans. Antennas Propag.* **1956**, *4*, 666–671. [CrossRef]
9. Feresidis, A.P.; Vardaxoglou, J. High gain planar antenna using optimised partially reflective surfaces. *IEE Proc. Microwaves Antennas Propag.* **2001**, *148*, 345–350. [CrossRef]
10. Kaymaram, F.; Shafai, L. Enhancement of microstrip antenna directivity using double-superstrate configurations. *Can. J. Electr. Comput. Eng.* **2007**, *32*, 77–82. [CrossRef]
11. Liu, Z.G. Fabry-Perot resonator antenna. *J. Infrared Millim. Terahertz Waves* **2010**, *31*, 391–403. [CrossRef]
12. Burokur, S.N.; Latrach, M.; Toutain, S. Theoretical investigation of a circular patch antenna in the presence of a left-handed medium. *IEEE Antennas Wirel. Propag. Lett.* **2005**, *4*, 183–186. [CrossRef]
13. Ziolkowski, R.W.; Heyman, E. Wave propagation in media having negative permittivity and permeability. *Phys. Rev. E* **2001**, *64*, 056625. [CrossRef] [PubMed]
14. Majid, H.A.; Abd Rahim, M.K.; Masri, T. Microstrip antenna's gain enhancement using left-handed metamaterial structure. *Prog. Electromagn. Res. M* **2009**, *8*, 235–247. [CrossRef]
15. Foroozesh, A.; Shafai, L. Investigation into the effects of the reflection phase characteristics of highly-reflective superstrates on resonant cavity antennas. *IEEE Trans. Antennas Propag.* **2010**, *58*, 3392–3396. [CrossRef]
16. Li, X.y.; Li, J.; Tang, J.j.; Wu, X.l.; Zhang, Z.x. High gain microstrip antenna design by using FSS superstrate layer. In Proceedings of the 2015 4th International Conference on Computer Science and Network Technology (ICCSNT), Harbin, China, 19–20 December 2015; IEEE: Piscataway, NJ, USA, 2015; Volume 1, pp. 1186–1189.
17. Enoch, S.; Tayeb, G.; Sabouroux, P.; Guérin, N.; Vincent, P. A metamaterial for directive emission. *Phys. Rev. Lett.* **2002**, *89*, 213902. [CrossRef] [PubMed]
18. Wu, B.I.; Wang, W.; Pacheco, J.; Chen, X.; Grzegorczyk, T.M.; Kong, J. A study of using metamaterials as antenna substrate to enhance gain. *Prog. Electromagn. Res.* **2005**, *51*, 295–328. [CrossRef]
19. Veselago, V.G. Electrodynamics of substances with simultaneously negative ϵ and μ. *Usp. Fiz. Nauk* **1967**, *92*, 517. [CrossRef]
20. Pendry, J.B.; Holden, A.; Stewart, W.; Youngs, I. Extremely low frequency plasmons in metallic mesostructures. *Phys. Rev. Lett.* **1996**, *76*, 4773. [CrossRef] [PubMed]
21. Pendry, J.B.; Holden, A.J.; Robbins, D.J.; Stewart, W. Magnetism from conductors and enhanced nonlinear phenomena. *IEEE Trans. Microw. Theory Tech.* **1999**, *47*, 2075–2084. [CrossRef]
22. Smith, D.R.; Padilla, W.J.; Vier, D.; Nemat-Nasser, S.C.; Schultz, S. Composite medium with simultaneously negative permeability and permittivity. *Phys. Rev. Lett.* **2000**, *84*, 4184. [CrossRef]
23. Yoo, K.; Mittra, R.; Farahat, N. A novel technique for enhancing the directivity of microstrip patch antennas using an EBG superstrate. In Proceedings of the 2008 IEEE Antennas and Propagation Society International Symposium, San Diego, CA, USA, 5–11 July 2008; IEEE: Piscataway, NJ, USA, 2008; pp. 1–4.
24. Kim, J.H.; Ahn, C.H.; Bang, J.K. Antenna gain enhancement using a holey superstrate. *IEEE Trans. Antennas Propag.* **2016**, *64*, 1164–1167. [CrossRef]
25. Liu, H.; Lei, S.; Shi, X.; Li, L. Study of antenna superstrates using metamaterials for directivity enhancement based on Fabry-Perot resonant cavity. *Int. J. Antennas Propag.* **2013**, *2013*, 209741. [CrossRef]
26. Gabriel, S.; Lau, R.; Gabriel, C. The dielectric properties of biological tissues: II. Measurements in the frequency range 10 Hz to 20 GHz. *Phys. Med. Biol.* **1996**, *41*, 2251. [CrossRef] [PubMed]
27. Merli, F.; Fuchs, B.; Mosig, J.R.; Skrivervik, A.K. The effect of insulating layers on the performance of implanted antennas. *IEEE Trans. Antennas Propag.* **2010**, *59*, 21–31.
28. Bahrami, H.; Mirbozorgi, S.A.; Ameli, R.; Rusch, L.A.; Gosselin, B. Flexible, polarization-diverse UWB antennas for implantable neural recording systems. *IEEE Trans. Biomed. Circuits Syst.* **2015**, *10*, 38–48. [CrossRef]
29. Blauert, J.; Kiourti, A. Theoretical modeling and design guidelines for a new class of wearable bio-matched antennas. *IEEE Trans. Antennas Propag.* **2019**, *68*, 2040–2049. [CrossRef]
30. Chen, W.C.; Lee, C.W.; Kiourti, A.; Volakis, J.L. A multi-channel passive brain implant for wireless neuropotential monitoring. *IEEE J. Electromagn. RF Microwaves Med. Biol.* **2018**, *2*, 262–269. [CrossRef]
31. Liu, Y.; Chen, Y.; Lin, H.; Juwono, F.H. A novel differentially fed compact dual-band implantable antenna for biotelemetry applications. *IEEE Antennas Wirel. Propag. Lett.* **2016**, *15*, 1791–1794. [CrossRef]
32. Xu, L.J.; Duan, Z.; Tang, Y.M.; Zhang, M. A dual-band on-body repeater antenna for body sensor network. *IEEE Antennas Wirel. Propag. Lett.* **2016**, *15*, 1649–1652. [CrossRef]

 sensors 

Communication

Integration and Prototyping of a Pulsed RF Oscillator with an UWB Antenna for Low-Cost, Low-Power RTLS Applications

Stefano Bottigliero * and **Riccardo Maggiora**

Department of Electronics and Telecommunications, Politecnico di Torino, 10129 Torino, Italy; riccardo.maggiora@polito.it
* Correspondence: stefano.bottigliero@polito.it

Abstract: The goal of this paper is to present a compact low-cost and low-power prototype of a pulsed Ultra Wide Band (UWB) oscillator and an UWB elliptical dipole antenna integrated on the same Radio Frequency (RF) Printed Circuit Board (PCB) and its digital control board for Real Time Locating System (RTLS) applications. The design is compatible with IEEE 802.15.4 high rate pulse repetition UWB standard being able to work between 6 GHz and 8.5 GHz with 500 MHz bandwidth and with a pulse duration of 2 ns. The UWB system has been designed using the CST Microwave Studio transient Electro-Magnetic (EM) circuit co-simulation method. This method integrates the functional circuit simulation together with the full wave (EM) simulation of the PCB's 3D model allowing fast parameter tuning. The PCB has been manufactured and the entire system has been assembled and measured. Simulated and measured results are in excellent agreement with respect to the radiation performances as well as the power consumption. A compact, very low-power and low-cost system has been designed and validated.

Keywords: elliptical dipole antenna; EM/circuit co-simulation; low-cost; low-power; power gating; RF oscillator; RTLS; ultrawide band antennas

Citation: Bottigliero, S.; Maggiora, R. Integration and Prototyping of a Pulsed RF Oscillator with an UWB Antenna for Low-Cost, Low-Power RTLS Applications. *Sensors* **2021**, *21*, 6060. https://doi.org/10.3390/s21186060

Academic Editor: Pedro Pinho

Received: 5 July 2021
Accepted: 7 September 2021
Published: 10 September 2021

1. Introduction

In recent years a great interest has been shown in UWB localization technology, demonstrated by the definition of the IEEE 802.15.4 standard for precision ranging [1]. The main reason is that its peculiar characteristics are suitable for high accuracy real time indoor localization [2]. In this paper we propose a compact, very low-power and low-cost solution for both the RF module of a transmitting tag that integrates a pulsed RF oscillator with a UWB antenna, and a digital pulse sequence generator that drives the RF oscillator circuit. The digital and RF modules are connected together to a rechargeable battery. The integration of the pulsed RF oscillator with the UWB antenna and the digital sequence generator is a major requirement in order to reduce the tag dimensions, manufacturing cost and power requirements. The main issue is to be able to tune the oscillator output with the antenna using the smallest PCB area possible. To achieve this result, it is important to understand the design implications when a real PCB is involved. Thanks to the EM/circuit co-simulation approach, we are able to evaluate the effects of the different PCB elements on the tag behavior.

Using a certain number of receiving sensors and a receiving computer it is possible to implement the localization engine for a multitude of such transmitting tags.

In the following Section 2 we will introduce the design details of both RF and digital modules. In Section 3 we will discuss the RF simulation setup and results. In Section 4 we will present measurement results of the manufactured prototype and in Section 5 we draw the conclusions.

2. Design

The transmitting tag is composed of two boards. The first board hosts the digital and power gating circuit while the second board generates the carrier frequency using an RF pulsed oscillator driven by an external signal. The RF oscillator output is provided to the linear vertically polarized elliptic dipole antenna. The tag's high level block diagram is shown in Figure 1 and the different blocks are described in the following.

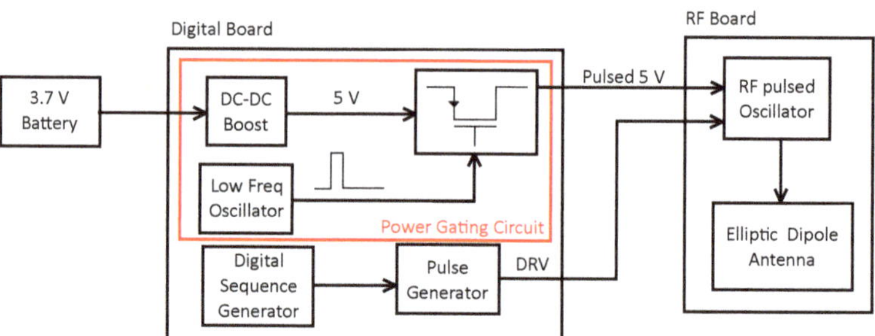

Figure 1. Tag's high level block diagram. The digital board generates the pulsed voltage supply and the modulating pulse sequence for the RF oscillator.

The RF pulsed oscillator circuit topology and its design methodology are presented in [3] where antenna and oscillator are on two separated printed boards. The design follows the method used for negative resistance oscillators where the Barkhausen criteria are satisfied so that the imaginary part of the input impedance, the one seen from the base of the transistor, is $Im(Z_{in}) = 0$ while the real part of the same impedance is $Re(Z_{in}) < 0$. To tune the resonance frequency is important to properly balance the reactance on the emitter to maximize the negative conductance at the base of the transistor. The oscillator operates in a common collector configuration; a command signal drives the emitter of the Infineon BFP740 SiGe-BJT transistor while the output is taken from the collector and sent to the antenna. The final RF pulsed oscillator schematic is shown in Figure 2.

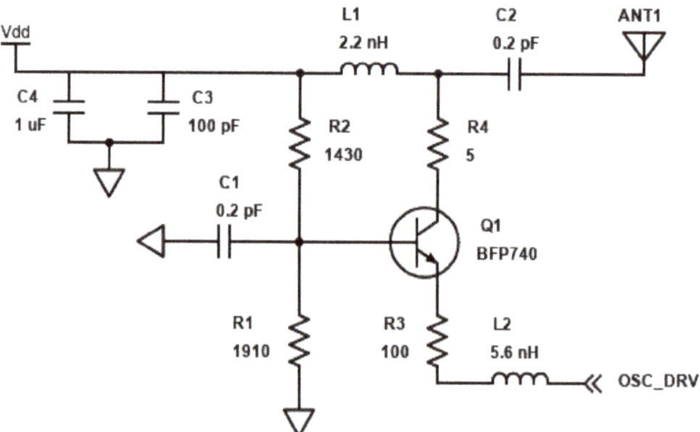

Figure 2. Schematic of the 7 GHz pulsed oscillator

The passive components values obtained during the design phase, are the starting point for the parameter tuning simulation phase to center the oscillating frequency at 7 GHz using a supply voltage of 5 V. Thanks to the co-simulation method, we are able to integrate the PCB, surface mount devices (SMDs) component and antenna contribution to the oscillator analysis, to estimate their effects, and to optimize the passive SMD components value accordingly. A RO4350B core 0.508 mm thick with very low losses (tanδ = 0.003) [4] has been adopted as dielectric substrate for the PCB hosting the RF pulsed oscillator and the antenna.

The results of the simulation phase are shown in Figure 2. In summary, to match the signal amplitude and frequency requirement, it was necessary to have a very small the capacitance at the base of the transistor and at the output of the circuit.

The antenna connected to the RF pulsed oscillator is a microstrip elliptic dipole antenna in linear vertical polarization. Microstrip antennas are inexpensive compared to ceramic chip ones currently on the market and have comparable performances [5]. An experimental study for UWB elliptic dipole antennas is presented in [6]. In this study only FR-4 and high dielectric constant materials were used.

In our case, we decided to adopt the same design methodologies but using different RF dielectric constant materials that allowed us to integrate the RF pulsed oscillator and the antenna on the same PCB and to reduce the overall dimensions. By changing the ratio between the minor semi-axis b and the major semi-axis a of the ellipses it is possible to extend the antenna bandwidth. Starting from the elliptic configuration, we optimized it. A unitary ratio between the semi-axis was sufficient to cover the required bandwidth of 500 MHz.

The geometrical parameters of the optimized antenna are reported in Figure 3.

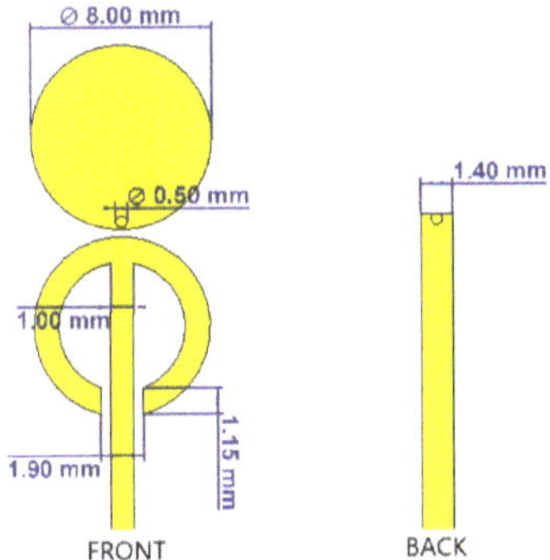

Figure 3. Antenna geometric parameters, front (**top**) view on the left and back (**bottom**) on the right.

The digital sequence generator circuit provided the transmitted sequence signal to the RF pulse generator that implemented the 2 ns pulse signal driving the RF pulsed oscillator. The tag transmitted a sequence of pulses modulated using the On-Off Keying (OOK) technique. The whole sequence was hardwired for each tag and it was 15 bits long where the first 7 bits represented a preamble common to all tags and the latter 8 bits were a unique tag ID number. The preamble took the specific values of a Barker 7 code [7]. We used a modified version of the code where there were no pulses in correspondence of −1 in the sequence, this allowed for a simpler receiver architecture without losing the benefit of Barker codes. In this way, the carrier frequency generated by the RF oscillator was OOK modulated so that, when a sequence bit was equal to "1", a pulse was transmitted, and when it was equal to "0", no pulse was transmitted. This modulation simplified the tags's hardware design dramatically, allowing the sequence generation circuit to be only the cascade of two 8 bit shift registers.

The separation in time between two subsequent pulses in a sequence was fixed by the 20 MHz (50 ns) clock signal used to time the shift registers. The sequence repetition frequency (SRF) was set by the power gating circuit (described in the following). The driving signal that modulated the carrier frequency was a 2 ns baseband pulse generated using the circuit shown in Figure 4. The implementation of short pulses may require very high speed and expensive hardware. Here, by using only discrete logic gates, we were able to maintain very low costs.

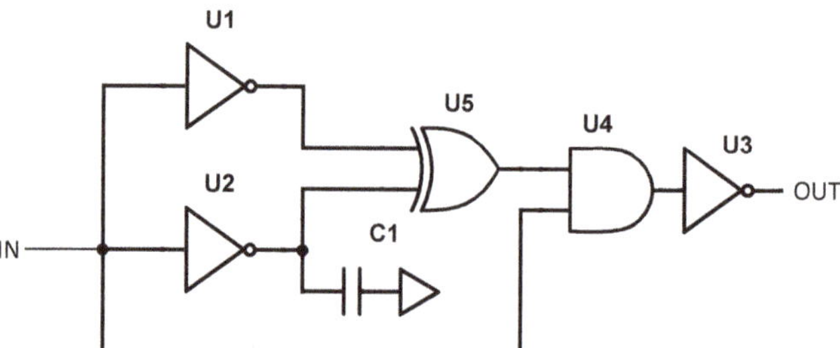

Figure 4. Low-cost 2 ns pulse generation circuit.

The transmitted sequence signal from the shift registers was provided at the input of two equal inverters where one of the two had an additional capacitive load at the output. This allowed us to increase the gate delay arbitrarily. A value of 33 pF for C1 was sufficient to add a delay of 2 ns. The pulse was obtained at the output of the XOR gate and its duration was proportional to the delay between the two inverters' outputs. The pulse sequence went into an AND gate together with the original sequence signal in order to filter out a second unwanted pulse exiting the XOR gate. The signal was inverted and provided to the RF pulsed oscillator emitter.

The power gating circuit was the most efficient solution to drastically reduce the power absorption and its schematic is shown in Figure 5. The supply voltage of 3.7 V was obtained from a rechargeable Lipo battery. The battery voltage was up-converted using a switching DC-DC boost converter to 5 V. The higher supply voltage allowed us to obtain a higher RF output signal.

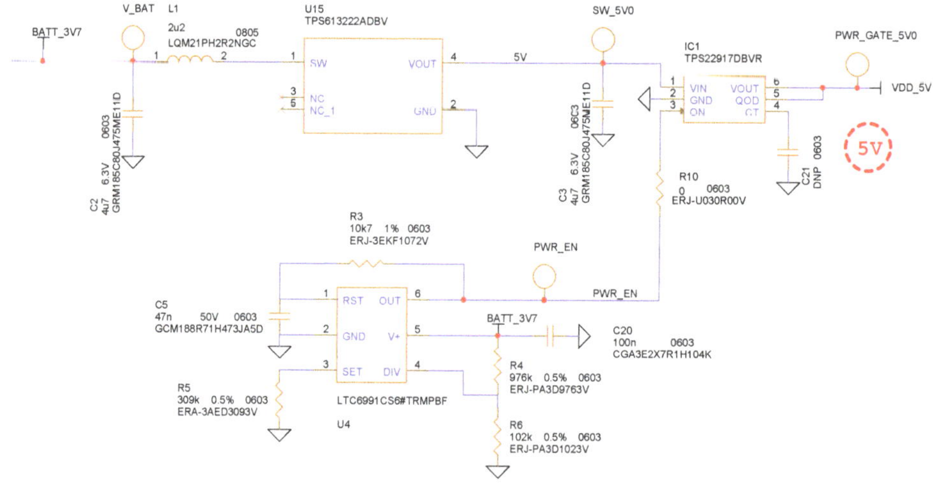

Figure 5. Power gating circuit implemented to reduce power consumptions. Only the LTC6991 low frequency oscillator and the DC-DC converter are always powered on.

The output of the DC-DC converter was provided to a TPS22917 [8] switch from Texas Intruments. The On-Off state of the transistor was controlled by the LTC6991 [9] low frequency oscillator from Analog Devices. This component was the key element of the power gating circuit and allowed us to create low frequency, low duty cycle waveforms. Setting the value of R5 it was possible to set the SRF; in this case we set the value to have 20 sequences per second. The product of R3 and C5 set the duration of the pulse that drove the switch.

The entire transmission of a single 15 bit sequence lasted 750 ns, but in order to take into account the charge and discharge time of the power gating circuit, we had to set the driving pulse duration to a minimum value of 470 μs. In these conditions, both the RF oscillator and the digital circuit were power supplied for only 0.94% of the time for an SRF of 20 Hz (corresponding to a sequence repetition interval of 50 ms). If the application allowed, the SRF could be reduced to less than a repetition per second, further reducing the power consumption.

3. Simulation

The simulations of the antenna and RF pulsed oscillator assembly were performed using CST Microwave Studio adopting the EM/circuit co-simulation method [10]. Examples of usage of this method were presented in [11,12]: in both cases, the co-simulation method allowed them to integrate non linear component and SMD components in the 3D model. The EM simulation was set to have a port for each component in the PCB and to generate the complete scattering matrix of the 3D model and the farfield results.

The 3D model of the PCB is shown in Figure 6.

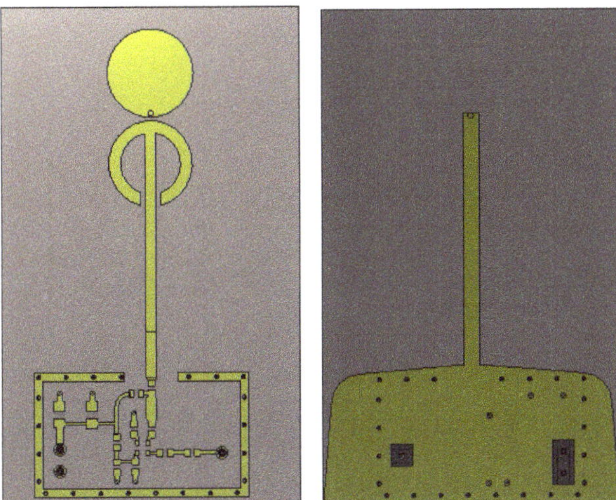

Figure 6. Top and bottom view of the PCB 3D model

The top view shows the dipole antenna, the oscillator circuit and a guard of ground vias while the bottom view shows the ground plane and the pin header used to power the board and to provide the driving signal to the RF pulsed oscillator. The EM simulation allowed us to have a full description of the PCB behavior and to see how it affected the tag functionality. For the complete tag simulation we connected the Gummen Pool SPICE model of the transistor [13] and of the SMD components to the PCB circuital n-port block and performed a transient simulation. In Figure 7 the complete schematic is shown. The part number and value of the simulated components are reported in Table 1.

Table 1. Details of the components used in the simulation.

Component	Value	Part Number	Manufacturer
C1, C2	0.2 pF	GCQ1555C1HR40BB01	Murata
C3	100 pF	GCG1885G1H101JA01D	Murata
C4	1 uF	GRT188C81A105KE13D	Murata
L1	2.2 nH	LQG15HH2N2B02D	Murata
L2	5.6 nH	LQG15HH5N6C02D	Murata
R1	1.91 kΩ	ERJ-2RKF1911X	Panasonic
R2	1.43 kΩ	ERJ-2RKF1431X	Panasonic
R3	100 Ω	ERJ-U02F1000X	Panasonic
R4	5 Ω	ERJ-U02F5R10X	Panasonic
Q1	BFP740	BFP740FH6327XTSA1	Infineon

The co-simulation allowed us to estimate the antenna radiation pattern and the shape of the output signal provided to the antenna. The radiation pattern simulated at 7 GHz is shown in Figure 10 with the blue dashed curve. The first two plots represent the $\phi = 90°$ and $\phi = 0°$ cuts while the third is the equatorial cut $\theta = 0°$ (as shown in Figure 9 where the z axis is parallel to the antenna polarization). The antenna main lobe was slightly tilted upwards and radiated almost uniformly in all ϕ directions.

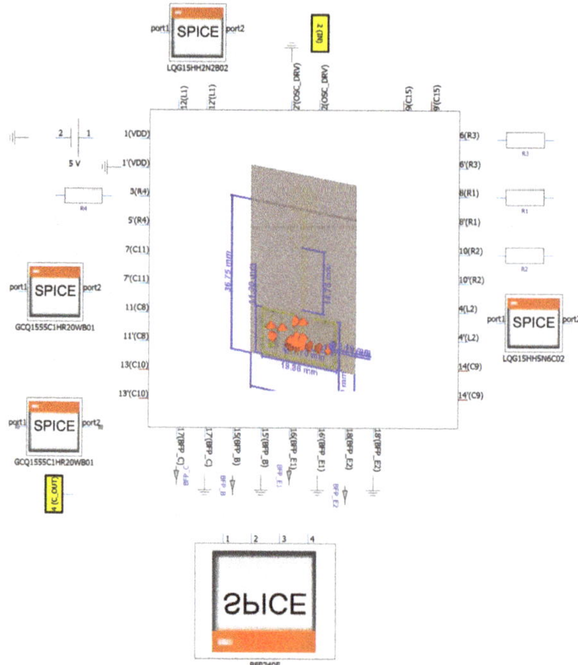

Figure 7. CST transient simulation schematic. The 3D model used in the EM simulation is instantiated as an N-port component.

The transient simulations results are shown in Figure 8. The blue curve represents the command signal, simulated as 2 ns square pulse with 100 ps rise and fall times and 4.5 V amplitude; the red curve is the pulsed oscillator output. The signal had a peak to peak amplitude of 1.5 V and reached the 90% of the maximum amplitude in 2–3 carrier frequency periods. The circuit behaved as intended generating a 2 ns pulse at 7 GHz.

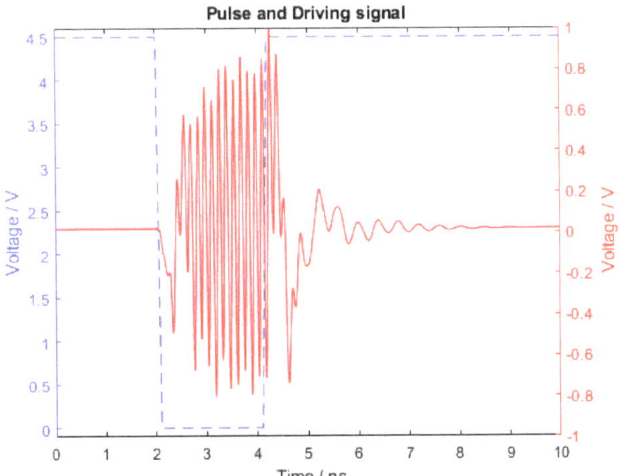

Figure 8. Voltage of the RF output across the output capacitor (red), and the driving signal (blue).

Furthermore, a functional simulation of the digital circuit was performed using PSpice to test the timing and feasibility of the power gating circuit.

4. Results

The tag was manufactured, assembled and measured. The final prototype is shown in Figure 9. It was possible to distinguish two different boards connected one on top of the other, the smaller one was the RF pulsed oscillator and antenna PCB while the other was the digital circuit board. The rechargeable battery was positioned on the bottom side of the digital circuit. The whole system dimensions were 75 mm × 55 mm × 10 mm making the whole tag smaller than a credit card. The tag did not require any programming since all parameters were hardwired.

To evaluate the radiation pattern, the tag was measured in the anechoic chamber of our institution. To perform this operation, the driving signal on the oscillator emitter circuit was fixed to ground setting the oscillator to work in a continuous wave (CW).

The measured radiation pattern is shown in Figure 10 with the red solid curve.

The results were post processed in Matlab using a cubic spline interpolation to reduce noise and normalized to the measured transmitted power of 7.5 dBm to properly compare to the simulated radiation patterns. The comparison between simulations and measurements showed an excellent agreement.

To test the pulsed behavior of the circuit, we connected the oscillator board to the digital control circuit. The signal radiated by the antenna was measured using a receiving antenna probe connected to an high frequency oscilloscope. The measurement setup was calibrated by comparing the output signal of an RF signal generator transmitting 0 dBm in two cases: (1) direct connection between the RF generator and the oscilloscope through a coaxial cable, and (2) an over the air configuration where the RF generator was connected to the tag antenna and spaced apart by a known distance equal to 1 mm from the probe antenna. The losses of the measurement setup were estimated to be equal to L = 9.5 dB. The tag signal measured with the receiving antenna probe at the same distance of 1 mm from the tag antenna is shown in Figure 11. The driving signal in this case was provided by the digital control circuit. The measured signal amplitude was comparable with the simulation results shown in Figure 8 once the calibrated losses, equal to 9.5 dB, were taken into account. The oscillation frequency was measured to be equal to 7 GHz.

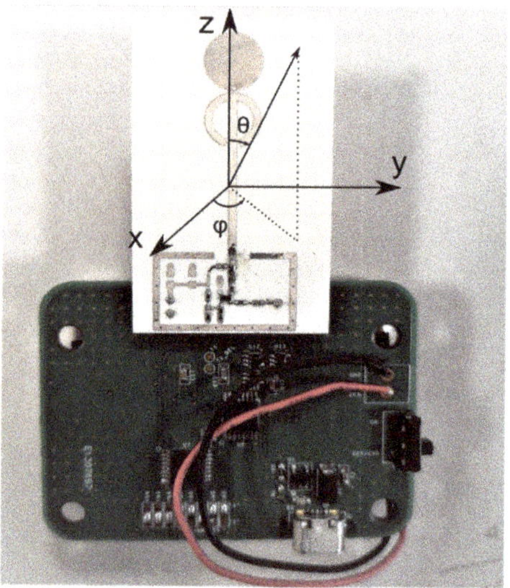

Figure 9. Assembled tag prototype with the oscillator and antenna board connected to the digital control circuit.

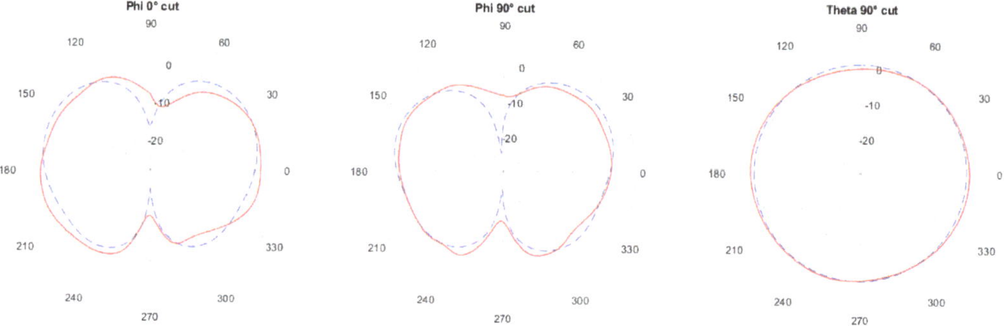

Figure 10. Comparison between the simulated radiation pattern (dashed line) and the measured one (red curve).

Again, the comparison between simulation and measurement results showed an excellent agreement.

The tag power absorption and battery duration were estimated. The tag was connected to a laboratory power supply set at 3.7 V and to a high precision series resistor. By measuring the voltage across the resistor using the oscilloscope, and dividing it by the sensing resistor value, we obtained the current absorption for a single sequence transmission. In Figure 12 the detail of the current absorbed during the 470 µs interval during which the voltage supply was provided to the circuit is shown. We computed the average current consumption over the entire duration of the voltage supply pulse and obtained the average power of a single sequence transmission as

$$P_{avg} = R \cdot I_{avg}^2 = 12.3 \text{ mW}. \quad (1)$$

The energy consumed was equal to the average power of a single sequence transmission multiplied by the voltage supply pulse duration, in our case, 470 µs.

$$E_{singleTX} = P_{avg} \cdot \tau = 6 \ \mu\text{Ws} \quad (2)$$

In our case the SRF was set to 20 repetitions per second leading to a total absorption per second of $E_{abs} = 120$ μWs.

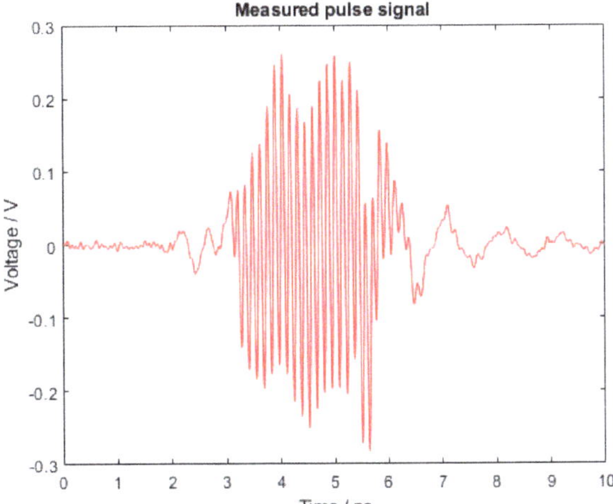

Figure 11. Voltage of the radiated pulse signal measured on the oscilloscope.

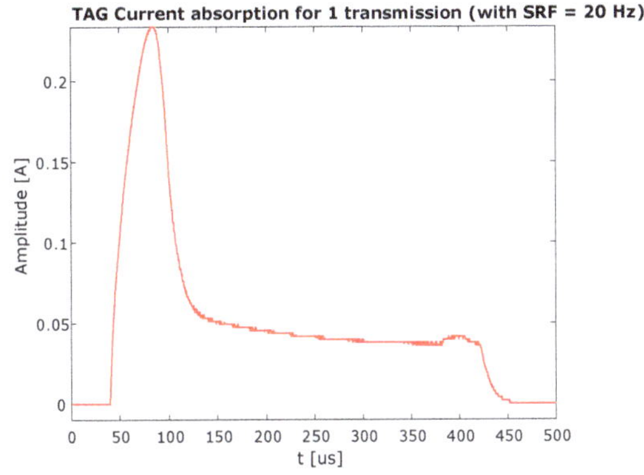

Figure 12. Measured waveform of the current absorbed from the battery, with focus on the 470 μs during the On state of the circuit. The waveform is obtained averaging 64 successive transmissions.

The prototype used a 3.7 V Lipo battery with 1800 mAh current rating, for a total of 6.6 Wh. Taking the ratio between the battery energy rating and the absorbed energy per second we obtained a rough estimation of the battery duration of around 6 years. The energy absorbed by the system during the off time was negligible. These results can be further improved for those RTLS applications that allow less than 20 transmissions per second allowing the battery life span to increase further.

5. Conclusions

In this paper we proposed the design of a compact, low-cost, low-power UWB tag for RTLS application. The RF pulsed oscillator that generates the 7 GHz carrier signal is integrated on a single PCB integrating a custom UWB antenna. The EM/circuit co-simulation method allowed us to evaluate the effects of the PCB on the oscillation frequency and on the antenna radiation pattern and to tune the behavior of the oscillator using the component's SPICE models. The designed solution is cost efficient both for components and bare boards and it is versatile in terms of channel frequency selection through components values tuning. The overall cost of the components and the PCBs is in the order of $10. The digital circuit designed to drive the oscillator is able to reduce the power absorption using power gating techniques drastically increasing the battery life.The power absorption of the entire system have been estimated by measuring the current absorbed from the battery over time and multiplying it by the battery voltage value. The tag has been manufactured to evaluate its performances in terms of radiation pattern, output signal amplitude and frequency and to compare them with simulated results.A comparison based on battery life, dimensions and cost between our tag and two high end industrial solutions [14,15] is reported in Table 2.

Table 2. Comparison between this work and two main industrial solutions.

System	Battery Life (Years)	Dimensions (mm)	Tag Cost (USD)
Sewio	2.8	46 × 55 × 17	30–100
Simatic	1	62 × 95 × 13	30–100
This work	6	55 × 75 × 10	10–20

The obtained results shows that with the adopted method we have been able to integrate the RF oscillator and the UWB antenna on a single board and, together with the power gating technique and low-cost design choices, we obtained an UWB tag for RTLS applications performances better than some already on the market.

Author Contributions: Methodology, S.B. and R.M.; Project administration, R.M.; Resources, R.M.; Software, S.B.; Supervision, R.M.; Validation, S.B. and R.M.; Visualization, S.B.; Writing—original draft, S.B. and R.M.; Writing—review & editing, S.B. and R.M. All authors have read and agreed to the published version of the manuscript.

Funding: This research received no external funding.

Institutional Review Board Statement: Not applicable.

Informed Consent Statement: Not applicable.

Data Availability Statement: Data can be provided upon request to stefano.bottigliero@polito.it.

Conflicts of Interest: The authors declare no conflict of interest.

References

1. Sedlacek, P.; Masek, P.; Slanina, M. An Overview of the IEEE 802.15.4z Standard and Its Comparison to the Existing UWB Standards. In Proceedings of the 29th International Conference Radioelektronika, Pardubice, Czech Republic, 16–18 April 2019.
2. Dabove, P.; Pietra, V.D.; Piras, M.; Jabbar, A.A.; Kazim, S.A. Indoor positioning using Ultra-wide band (UWB) technologies: Positioning accuracies and sensors' performances. In Proceedings of the 2018 IEEE/ION Position, Location and Navigation Symposium (PLANS), Monterey, CA, USA, 23–26 April 2018; pp. 175–184. [CrossRef]
3. Toccafondi, A.; Zampilli, D.; Giovampaola, C.D.; Tesi, V. Low-power UWB transmitter for RFID transponder applications. In Proceedings of the 2012 IEEE International Conference on RFID-Technologies and Applications (RFID-TA), Nice, France, 5–7 November 2012; pp. 234–238.
4. Rogers Corporation Website, Substrate Datasheet. Available online: https://rogerscorp.com/advanced-connectivity-solutions/ro4000-series-laminates/ro4350b-laminates (accessed on 14 April 2021).
5. Taoglas Ceramic Chip UWB Antenna Datasheet. Available online: https://cdn.taoglas.com/datasheets/UWC.01.pdf (accessed on 14 April 2021).
6. Lee, C.-C. An Experimental Study of the Printed-Circuit Elliptic Dipole Antenna with 1.5–16 GHz Bandwidth. *Int. J. Commun. Netw. Syst. Sci.* **2008**, *1*, 295–300. [CrossRef]
7. Rosli, S.J.; Rahim, H.; Ngadiran, R.; Rani, K.N.A.; Ahmad, M.I.; Hoon, W.F. Design of binary coded pulse trains with good autocorrelation properties for radar communications. *MATEC Web Conf.* **2018**, *150*, 6016. [CrossRef]
8. Texas Instruments Website. Available online: https://www.ti.com/product/TPS22917 (accessed on 28 May 2021).
9. Analog Devices Website. Available online: https://www.analog.com/en/products/ltc6991.html (accessed on 28 May 2021).
10. Adrian Scott and Vratislav Sokol "True Transient 3D EM/Circuit CoSimulation Using CST STUDIO SUITE" Article from CST-Computer Simulation Technology AG, Page 7, October 2008. Available online: www.mpdigest.com (accessed on 2 June 2021).

11. Nandyala, C.; Litz, H.; Hafner, B.; Kalayciyan, R. Efficient use of circuit & 3D-EM simulation to optimize the automotive Bulk Current Injection (BCI) performance of Ultrasonic Sensors. In Proceedings of the 2020 International Symposium on Electromagnetic Compatibility—EMC EUROPE, Rome, Italy, 23–25 September 2020; pp. 1–4. [CrossRef]
12. Négrier, R.; Lalande, M.; Joel, A.; Bertrand, V.; Couderc, V.; Pecastaing, L.; Ferron, A. Improvement of an UWB impulse radiation source by integrating photoswitch device. In Proceedings of the 2014 European Radar Conference, Rome, Italy, 8–10 October 2014. [CrossRef]
13. Infineon Website Providing Transistor Model. Available online: https://www.infineon.com/cms/en/product/rf-wireless-control/rf-transistor/ultra-low-noise-sigec-transistors-for-use-up-to-12-ghz/bfp740/ (accessed on 11 June 2021).
14. Sewio Website, UWB Tag Product Page. Available online: https://docs.sewio.net/docs/tag-leonardo-imu-personal-30146967.html (accessed on 23 August 2021).
15. Simatic RTLS Systems, UWB Transponder Page. Available online: https://support.industry.siemens.com/cs/pd/1418186?pdti=td&dl=en&pnid=25277&lc=en-WW (accessed on 23 August 2021).

Article

Conformal Design of a High-Performance Antenna for Energy-Autonomous UWB Communication

Shobit Agarwal [1], Diego Masotti [1,*], Symeon Nikolaou [2] and Alessandra Costanzo [1]

1 Department of Electrical, Electronic, and Information Engineering, Università di Bologna, 40136 Bologna, Italy; shobit.agarwal2@unibo.it (S.A.); alessandra.costanzo@unibo.it (A.C.)
2 Frederick Research Center, Frederick University, 1036 Nicosia, Cyprus; s.nikolaou@frederick.ac.cy
* Correspondence: diego.masotti@unibo.it

Abstract: In view of the need for communication with distributed sensors/items, this paper presents the design of a single-port antenna with dual-mode operation, representing the front-end of a future generation tag acting as a position sensor, with identification and energy harvesting capabilities. An Archimedean spiral covers the lower European Ultra-Wideband (UWB) frequency range for communication/localization purposes, whereas a non-standard dipole operates in the Ultra High Frequency (UHF) band to wirelessly receive the energy. The versatility of the antenna is guaranteed by the inclusion of a High Impedance Surface (HIS) back layer, which is responsible for the low-profile stack-up and the insensitivity to the background material. A conformal design, supported by 3D-printing technology, is pursued to check the versatility of the proposed architecture in view of any application involving its deformation and tracking/powering operations.

Keywords: UHF antennas; ultra wideband antennas; conformal antennas; radio frequency identification

Citation: Agarwal, S.; Masotti, D.; Nikolaou, S.; Costanzo, A. Conformal Design of a High-Performance Antenna for Energy-Autonomous UWB Communication. *Sensors* **2021**, *21*, 5939. https://doi.org/10.3390/s21175939

Academic Editor: Giovanni Andrea Casula

Received: 2 August 2021
Accepted: 1 September 2021
Published: 3 September 2021

1. Introduction

With the rapid increase of so-called "smart dust" in the framework of the Internet of Things, i.e., small devices distributed in the environment able to perform basic sensing operations, there is an increased interest in zero-power pervasive computing [1]. In some cases, this operation can be performed by using inductive or resonant coupling [2,3]. However, this technique is applicable when there is a short distance between the charger and the device, and mainly static objects are involved. When these limitations need to be removed, i.e., moving objects far from the radiofrequency (RF) source have to be charged (such as in logistic applications), Energy Harvesting (EH) and/or Wireless Power Transfer (WPT) are the candidate technologies to make this feasible, because they can extend the battery lifetime or completely avoid its use. This can be done, for instance, by adopting Ultra High Frequency (UHF) combined with Radio Frequency Identification (RFID) technology, if the identification of the device is of primary importance important. Moreover, if we refer to applications with moving objects, operation as identification and localization both become, at the same time, of strategic importance: typically these functionalities are offered by separate wireless systems. The purpose of this paper is to present an advanced antenna solution representing the front-end of a future generation RFID tag able to effectively combine all the aforementioned needs, hence, to be an autonomous position sensor. For these purposes, the UHF band is exploited with the twofold goal of having a tag compatible with the existing RFID tag generations and being energy autonomous by means of the exploitation of dedicated RF energy showers [4]. As per the localization and communication operations, the antenna resort to the well-established Impulse Radio Ultra Wide Band (IR-UWB) technology, which has already demonstrated its effectiveness especially in indoor environment, by reaching sub-meter localization precision through the signal backscattered by the tag itself despite the ultralow power signals involved [4–6].

The proposed solution takes its inspiration from [7], where the authors designed a hybrid UWB-UHF antenna that has a good matching response, but with unsatisfactory performance from the axial ratio point of view. Further, designing an antenna which is insensitive to the background material is of paramount importance in case of the envisaged application as a tag of future generation. This makes an antenna that can be used wherever needed without changing the response according to the background material. For these reasons, the present work starts from the preliminary results of [8] where a first idea to shield the antenna was proposed, but for the UWB band only. Here, the shielding effect is extended for both the bands and a prototype realization is presented. To the authors' knowledge, this is the first HIS able to guarantee both the circular polarization and low back radiation in a frequency band (in the literature just one of the two results is achieved at a time) while also acting as a shield at a completely different frequency (the UHF, in this case). The layout is then slightly changed because of conformal realization of the antenna: this additional step is made to test the versatility of the proposed architecture for future envisaged applications where a deformation is mandatory. As a final step, a planar balun is also added in order to guarantee a safe transition from the balanced antenna to the future diplexing network in microstrip technology responsible for the management of simultaneous UWB communication and UHF energy harvesting.

This paper is organized as follows. Related work is presented in Section 2. Section 3 discusses the design of circularly polarized hybrid antenna followed by the HIS structure design presented in Section 4. Further, modeling and fabrication of the proposed hybrid antenna with a suspended HIS structure and 3-D printed posts is presented in Section 5. Later, a dual-sided conformal design is discussed in Section 6 followed by the complete antenna structure with the inclusion of a balun for impedance transformation in Section 7. The article concludes in Section 8.

2. Related Work

The effective use of IR-UWB technology for the localization and tracking of moving items was first theoretically proved almost one decade ago [9,10], and more recently experimentally demonstrated [5,6,11]. For this reason, many solutions of new tags exploiting the UWB backscattering mechanism have been proposed in the recent past. Many of these solutions envisage the coexistence of both UWB and UHF standards, mostly for compatibility reasons, with the already existing RFID tags operating in the UHF band.

To cite some of them, in [3], a planar antenna is presented for dual-mode passive tag systems. The presented antenna operates in both UHF-RFID band and in the UWB band. A typical UHF-RFID antenna is designed on the top side of the substrate and the slot-loaded antenna for UWB communication is placed on the opposite side. The designed antenna can thus be used for item identification in the two bands and for precise tracking in indoor environments. In [12], a semi-passive UWB RFID system is designed. The tag consists of an RFID antenna operating at 2.4 GHz and a UWB antenna. The 2.4 GHz antenna is used, in this case, to wake up the tag, thus enhancing the battery lifetime. Similarly to [12], in [13] two co-located antennas are used for the dual-mode operation, but an 868 MHz signal is adopted for both waking up the tag and for ensuring compatibility with UHF RFID systems. Another hybrid UHF-UWB tag with two antennas on the same substrate is proposed in [14], where the dual-frequency behavior is just for compatibility with existing UHF RFID tags, not for energy harvesting purposes.

In addition, ref. [15] describes a planar hybrid UHF/UWB single antenna whose main body is used as an antenna at UHF to receive energy from a standard RFID reader, whereas a slot antenna obtained from the previous one, but with a separate port, transmits the UWB signal for communication. However, maintaining a low mutual coupling between the two ports was indeed a tough challenge. A planar inverted-F antenna (UHF band) with an ellipse shaped dipole (UWB band) is reported in [16] that has only identification and localization capabilities. To improve the isolation between the two ports, the design includes a two-port layout with inductive loading, and a complex differential feeding solution in the UWB band. A quarter elliptical patch along with two U-shaped striplines are used as antennas, but only for indoor positioning purposes in [17]: the proposed solution is compact, but it is incapable of mitigating high back radiation.

However, all the aforementioned designs require multi-port excitation and just a few of them are energy-aware, even if not battery-less, solutions. As an additional drawback, they do not support circular polarization (CP) in the UWB band. From the latter point of view, the literature offers CP solutions, as in [18,19], by they still are two-ports layouts with no energy autonomy functionality. In [18] the spiral UWB antenna uses two 100 Ω-RF resistors to achieve the needed axial ratio, and a planar loop loaded by meandered lines achieves CP also in the UHF band. The CP of the dual-band reader antenna in [19] is achieved by varying the shape of different patches acting as electric and magnetic dipoles in the UWB band, and through a complex feeding network piloting four suspended F-antennas in the UHF band.

Finally, in [7,20], the authors designed the first single-port hybrid UWB-UHF RFID antenna system for both energy harvesting and localization applications with backward compatibility with existing UHF RFID systems. A two-armed dipole antenna is used in the UHF band, and a typical CP UWB Archimedean spiral antenna is designed for localization and tracking purposes. A diplexer is also placed behind the spiral to guarantee high isolation between the two operating frequency bands. As a limitation, the designed antenna provides strong back radiation because of the absence of the ground plane, thus resulting strongly dependent on the background material where it lies on. The present paper tries to bridge this gap, with an improved and advanced antenna layout, also taking a look at the effect of a conformal design on the radiation performance. To the best of our knowledge, there are no solutions of dual-mode antennas in the UHF/UWB bands encompassing all the cited functionalities, i.e., high-performance dual-band operation that can be easily equipped with an energy harvesting unit, single-port layout, insensitivity to the back-ground material, and robustness with respect to deformation.

3. Antenna Design

A schematic representation of the antenna, referred to as Iteration 1 design in this article, with all the parameters is shown in Figure 1, and the optimized values of the design parameters are listed in Table 1. A paper substrate ($L_{S1} \times W_{S1}$) having dielectric constant, $\epsilon_r = 2.85$, thickness, $h = 5$ mm, and tan $\delta = 0.053$ is first used to design the antenna. From Figure 1, it can be observed that the presented structure is composed of two different antennas. An archimedean spiral antenna that operates in the lower European UWB frequency range (3.1–4.8 GHz) and responsible for communication/localization purposes. Meanwhile, the long dipole antenna, made of the meandered traces of the spiral plus the straight extensions (highlighted in orange in Figure 1a) antenna is operating at UHF band (868 MHz) and also considered as a receiving antenna to receive the electromagnetic (EM) energy able to guarantee its energy autonomy. Regarding the UWB antenna, the design rules follow the standard auto-complementary antenna rules [21], i.e., the copper and empty traces share the same width. The width and the number of turns have been optimized to reach an AR superior than in [7]. An increment in the width of the spiral and of the number of turns was recorded. Regarding the UHF antenna, its physical length is the result of the straight lines and of the spiral loops. In terms of radiation, the close loops cancel each other and the resulting far-field turns out to be horizontally polarized, as the main longer branches prolong the spiral arms. Moreover, the length of the dipole is greater than the standard half-wavelength, i.e., it is 2.5λ long, as can be evinced by the multiple resonances of the reflection coefficient plot.

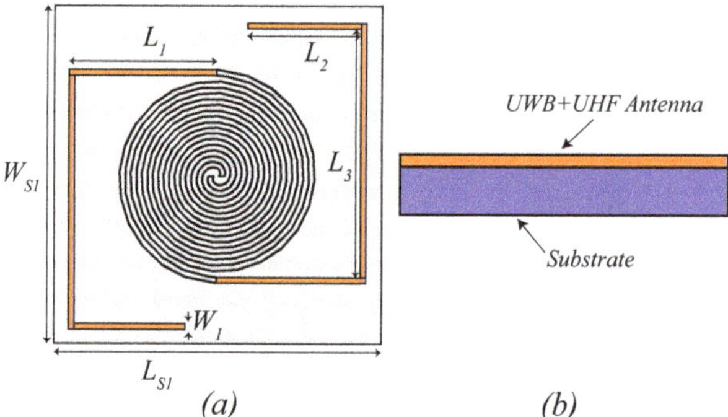

Figure 1. Iteration 1 design (**a**) Top view (**b**) side view.

Table 1. Parameters of Iteration 1 design.

Parameter	Value (mm)
L_{s1}	130.20
W_{s1}	130.20
L_1	63.60
W_1	2.40
L_2	48.20
L_3	103.00

The simulated results of the Iteration 1 design are shown in Figure 2.

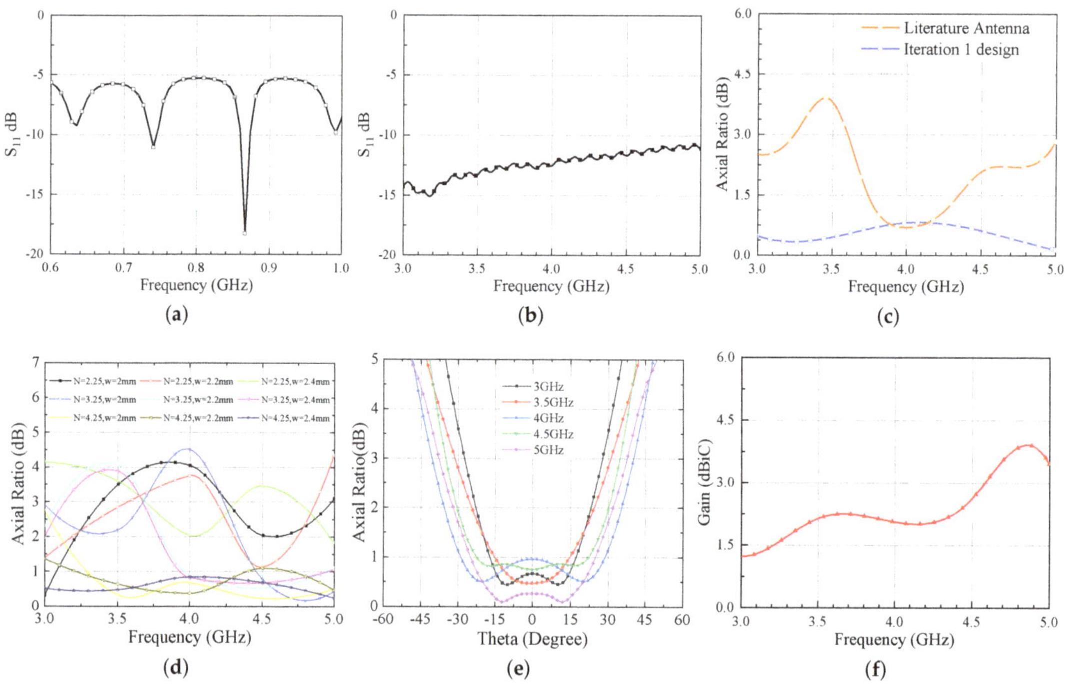

Figure 2. Simulated responses of hybrid UWB antenna (**a**) S_{11} for UHF band (normalized to 12 Ω) (**b**) S_{11} for UWB band (normalized to 120 Ω) (**c**) Axial Ratio v/s Freq (**d**) parametric study of AR optimization (**e**) Axial Ratio v/s Theta (**f**) Gain.

It is evident from Figure 2a,b that the antenna is operating well within the UHF and UWB frequency bands. The dipole antenna bandwidth (considering $|S_{11}|$ at the antenna balanced port ≤ -10 dB) is evaluated with respect to a 12 Ω resistance, because a rectifier, whose input resistance is typically in the range 10–15 ohm at low power levels, is foreseen as a load in this frequency band. Conversely, the UWB bandwidth is computed with respect to the standard impedance of an Archimedean spiral (120 Ω) [21]. In a spiral antenna, the critical parameters that contribute the most to achieve the circular polarization are the number of turns (N) and the width (w) of the traces [22]. Considering this, an extensive parametric study on the design presented in [7] is performed to achieve an improved circular polarization and the optimized response is shown in Figure 2c. Additionally, from Figure 2d, it can be seen that the most effective parameter is N and that a wider trace (hence a larger antenna footprint) is beneficial for the axial ratio, which is below 1 dB within the UWB range for N = 4.25 and w = 2.4 mm. Regarding the dipole length, it can be obviously tuned according to the desired matching condition. In the present application, the antenna length corresponds to the resonant condition (i.e., imaginary part of the antenna input port impedance equal to zero), but of course a proper tuning of the length can be exploited for conjugate matching condition fulfilment with the needed RFID chip impedance [7]. Additionally, the variation of AR w.r.t. to theta is shown in Figure 2e to gain an insight of its variation within the beamwidth. Figure 2f, demonstrates the gain performance of the Iteration 1 antenna. One can observe from the figure that the gain of the antenna slightly increases with the frequency.

Next, the radiation performance of the antenna is studied as shown in Figure 3. The performance is shown at UHF frequency (868 MHz) along with frequencies in UWB range (3–5 GHz). It is evident from the results that the antenna radiates in $\theta = 0\,^{\circ}$ and $\theta = 180\,^{\circ}$ directions. This happens due to the fact that the structure is not backed by the ground plane that deteriorates the radiation properties of this antenna. Next, a plausible solution to mitigate the back radiation using a high impedance surface is discussed.

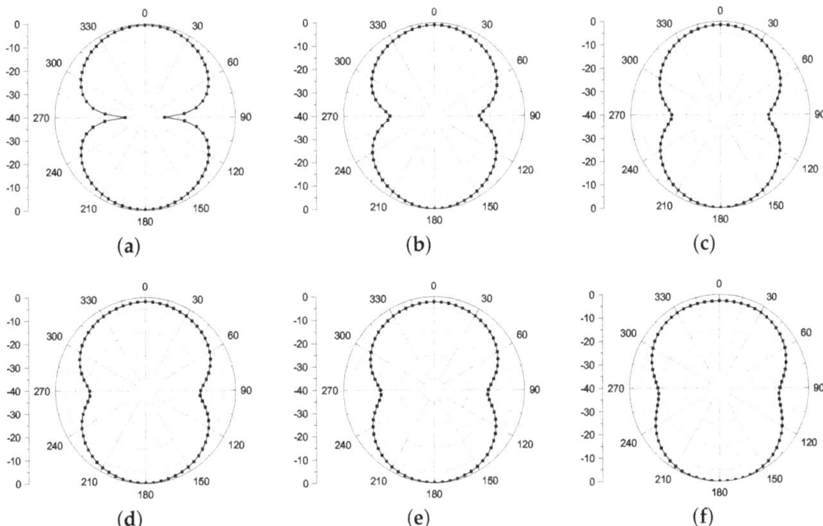

Figure 3. Simulated radiation patterns of Iteration 1 design (xz plane) at (**a**) 868 MHz (**b**) 3 GHz (**c**) 3.5 GHz (**d**) 4 GHz (**e**) 4.5 GHz (**f**) 5 GHz.

4. Design of High Impedance Surface

A high impedance surface (HIS), in general, is a combination of periodically arranged metallic patches placed over a grounded dielectric substrate [23]. HIS can be exploited to serve as a Perfect Magnetic Conductor (PMC). Unlike PEC, a PMC does not introduce any phase shift to the reflected waves; i.e., the incident and reflected wave remain coherent. The HIS structure for UWB spiral antenna designed on a 1-mm-thick paper substrate has already been reported in [8] (similar to the structure shown in Figure 4a). However, due to fabrication hindrance of achieving 1 mm paper thickness precision, the HIS is re-optimized on the Rogers/RT duroid 6002 (ε_r = 2.94, tan δ = 0.0012, h = 0.508 mm) substrate. Furthermore, a novel inverted-L shaped HIS (to mitigate back radiation of UHF dipole) is also designed on the same substrate. A basic difference between rectangular and circular HIS is observed in the simulation setup. In rectangular HIS, the simulation can be performed using a unit cell. Only for the circular HIS must a unit ring be considered. A schematic of the complete layer and circular HIS is shown in Figure 4a,b, respectively. The optimized parameters of the HIS layer are summarized in Table 2 where g_r and g_w are the distance between the rings and gap between the patches within a ring, respectively. The response of the circular HIS is obtained in accordance with [24].

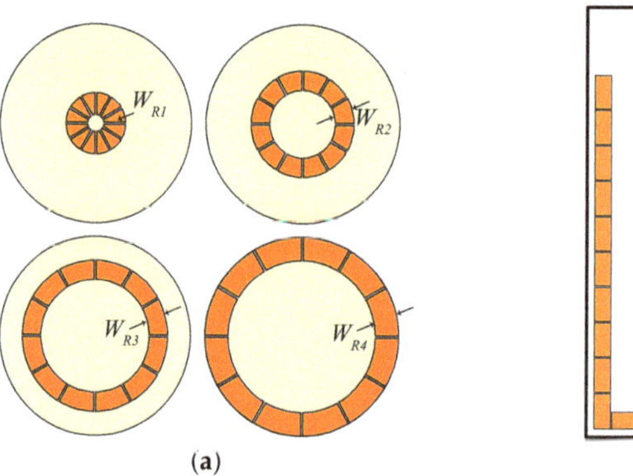

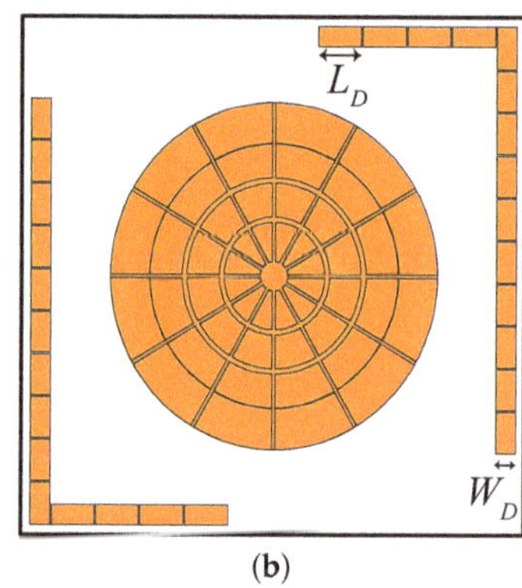

Figure 4. (**a**) Four HIS rings (innermost to outermost) and (**b**) complete HIS structure.

Table 2. Optimized parameters of HIS layer.

Parameter	Value (mm)
W_{R1}	10.10
W_{R2}	9.49
W_{R3}	9.42
W_{R4}	9.49
g_r	0.6
g_w	0.4
W_D	5.5
L_D	9.8

More number of HIS rings can be can be utilized for achieving wideband performance. For this purpose, all the rings must have a similar phase response within the frequency band of interest [25]. In this work, four such rings are designed (Figure 4a) and their phase response is optimized for the UWB band. The optimized phase response of the four unit rings is shown in Figure 5 having a linear phase variation within the operating frequency range; then, in the following simulations, the HIS is considered as a whole (no more as a combination of four separated unit rings). A symmetric structure is maintained along both planes by introducing same number of patches, which in turn helps in improving circular polarization in UWB band. The surface phase response can be tuned within the desired frequency band by changing mostly the width (W_{R1} – W_{R4}) in Figure 4. With increasing value of the widths, the phase response of Figure 5 shift towards lower frequency ranges.

Also, the response shifts a bit towards right with increasing gap value between the patches. Moreover, the number of cells also plays an important role to have a linear phase response. This can be explained with mitigation of the curvature effect with increasing numbers of the patches within a ring.

Further, as shown in Figure 4b, a conventional inverted-L shaped HIS is designed [26], optimized, and inserted behind the dipole antenna operating at UHF band just to act as a shield, since linearly (horizontally in this case) polarized field is radiated at this frequency. Although the back radiation at the UHF range is not completely cancelled with L-shaped HIS (as can be seen in the next Figure 9a), it enhances the efficiency to 76% from 57% at the UHF band. Therefore, it can be concluded that HIS is playing a vital role to maintain a low profile stack up with satisfactory performance for both spiral and dipole antennas.

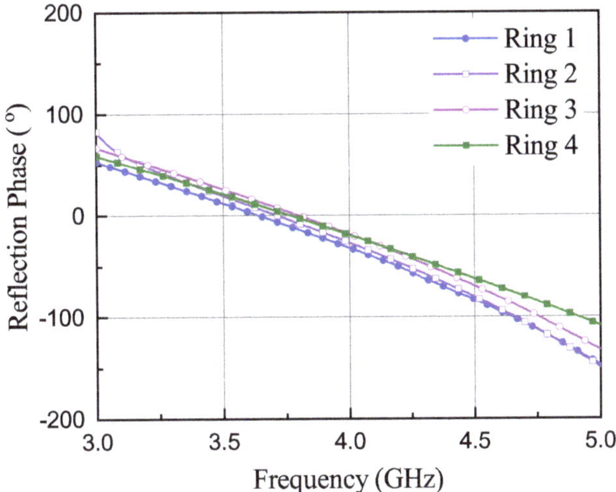

Figure 5. Phase response of four circular HIS rings.

5. Antenna with Suspended HIS and 3-D Printed Posts

As mentioned earlier, due to fabrications perplexities of thick paper substrate, the iteration 1 antenna is also re-designed on the Rogers/RO 6002 substrate with 1.5 mm thickness. Since the parameters of the substrate material are changed, the antenna need to be re-tuned for the desired performance. A schematic of the antenna (referred as Iteration 2 design in the article) is shown in Figure 6a,b, the top view and the perspective view, respectively. The corresponding parameters value of the antenna and the HIS are summarized in Table 3.

Table 3. Optimized Parameters of Iteration 2 design and HIS.

Parameter	Value (mm)	Parameter	Value (mm)	Parameter	Value (mm)	Parameter	Value (mm)
W_S	120	L_1	58.4	W_{R1}	10.10	g_r	0.6
W_g	140	L_2	52.2	W_{R2}	9.49	g_w	0.4
L_S	120	L_3	94.6	W_{R3}	9.42	W_D	5.5
L_g	140	W_1	2.2	W_{R4}	9.49	L_D	9.8

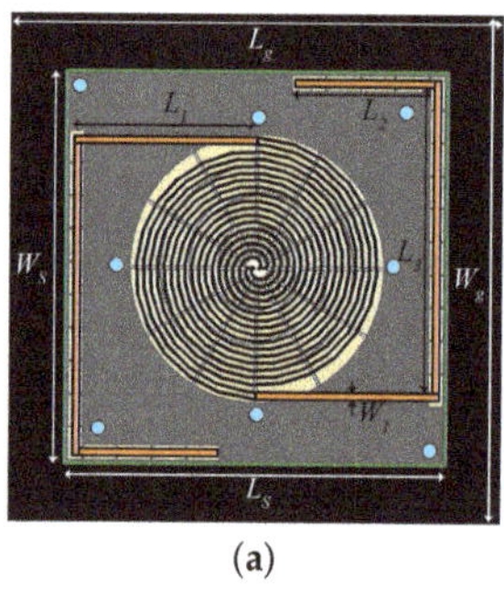

(a)

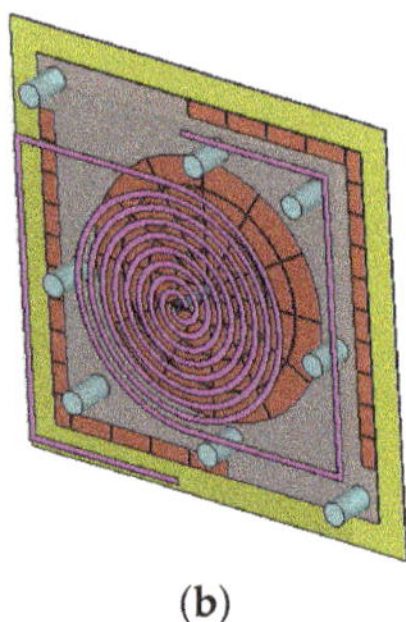

(b)

Figure 6. Antenna with suspended HIS and 3D printed posts: (**a**) front view; (**b**) perspective view (with transparent antenna substrate for ease of visualization). (Iteration 2 design).

A cross section view of the antenna is shown in Figure 7b. From the figure, it can be observed that Iteration 2 design comprises 6-layers viz. hybrid antenna, substrate, suspended air gap, HIS layer, substrate, and the ground plane. To enhance the gain performance of the antenna, which was previously affected by the lower substrate height, an air gap of 13 mm is introduced between the antenna and the HIS layer. To support the top layer, eight 3-D printed posts are utilized. The supporting posts (in light blue in the figure) are designed using 3D printing technology and PLA filament having $\varepsilon_r = 3.5$ and $\tan\delta = 0.04$. The radius of the post, R_{post}, is 4 mm.

Furthermore, a novel feeding mechanism to feed the hybrid antenna is proposed in this work. A schematic of this feeding structure is shown in Figure 7a. It can be observed from the figure that one balanced bi-filar line is inserted within a PLA cylinder with radius of 2 mm. The distance between both lines is 1.2 mm that ensures a characteristics impedance of 120 Ω which is in accordance with standard impedance of the spiral antenna. The PLA cylinder serves two purposes: serving as the dielectric material for the bi-filar lines, and providing central support to the top layer (as shown in Figure 7b).

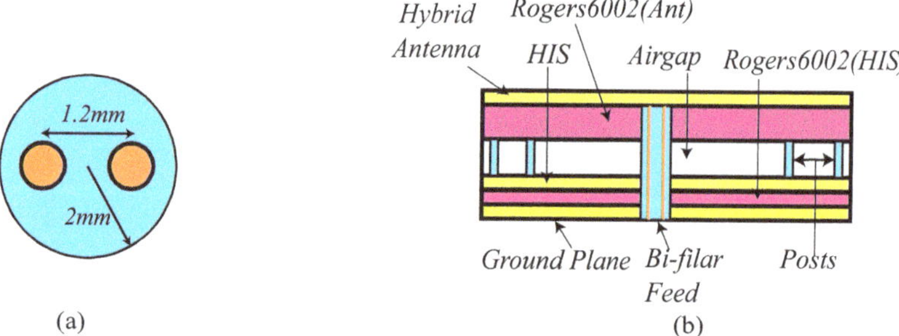

Figure 7. Bi-filar feeding (**a**) top view (**b**) Cross section view with antenna (not to scale).

The simulated results of the Iteration 2 design are shown in Figure 8. Similar to the previous design, the UHF antenna bandwidth is calculated assuming resistive part of a voltage-doubler rectifier for low incident power which is 12 Ω in this case and is displayed in Figure 8a. On the other hand, the UWB bandwidth is estimated with standard spiral antenna impedance [21] and shown in Figure 8b. From the results, it is evident that both antennas are operating well within the intended frequency band. However, a diplexing network would serve as a key element which can separate these two signals in the final realization of the next generation RFID tag. The design of this network will follow the rules already established in [7] and will be a future research activity.

Moving next, the axial ratio and gain results of Iteration 2 design are shown in Figure 8c–e, respectively. From Figure 8c,d, one can observe that the axial ratio of the antenna is below 3 dB which ensures circular polarization within the entire UWB range. Further, the suspended configuration improves the gain performance of the antenna. The simulated gain is ≥8 dBi for the UWB range and ∼4.8 dBi for the UHF band.

Since the HIS is serving as a PMC, the radiation characteristics of the Iteration 2 design are expected to be improved. The simulated E-field radiation patterns of the antenna are shown in Figure 9 at six different frequencies covering UHF and UWB frequencies. It is evident from the figures that after inclusion of HIS, the back radiation is improved by 10 dB for UHF band and more than 20 dB for the UWB band.

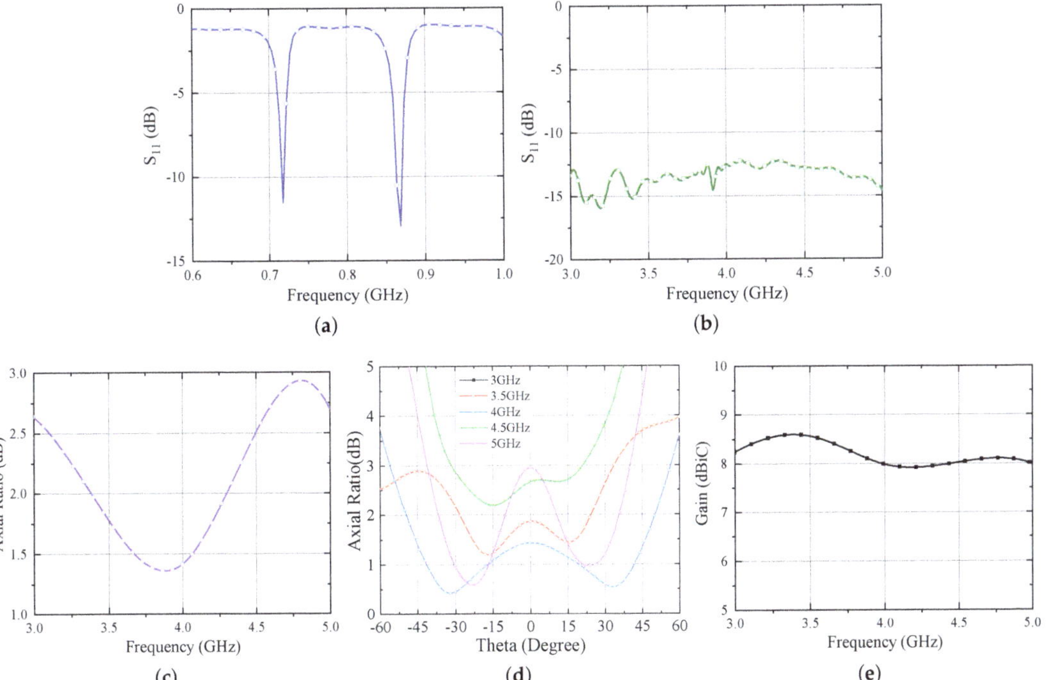

Figure 8. Simulated responses of hybrid UWB antenna (**a**) S_{11} for UHF band (normalized to 12 Ω) (**b**) S_{11} for UWB band (normalized to 120 Ω) (**c**) Axial Ratio v/s Freq (**d**) Axial Ratio v/s Theta (**e**) Gain.

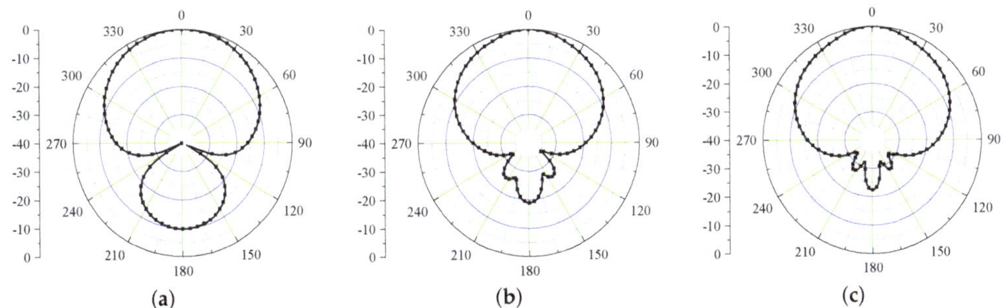

Figure 9. *Cont.*

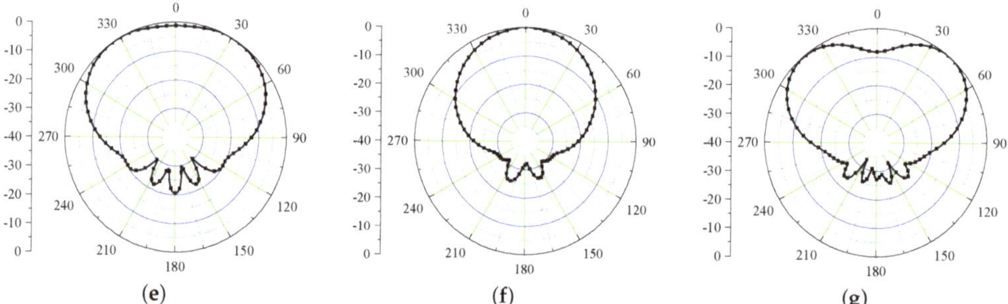

Figure 9. Simulated radiation patterns of Iteration 2 design (xz plane) at (**a**) 868 MHz (**b**) 3 GHz (**c**) 3.5 GHz (**d**) 4 GHz (**e**) 4.5 GHz (**f**) 5 GHz.

Fabrication and Measurement

To prove the veracity of the design, a prototype of the Iteration 2 design is fabricated and reflection coefficient characteristics were measured using Keysight PNA Network Analyzer E8368B (Keysight Technologies, Santa Rosa, CA, USA). The fabricated prototype is shown in Figure 10.

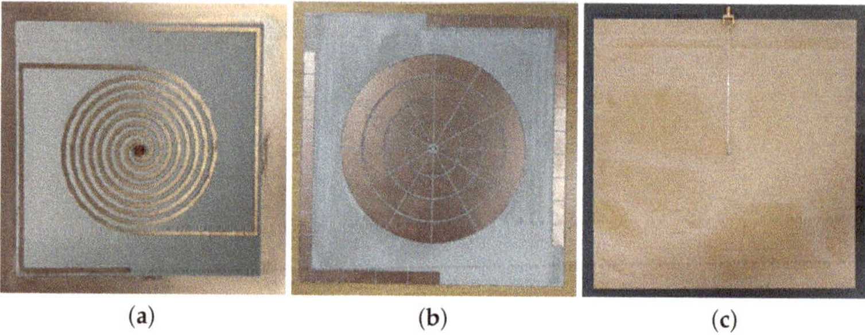

Figure 10. Fabricated prototype (**a**) Hybrid antenna layer (**b**) HIS layer (**c**) Ground back FR-4 layer with feed line.

Considering the difficulty to measure the response directly using balanced bi-filar lines by directly attaching an instrument, one of the two conductors of the bi-filar line is connected to a 120 Ω microstrip line (Figure 10c). To facilitate this, a FR4 substrate ($\varepsilon_r = 4.4$, $h = 1.6$ mm) layer is added behind the ground plane to host the microstrip line and to provide support to the SMA connector used to excite the 120 Ω feed line. The other conductor of the bi-filar line is connected to the ground plane. This abrupt transition has been realized just to have a first measurement capable of demonstrating the correctness of the complex design at this stage. Of course, the correct solution to this issue would be the realization of a planar balun able also to transform the impedance level from 120 Ω to 50 Ω. This additional step has been performed later and is presented in Section 7. Unfortunately, the corresponding fabrication was not possible in reduced time because of the pandemic situation. The simulated and experimental results (both normalized to 120 Ω) are shown in Figure 11. From the figure, it can be seen that the results are in quite good agreement;of course, they are different from the earlier presented results because of the unconventional transition from bi-filar to microstrip line adopted in the measurement.

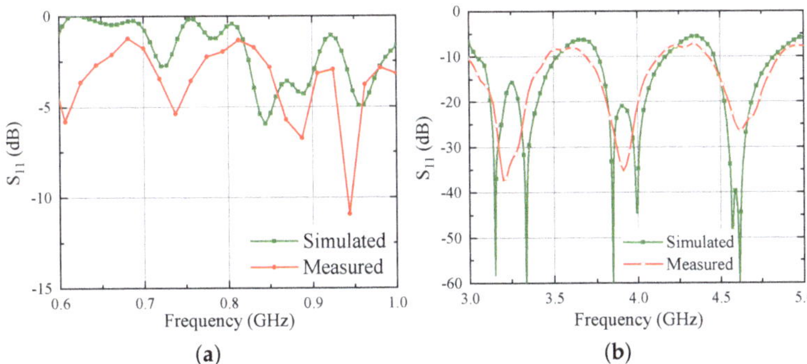

Figure 11. Simulated and measured reflection coefficient of the hybrid UWB antenna (**a**) S_{11} for UHF band (normalized to 120 Ω) (**b**) S_{11} for UWB band (normalized to 120 Ω).

6. Dual Side Conformal Antenna

In this section, a conformal design of the suspended antenna is discussed. Despite the usage of rigid substrates, this investigation is motivated by the fact that a future realization of the antenna (with the diplexer, the energy harvesting circuitry and the UWB backscatter modulator) is foreseen with flexible materials in order to be easily located in a wide range of placements and for a wide selection of applications where localization/tracking and energy autonomy are needed. One envisaged possibility is on the fuselage of a drone for charging it wirelessly while flying. All the parameters of this antenna are identical to the Iteration 2 design discussed in the previous section. A schematic of the conformal antenna is shown in Figure 12. As shown in the figure, the bending is performed around a hemisphere with a radius of 100 mm.

Here one thing to note is that, during bending of the stacked antenna, it was observed that the inverted-L shaped HIS was no more behind the UHF antenna as expected. As a consequence of this, the effect of the HIS at the UHF frequency was completely lost. To mitigate this problem, the size of the HIS layer is scaled up by a factor of 1.05 which results in the total size of 126 × 126 mm^2. The simulated responses of the conformal antenna are shown in Figure 13a–e.

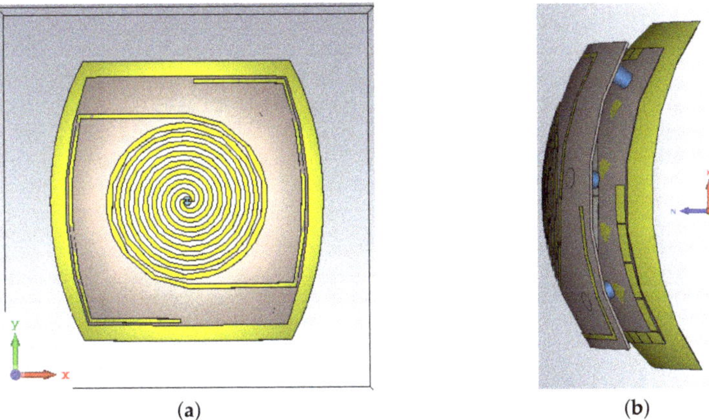

Figure 12. Conformal antenna design (**a**) Top view (**b**) side view.

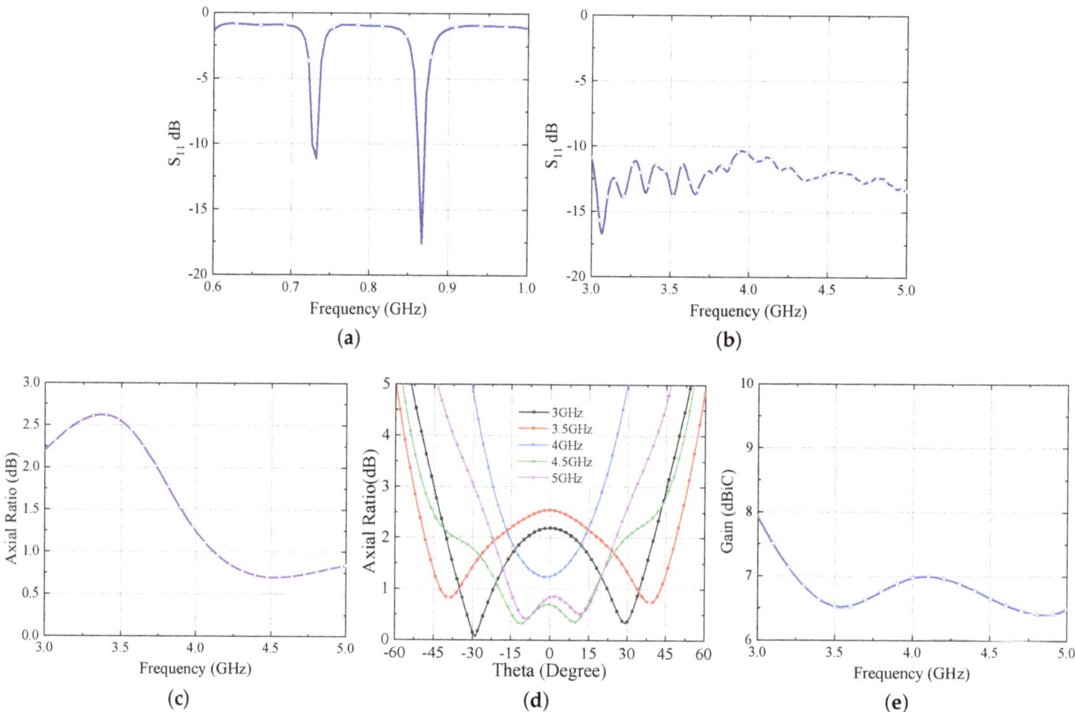

Figure 13. Simulated responses of hybrid UWB antenna (**a**) S_{11} for UHF band (normalized to 12 Ω) (**b**) S_{11} for UWB band (normalized to 120 Ω) (**c**) Axial Ratio v/s Freq (**d**) Axial Ratio v/s Theta (**e**) Gain.

One can observe from the results that the hybrid antenna is operating both in the UHF and UWB band, as expected. Moreover, the axial ratio performance of the antenna is still intact and it provides circular polarization for the entire UWB range despite the significant bending of the antenna layer. Figure 13e displays the gain of the conformal antenna. By comparing this with the Iteration 2 design, the gain of the antenna reduces because of the bending; however, it is still high enough for the communication purposes. As mentioned earlier, after bending, the UHF antenna becomes more vulnerable as it is present at the edges of the substrate; hence, a significant reduction is observed in the gain at UHF band and is 1 dBi at 868 MHz.

Moving further, the simulated radiation characteristics of the conformal antenna at UHF and within UWB frequency range are shown in Figure 14. Here, the major back radiation is observed at the UHF band (a front-to-back ratio of 3 dB is observed). This is due to the fact that, as the antenna is bent in both the planes, the UHF antenna which is already at the edge of the substrate becomes vulnerable to the back radiation. It is worth noticing that this performance is achieved thanks to the slight increase of the HIS previously described. At UWB frequencies (Figure 14b–f), the back radiation is below −15 dB.

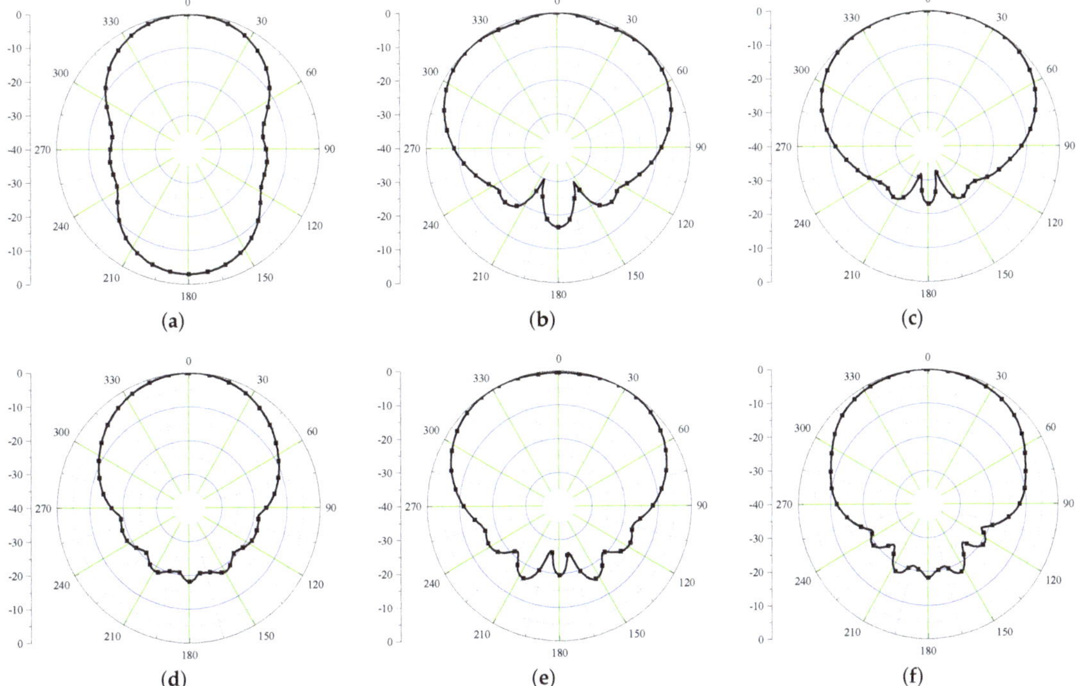

Figure 14. Simulated radiation patterns of conformal antenna (xz plane) at (**a**) 868 MHz (**b**) 3 GHz (**c**) 3.5 GHz (**d**) 4 GHz (**e**) 4.5 GHz (**f**) 5 GHz.

7. Antenna with Balun

As discussed earlier, to feed the spiral antenna using the bifilar line feeding structure reported in this article, a balun is a strategic part to be developed for a direct connection of the dual-mode antenna to a 50 ohm instrumentation. One such balun is designed according to the rules given in [27] on a 0.508-mm-thick Rogers/RO 6002 substrate (ε_r = 2.94, tan δ = 0.0012) as shown in Figure 15. It consists of an un-grounded planar 120 Ω bi-filar line, as a prosecution of the vertical one whose top side is the port 1, followed by an open stub (partially grounded) and a microstrip line (grounded) plus a microstrip impedance step to reach the 50 Ω at the output port (port 2). The parameters of the balun are summarized in Table 4.

Table 4. Optimized dimensions of the proposed metasurface.

Parameter	Value (mm)	Parameter	Value (mm)	Parameter	Value (mm)	Parameter	Value (mm)
BL_1	5	BG_1	20	BL_{in}	8	L_{gb}	60
BL_2	12.4	BG_2	9.3	W_b	2.4	W_{gb}	60
BL_3	14.4	BG_3	27	BG_4	15	W_s	1.4

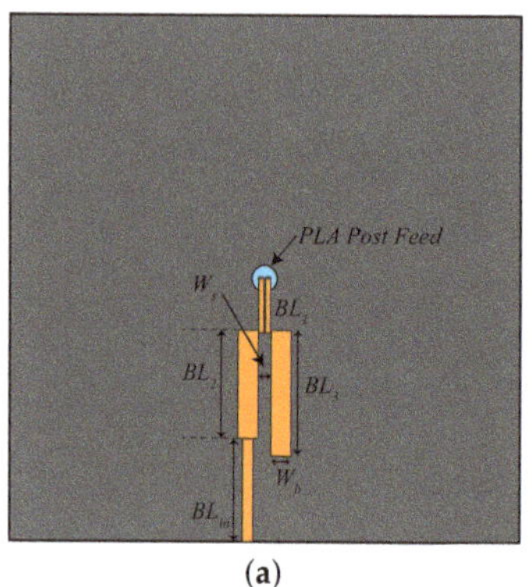

(a)

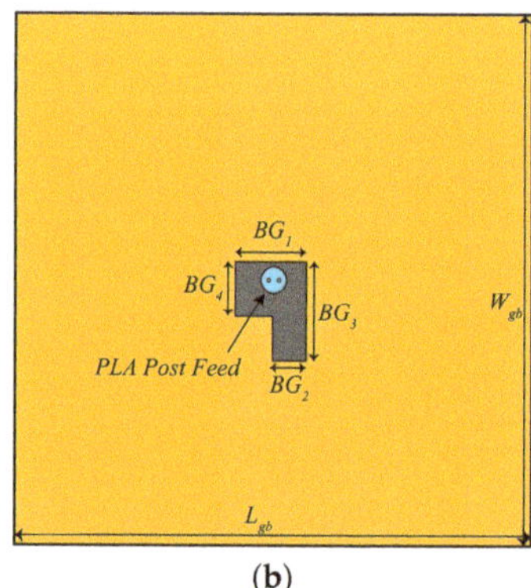

(b)

Figure 15. Balanced to Unbalanced (Balun) design (**a**) Top view (**b**) ground plane.

The simulated results of the balun structure shown in Figure 15 are displayed in Figure 16. As shown in Figure 16a, the transmission parameter between port 1 and 2 varies between −1.2 dB and −1.5 dB within the UWB band. Slightly worse, but still acceptable, performance are achieved at 868 MHz. Of course, the balun design is the result of a delicate trade-off between the performance in the two distant frequency ranges. Next, this balun is deployed behind the Iteration 2 design. One can note that the removal of the ground plane has been limited as much as possible in order not to affect the role of the HIS, even if this cannot be completely assured, as described later.

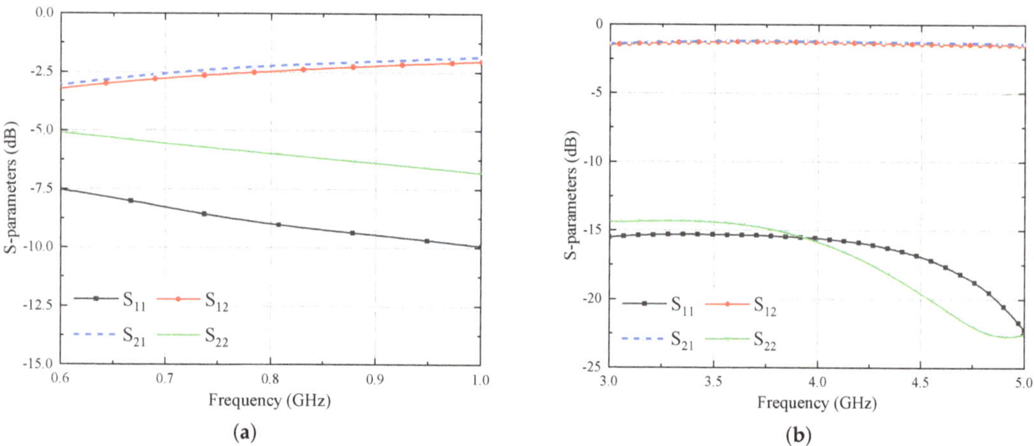

Figure 16. Simulated S-parameter results of planar balun at (**a**) UHF band (**b**) UWB band.

The schematic of the balun inserted at the bottom side of Iteration 2 design is similar to Figure 15. However, due to the large ground plane of the Iteration 2 antenna (140 mm × 140 mm), the value of 50 Ω feed line length (BL_{in}) is increased to 52.6 mm. In addition, the ground plane and the substrate size is increased to 140 mm × 140 mm. The simulated results of the antenna integrated with the balun are shown in Figure 17.

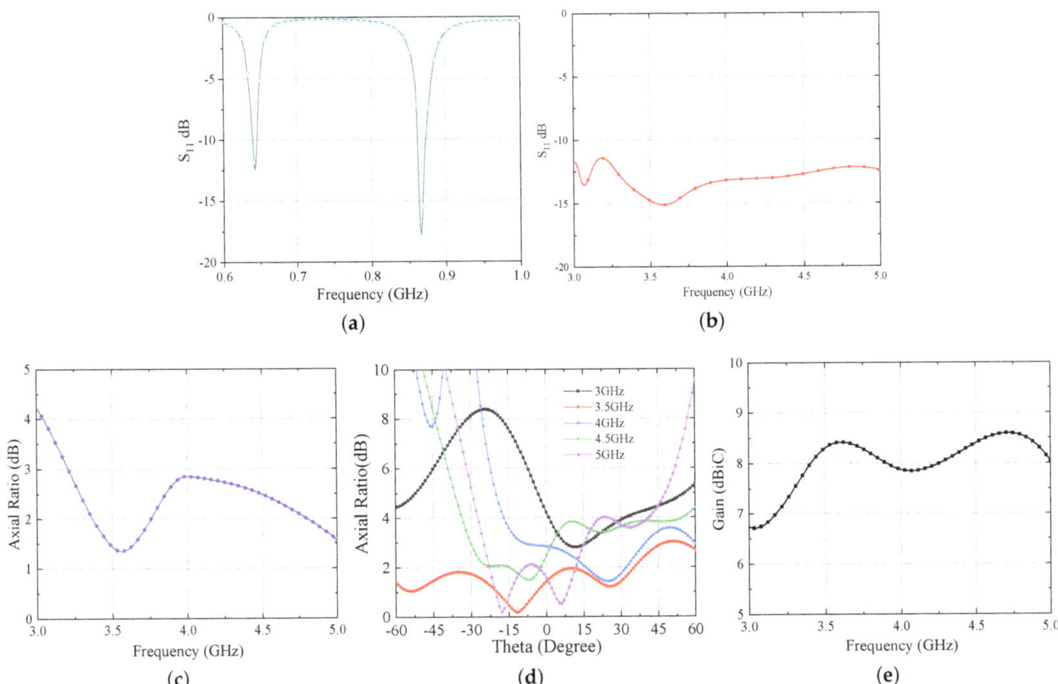

Figure 17. Simulated responses of hybrid UWB antenna (**a**) S_{11} for UHF band (normalized to 12 Ω) (**b**) S_{11} for UWB band (normalized to 120 Ω) (**c**) Axial Ratio v/s Freq (**d**) Axial Ratio v/s Theta (**e**) Gain.

From the figures, it can be observed that the antenna is operating well within the frequency bands of interest. However, the axial ratio of the antenna is compromised a little at the beginning of the UWB band: it remains below 3 dB from 3.22 GHz and provides circular polarization up to 5 GHz. The gain of the antenna is also good with a peak value of 8.6 dBi at 4.7 GHz with an average gain of more than 7 dBi in the entire UWB band, and 3 dBi at the UHF frequency. The improved gain performance of the present multi-mode antenna with respect to previous realizations, lead the authors to state that 10 m of distance from the reader/source for both tracking/localization [5] and energy transmission [28] in the UWB and UHF bands, respectively, could be exceeded. This will be part of our future research activity.

Next, the radiation characteristics of the antenna are computed at six different frequencies as shown in Figure 18.

From the figures, it can be concluded that after insertion of the Balun, the performance of the antenna is intact and is able to mitigate the back radiation at UHF as well as UWB band. However, it can be seen that after insertion of the balun the radiation patterns are slightly slanted. This happened due to the asymmetric removal of the ground plane behind the HIS to ascertain the balun performance. It is also worth noting that, because of the compact layout of the balun, its behavior, not shown for the sake of brevity, is not significantly affected while bending the whole system.

A comparative study of the proposed antenna was carried out with the existing literature and the results are summarized in Table 5. From the table, it is evident that the main advantages of the proposed antenna are its single-port nature that guarantees an easier feeding network, as well as a more compact layout, the circular polarization within the UWB band, and its platform independent behavior thanks to the HIS layer. Further, the robustness of the proposed antenna with respect to dual side bending, not always studied in the literature, is quite encouraging. Therefore, from the presented data, it can be concluded that the proposed antenna represents a step forward if compared with the existing literature designs.

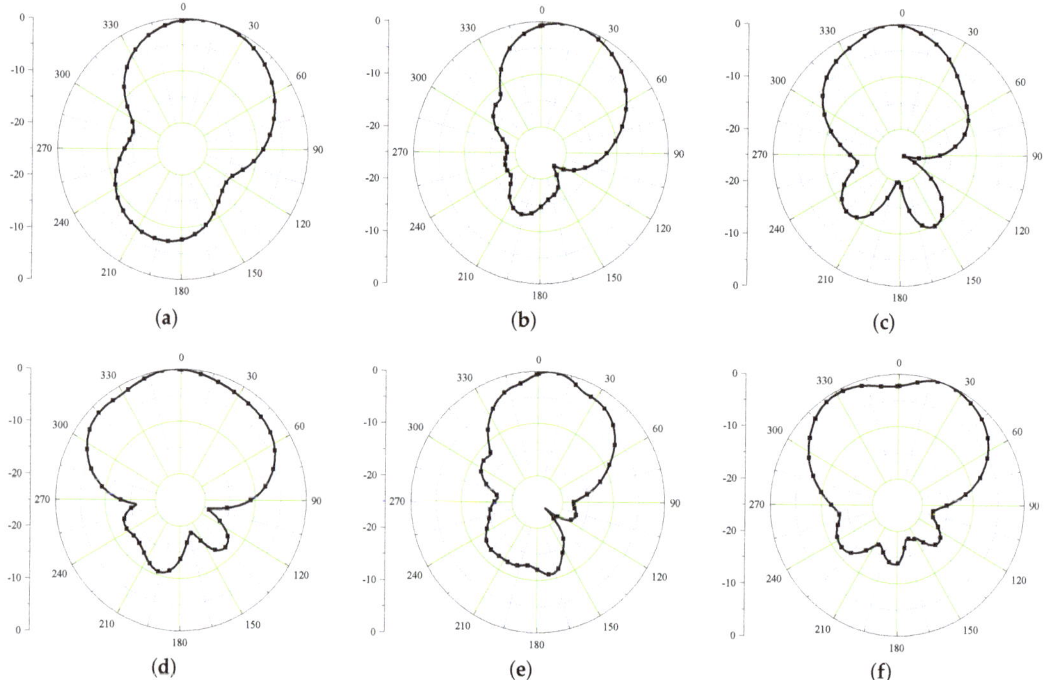

Figure 18. Simulated radiation patterns of conformal antenna (xz plane) at (**a**) 868 MHz (**b**) 3 GHz (**c**) 3.5 GHz (**d**) 4 GHz (**e**) 4.5 GHz (**f**) 5 GHz.

Table 5. Comparison of the proposed antenna with the existing literature.

Ref.	Separate Antennas	No. of Antenna Ports	Circular Polarization (in UWB Band)	Insensitive to Background Material	Robustness wrt Deformation	Energy Harvesting Arrangement
[3]	yes	2	no	no	N.A.	no
[7]	no	1	yes	no	N.A.	yes
[15]	no	2	no	no	N.A.	yes
[16]	yes	2	no	yes	N.A.	no
[17]	yes	2	no	no	N.A.	no
[18]	yes	2	yes	yes	N.A.	no
[19]	yes	2	yes	yes	N.A.	no
This work	no	1	yes	yes	yes	yes

8. Conclusions

A circularly polarized hybrid conformal antenna serving dual purposes of energy autonomy and UWB communication is presented in this article. A suspended high impedance surface is designed and deployed behind the antenna offering, for the first time, a simultaneous control of the back-radiation mechanism at highly separated frequencies, and of the circular polarization in the UWB band. The exploitation of 3D-printed posts plays a strategic role for both the mechanical stability and the original bi-filar feeding strategy of the proposed multilayer architecture. Furthermore, a conformal design of the proposed hybrid antenna is studied for future technological systems, drones for example, that would require wireless charging while flying: despite the bidirectional bending, the structure reveals itself to be quite robust to mechanical changes. The proposed radiating system will be able to play a crucial role as an RFID tag of the next generation as soon as it is equipped with the diplexing network, the harvesting section, and the back-scatter modulator.

Author Contributions: Conceptualization, S.A., D.M., A.C.; methodology, D.M., A.C; software, S.A., D.M.; validation, S.N.; formal analysis, S.A., D.M., A.C.; investigation, S.A., D.M., A.C.; resources, S.A., D.M., S.N., A.C.; data curation, S.A., D.M., A.C.; writing—original draft preparation, S.A., D.M.; writing—review and editing, S.A., D.M., A.C.; visualization, S.A., D.M., A.C.; supervision, D.M., A.C., S.N.; project administration, D.M., A.C., S.N.; funding acquisition, A.C., S.N. All authors have read and agreed to the published version of the manuscript.

Funding: This work was co-funded by the European Regional Development Fund and the Republic of Cyprus through the Research and Innovation Foundation, under the project EXCELLENCE/1918/365—(ICARUS).

Institutional Review Board Statement: Not applicable.

Informed Consent Statement: Not applicable.

Data Availability Statement: Not applicable.

Conflicts of Interest: The authors declare no conflict of interest.

References

1. Costanzo, A.; Masotti, D. Wirelessly powering: An enabling technology for zero-power sensors, IoT and D2D communication. In Proceedings of the 2015 IEEE MTT-S International Microwave Symposium, Phoenix, AZ, USA, 17–22 May 2015; pp. 1–4.
2. Scorcioni, S.; Larcher, L.; Bertacchini, A.; Vincetti, L.; Maini, M. An integrated RF energy harvester for UHF wireless powering applications. In Proceedings of the 2013 IEEE Wireless Power Transfer (WPT), Perugia, Italy, 15–16 May 2013; pp. 92–95.
3. Cruz, C.C.; Costa, J.R.; Fernandes, C.A. Hybrid UHF/UWB antenna for passive indoor identification and localization systems. *IEEE Trans. Antennas Propag.* **2012**, *61*, 354–361. [CrossRef]
4. Decarli, N.; Guerra, A.; Guidi, F.; Chiani, M.; Dardari, D.; Costanzo, A.; Fantuzzi, M.; Masotti, D.; Bartoletti, S.; Dehkordi, J.S.; et al. The GRETA architecture for energy efficient radio identification and localization. In Proceedings of the 2015 International EURASIP Workshop on RFID Technology (EURFID), Rosenheim, Germany, 22–23 October 2015; pp. 1–8.
5. Costanzo, A.; Dardari, D.; Aleksandravicius, J.; Decarli, N.; Del Prete, M.; Fabbri, D.; Fantuzzi, M.; Guerra, A.; Masotti, D.; Pizzotti, M.; et al. Energy autonomous UWB localization. *IEEE J. Radio Freq. Identif.* **2017**, *1*, 228–244. [CrossRef]
6. Decarli, N.; Del Prete, M.; Masotti, D.; Dardari, D.; Costanzo, A. High-accuracy localization of passive tags with multisine excitations. *IEEE Trans. Microw. Theory Tech.* **2018**, *66*, 5894–5908. [CrossRef]
7. Fantuzzi, M.; Masotti, D.; Costanzo, A. A Novel Integrated UWB–UHF One-Port Antenna for Localization and Energy Harvesting. *IEEE Trans. Antennas Propag.* **2015**, *63*, 3839–3848. [CrossRef]
8. Agarwal, S.; Costanzo, A.; Masotti, D. Dual-Purpose Metasurface for Background Insensitive UWB Tag. In Proceedings of the 2021 15th European Conference on Antennas and Propagation (EuCAP), Dusseldorf, Germany, 22–26 March 2021; pp. 1–5.
9. Dardari, D.; d'Errico, R.; Roblin, C.; Sibille, A.; Win, M.Z. Ultrawide bandwidth RFID: The next generation? *Proc. IEEE* **2010**, *98*, 1570–1582. [CrossRef]
10. Costanzo, A.; Mastri, F.; Masotti, D.; Rizzoli, V. Circuit-level nonlinear/EM co-simulation and co-design of UWB receivers. In Proceedings of the 2011 IEEE International Conference on Ultra-Wideband (ICUWB), Bologna, Italy, 14–16 September 2011; pp. 425–429.
11. Zhou, Y.; Law, C.L.; Xia, J. Ultra low-power RFID tag with precision localization using IR-UWB. In Proceedings of the 2011 IEEE MTT-S International Microwave Symposium, Baltimore, MD, USA, 5–10 June 2011; pp. 1–4.
12. Ramos, A.; Lazaro, A.; Girbau, D. Semi-passive time-domain UWB RFID system. *IEEE Trans. Microw. Theory Tech.* **2013**, *61*, 1700–1708. [CrossRef]
13. d'Errico, R.; Bottazzi, M.; Natali, F.; Savioli, E.; Bartoletti, S.; Conti, A.; Dardari, D.; Decarli, N.; Guidi, F.; Dehmas, F.; et al. An UWB-UHF semi-passive RFID system for localization and tracking applications. In Proceedings of the 2012 IEEE International Conference on RFID-Technologies and Applications (RFID-TA), Nice, France, 5–7 November 2012; pp. 18–23.
14. Pigeon, M.; D'Errico, R.; Delaveaud, C. UHF-UWB tag antenna for passive RFID applications. In Proceedings of the 2013 7th European Conference on Antennas and Propagation (EuCAP), Gothenburg, Sweden, 8–12 April 2013; pp. 3968–3972.
15. Ziai, M.A.; John, C.B. UWB/UHF RFID tag. In Proceedings of the 2015 Loughborough Antennas & Propagation Conference (LAPC), Loughborough, UK, 2–3 November 2015; pp. 1–3.
16. An, W.; Shen, Z.; Wang, J. Compact low-profile dual-band tag antenna for indoor positioning systems. *IEEE Antennas Wirel. Propag. Lett.* **2016**, *16*, 400–403. [CrossRef]
17. Shan, X.; Shen, Z. Miniaturized UHF/UWB tag antenna for indoor positioning systems. *IEEE Antennas Wirel. Propag. Lett.* **2019**, *18*, 2453–2457. [CrossRef]

18. Gao, X.; Shen, Z. UHF/UWB tag antenna of circular polarization. *IEEE Trans. Antennas Propag.* **2016**, *64*, 3794–3802. [CrossRef]
19. Zhang, N.; Li, X.; Zhu, H.; Gao, G.; Qi, Z. Compact and Circular Polarization UHF/UWB RFID Reader Antenna. In Proceedings of the 2019 IEEE Asia-Pacific Microwave Conference (APMC), Singapore, 10–13 December 2019; pp. 1337–1339.
20. Fantuzzi, M.; Masotti, D.; Costanzo, A. Simultaneous UHF energy harvesting and UWB-RFID communication. In Proceedings of the 2015 IEEE MTT-S International Microwave Symposium, Phoenix, AZ, USA, 17–22 May 2015; pp. 1–4.
21. Wiesbeck, W.; Adamiuk, G.; Sturm, C. Basic properties and design principles of UWB antennas. *Proc. IEEE* **2009**, *97*, 372–385. [CrossRef]
22. Balanis, C.A. *Antenna Theory: Analysis and Design*; John Wiley & Sons: New York, NY, USA, 2016.
23. Sievenpiper, D.F. High-Impedance Electromagnetic Surfaces. Ph.D. Thesis, University of California, Los Angeles, CA, USA, 2000.
24. Sarrazin, J.; Lepage, A.C.; Begaud, X. Circular high-impedance surfaces characterization. *IEEE Antennas Wirel. Propag. Lett.* **2012**, *11*, 260–263. [CrossRef]
25. Amiri, M.A.; Balanis, C.A.; Birtcher, C.R. Analysis, Design, and Measurements of Circularly Symmetric High-Impedance Surfaces for Loop Antenna Applications. *IEEE Trans. Antennas Propag.* **2016**, *64*, 618–629. [CrossRef]
26. Clavijo, S.; Diaz, R.E.; McKinzie, W.E. Design methodology for Sievenpiper high-impedance surfaces: An artificial magnetic conductor for positive gain electrically small antennas. *IEEE Trans. Antennas Propag.* **2003**, *51*, 2678–2690. [CrossRef]
27. Lim, T.B.; Zhu, L. Compact microstrip-to-CPS transition for UWB application. In Proceedings of the 2008 IEEE MTT-S International Microwave Workshop Series on Art of Miniaturizing RF and Microwave Passive Components, Chengdu, China, 14–15 December 2008; pp. 153–156.
28. Fabbri, D.; Berthet-Bondet, E.; Masotti, D.; Costanzo, A.; Dardari, D.; Romani, A. Long range battery-less UHF-RFID platform for sensor applications. In Proceedings of the 2019 IEEE International Conference on RFID Technology and Applications (RFID-TA), Pisa, Italy, 25–27 September 2019; pp. 80–85.

MDPI

Article

Development and Validation of an ISA100.11a Simulation Model for Accurate Industrial WSN Planning and Deployment

Zoltan Padrah [1,2], Andra Pastrav [2], Tudor Palade [2], Ovidiu Ratiu [1,2] and Emanuel Puschita [2,*]

1 Control Data Systems S.R.L., Liberty Technology Park, 21 Garii Street, 400267 Cluj-Napoca, Romania; Zoltan.Padrah@cds.ro (Z.P.); Ovidiu.Ratiu@cds.ro (O.R.)

2 Communications Department, Technical University of Cluj-Napoca, 28 Memorandumului Street, 400267 Cluj-Napoca, Romania; Andra.Pastrav@com.utcluj.ro (A.P.); Tudor.Palade@com.utcluj.ro (T.P.)

* Correspondence: Emanuel.Puschita@com.utcluj.ro; Tel.: +40-744-760-356

Abstract: During the planning, design, and optimization of an industrial wireless sensor network (IWSN), the proposed solutions need to be validated and evaluated. To reduce the time and expenses, highly accurate simulators can be used for these tasks. This paper presents the development and experimental validation of an ISA100.11a simulation model for industrial wireless sensor networks (IWSN). To achieve high simulation accuracy, the ISA100.11a software stack running on two types of certified devices (i.e., an all-in-one gateway and a field device) is integrated with the ns-3 simulator. The behavior of IWSNs is analyzed in four different types of test scenarios: (1) through simulation using the proposed ISA100.11a simulation model, (2) on an experimental testbed using ISA100.11a certified devices, (3) in a Gateway-in-the-loop Hardware-in-the-loop (HIL) scenario, and (4) in a Node-in-the-loop HIL scenario. Moreover, the scalability of the proposed simulation model is evaluated. Several metrics related to the timing of events and communication statistics are used to evaluate the behavior and performance of the tested IWSNs. The results analysis demonstrates the potential of the proposed model to accurately predict IWSN behavior.

Keywords: industrial WSN; ISA100.11a model; ns-3; WSN

Citation: Padrah, Z.; Pastrav, A.; Palade, T.; Ratiu, O.; Puschita, E. Development and Validation of an ISA100.11a Simulation Model for Accurate Industrial WSN Planning and Deployment. *Sensors* **2021**, *21*, 3600. https://doi.org/10.3390/s21113600

Academic Editor: Carles Gomez

Received: 23 April 2021
Accepted: 19 May 2021
Published: 21 May 2021

Publisher's Note: MDPI stays neutral with regard to jurisdictional claims in published maps and institutional affiliations.

1. Introduction

Wireless sensor networks (WSNs) have become increasingly popular in the industrial domain as they collect and transmit data about the environment and provide a series of advantages that make them easier to expand than the wired networks: configuration flexibility, support for mobility, and reduced infrastructure weight [1].

This paper focuses on a particular category of WSNs, namely the industrial ones. Industrial WSNs (IWSNs) are addressed by several standards, including ISA100.11a [2] (IEC-62734 [3]), WirelessHART (IEC-62591) [4], ZigBee [5], Bluetooth Mesh Networking [6], Wireless Networks for Industrial Automation-Factory Automation (WIA-FA) (IEC-62948) [7,8], and Process Automation (WIA-PA) (IEC-62601) [9].

Problem statement. IWSNs are installed in environments having strict security, control, and safety measures. Such IWSNs are typically located on chemical plants, refineries, and maritime platforms, and their deployment is an iterative process including several steps that need to be carefully prepared and precisely executed. As such, during the deployment lifecycle of an IWSN, which includes defining objectives, factory survey, selecting candidates, designing a solution, and deploying and monitoring the network, a detailed analysis of the IWSN needs to be performed [1,10]. Such analyses become time-consuming and expensive when they rely on acquiring and managing sets of physical equipment and performing tests in specific locations. To overcome this issue, some of the tasks can be performed using highly accurate simulators and/or Hardware-in-the-loop (HIL) testbeds. In system planning activities, simulators are used to validate the candidate solution. Field device selection and network integration can also be validated using simulators and/or

HIL testbeds. Moreover, realistic simulations are useful in predicting and evaluating the operational performance of IWSNs without deploying physical devices in testing scenarios. Therefore, simulations represent a fast and cost-effective way to determine the behavior of the network beforehand and are highly recommended for IWSN planning, design, deployment, and monitoring [1].

Ideally, simulations should be so realistic and accurate that a simulated IWSN and the physically deployed replica should behave identically under all circumstances arising during the lifetime of the IWSN.

Scope of the paper. In this context, this paper aims to propose and evaluate an enhanced IWSN simulation model that integrates the ISA100.11a communication stack functionalities with the ns-3 network simulator [11].

Methodology. The following steps were carried out to integrate the accurate ISA100.11a simulation model with ns-3:

- Porting a proprietary ISA100.11a software stack that runs on certified commercial devices to the same operating system as the one running the ns-3 simulator;
- Defining the interfaces between the ported ISA100.11a software stack and the ns-3 environment;
- Integrating the ported ISA100.11a software stack with the ns-3 simulation environment;
- Functional testing of the ISA100.11a simulation model for IWSN network formation and data collection.
- Iterative refinement of the ISA100.11a integrated simulation model for accurate modeling of a physical operational IWSN using HIL scenarios.

Test and validation. The proposed ISA100.11a simulation model is used to implement a basic IWSN with an infrastructure device and a field device. To validate the accuracy of the implemented model, we compare the behavior and performance of the simulated network with those obtained in the following hardware-based setups that replicate the same network:

- Physical network—in this scenario, the infrastructure device and the field device are real hardware devices, i.e., CDS VR950 gateway and CDS VS210 node.
- Gateway-in-the-loop—in this scenario, the physical infrastructure device is used to manage the simulated ISA100.11a network.
- Node-in-the-loop—in this scenario, the simulated infrastructure device is used to manage a physical ISA100.11a network.

Moreover, the scalability of the simulation model is tested by means of 5-node and 10-node networks both physically deployed and simulated. By scalability of the simulator, we mean its capability to correctly simulate the IWSN behavior as the number of the ISA100.11a devices is increased. In each scenario, the network performance is evaluated in terms of specific network formation and operation metrics.

Originality and contributions. The originality of this work consists of the highly accurate integration of the ISA100.11a software communication stack with ns-3. As such, in a unique approach, the same software stack found on the ISA100.11a compliant hardware devices is used in both the simulated scenarios and the experimental testbed replicas.

This paper is an expanded version of the conference paper [12] presented at the 2020 International Workshop on Antenna Technology (iWAT 2020). This work has been significantly extended, the major contributions being as follows:

- Contribution 1: A more thorough state-of-art is included, presenting the available IWSN simulation solutions and HIL approaches, highlighting the need for a reliable simulation tool.
- Contribution 2: An enhanced set of metrics is used to evaluate the performance of the proposed simulation model, in comparison with the hardware-based scenarios.
- Contribution 3: An extended set of experiments to test the functionalities of the gateway and node is performed by means of the two HIL scenarios (i.e., Gateway-in-the-loop and Node-in-the-loop). The HIL scenarios show that the ISA100.11a devices

implemented by the simulator are the equivalents of the physical devices. In addition to showing that all the functionalities of physical devices have been successfully integrated with the simulator, the metrics collected from the HIL scenarios allow us to quantify the similarity in performance between the physical and simulated devices.
- Contribution 4: An extended set of experiments including real infrastructure and up to ten real field devices validating the accuracy and scalability of the proposed IWSN simulated model.

The rest of the paper is organized as follows. Section 2 indicates the key aspects of a realistic IWSN model. Section 3 presents existing IWSN simulation tools and indicates the main requirements of a reliable simulation environment. Section 4 briefly presents the ISA100.11a standard. Section 5 describes the implementation of the proposed ISA100.11a model and its integration with ns-3. Section 6 presents the test scenarios, while Section 7 highlights the performance evaluation results. Finally, Section 8 concludes the paper, emphasizing the potential of the implemented ISA100.11a model.

2. Industrial WSN Accurate Modeling

An accurate/realistic IWSN model is an IWSN model with externally observable behavior matching the behavior of a physical IWSN. This accuracy translates into an identical start-up sequence, communication characteristics, and response to changes in the physical topology or radio propagation conditions. Specifically, we consider the following aspects of a realistic IWSN modeling:

- The external interface and used communication protocol are the same for both the simulated and for the physical IWSN. This includes the case where an unmodified external control system receiving data from a physical IWSN can use a simulated IWSN as a source of the same data, with minimal configuration changes (e.g., changing the IWSN's address in the control system's configuration). From the control system's point of view, the physical and simulated IWSN are the same.
- The start-up sequence and timing (synchronization) for a simulated and physical IWSN are similar. This includes the moments when devices become operational and when the first collected data samples arrive at the external interface of the IWSN.
- The QoS evaluation parameters (e.g., data throughput, transmission delays, packet size of the collected and transmitted data) are the same for both simulated and physical IWSNs.
- The provisioning data used for configuring a simulated IWSN can be trivially applied in the deployment of a physical IWSN.
- For a given set of field device positions, the network topology of a simulated IWSN is identical to that of a physical IWSN.
- After deployment, the physical and simulated IWSNs can be configured at run time by means of the external IWSN interface. The latency and bandwidth of message exchanges in this situation is identical between simulated and physical IWSN. For example, such additional configuration might be performed by the external control system for fine-tuning the data transmission period of a specific field device.
- In case there are changes in the radio propagation conditions or in the physical topology of the IWSN, the simulated and the physical IWSN react identically.

Through quick and efficient modeling, a realistic IWSN simulator helps solving various problems in different stages of the IWSN's development, as follows.

In the planning phase, the IWSN simulator can be used to select the physical locations where field devices should be placed and to evaluate the feasibility of different network topologies. Moreover, it helps estimating the communication bandwidth and latency of a given candidate solution.

In the implementation phase, the provisioning data can be created inside the realistic simulator and exported to the field devices to be deployed. The start-up sequence and the integration of the IWSN with external systems can be implemented and validated using the realistic simulator.

In the deployment phase, for fine-tuning an IWSN, configuration changes can be validated in a realistic IWSN simulator. In addition, in case of changes in propagation conditions, the realistic simulator can predict the effects on the IWSN performance.

3. Review of IWSN Simulation Tools and Validation Methods

This section provides a concise and precise description of the experimental results, their interpretation, as well as the experimental conclusions that can be drawn.

In terms of accuracy, the widely used WSN and IWSN simulators [13] can be divided into two categories:

- Discrete-event network simulators that rely on generating events at certain moments of time: events that can, in turn, generate additional ones during the simulation. As such, a trade-off between computational efficiency and simulation accuracy is achieved. Such a simulator is OMNeT++ [14] together with specific frameworks usable with it such as Castalia [15], INET framework [16], MiXiM [17], InetManet [18], Riverbed Modeler (former OPNET Modeler) [19], QualNet simulator [20], TOSSIM [21], and ns-3 [11]).
- Cycle-accurate simulators that simulate each CPU cycle of a wireless device. These simulators require more computational power but are highly accurate, and they allow for the same binary files to run in the simulator and on real devices. Such simulators are Avrora [22] and Cooja [23].

3.1. Latest IWSN Simulation Solutions

The simulators listed above have been used to carry out several studies in the domain of IWSNs. We summarize here the most relevant works, emphasizing the accuracy of the simulations. We analyze the level of detail implemented in the simulations (as a greater level of detail allows more accurate simulations) and the metrics used during the simulator evaluation.

The work in [24] presents a WirelessHART implementation in OMNET++ for experiments concerning security. The model implements the full protocol stack and the Network Manager, using InetManet [18] features for the physical layer. This shows that implementing a full IWSN protocol stack in a simulator is feasible.

In [25], the authors modify a model of an industrial process (the widely used Tennessee Eastman Process Control Challenge Problem [26]) by applying wireless communication links in the model between sensors, actuators, and the process controller. Simulation results show that the process can operate by using wireless communication links. The effect of various wireless configurations on the operation of the process is evaluated. The used simulator is OMNET++ and models from INET and MiXiM frameworks are used for address resolution protocol (ARP) and wireless channel template, respectively. The simulation employs the WirelessHART protocol, and it focuses on the Physical (PHY) and Media Access Control (MAC) layers. The archived version of the source code of simulations is available online at [27]. The evaluated performance metrics are the deviation of process variables from their nominal values and the expected time for which the plant can operate normally. The packet error rate on the RF communications and the placement of the Access Points have been modified during simulations. This simulator does not implement a complete WirelessHART protocol stack nor the functionalities of a WirelessHART gateway.

In [28], Castalia and Pymote are used for simulating ISA100.11a, WirelessHART, and ZigBee networks. The metrics used for performance evaluations are throughput, number of packets transmitted, lost, and received, and device energy consumption. During simulations, the readily available generic models have been used (i.e., the IEEE 802.15.4 MAC and PHY layer models for ISA100.11a simulations and a generic "Throughput Test" application). Not all layers of the communication stack have been implemented, nor the functionalities of an ISA100.11a or WirelessHART gateway.

The work in [29] presents an ISA100.11a model for ns-3, implementing the physical and data link layers of the standard and a simplified application layer. The corresponding source code is available online at [30].

Avrora [22] is a cycle-accurate simulator for WSNs. It can handle as many as 25 nodes in real time. It is targeted for high timing accuracy with increased performance compared to ATEMU [31]. It achieves this by extracting fine-grained parallelism inside the simulation of a WSN.

A mathematical model for energy consumption of wireless sensor nodes is presented in [32]. It considers the energy consumption of communications, acquisition, and processing.

The importance of simulating the effects of CPU load on the IWSN is assessed in [33]. The ns-2 simulator [34] is integrated with the RTSim [35] software to accurately simulate wireless sensor nodes. The IEEE 802.15.4 standard is used for evaluation.

A discussion on the deployment of ISA100.11a and WirelessHART IWSNs in a refinery is presented in [36]. For assisting the deployment, a new simulator called RF Propagation Simulator (RFSim) has been created and used. It predicts the quality of the RF signal on the premises of the industrial installation. The simulation results are compared with onsite measurements and look promising. The authors note that most of the published work on sensor network deployment limits itself to 1D or 2D environment, and 3D environments are considered an open issue.

A complete WirelessHART stack and gateway implementation for the ns-2 simulator [34] is presented in [37]. To validate the model, the authors set up similar networks with real hardware and within the simulator. The performances of the two networks (i.e., simulated and real) are evaluated in terms of reliability in the network (i.e., failed transmission ratio and average of received signal level), communication scheduling and network throughput, real-time data transmission (i.e., end-to-end delay and interval between consecutively received packets), and energy consumption in the network. Management efficiency is evaluated in terms of overhead and delay for performance during joining and service request procedures. The simulator performance in multi-hop mesh networks is also assessed, highlighting the response of the network in case of node and link failure. Moreover, the data delivery ratio is evaluated for three lossy networks. The experiments showed that the simulated results are similar to the results obtained in real networks and that the proposed model is versatile and usable in diverse scenarios.

To enhance the simulation results, hardware-related aspects can be integrated into the simulation scenarios. This concept is known as Hardware-in-the-loop.

The TOSSIM simulator is integrated with physical wireless nodes at the radio communication level in [38]. The work demonstrates the feasibility of a WSN consisting of both simulated and physical nodes. This is achieved by using a pair of physical Dual Base Stations as a bridging device between the simulated and physical environments. The simulation runs in real time. The approach allows unmodified nodes to communicate with purely simulated nodes, but the number and location of the physical nodes are being constrained by the number and capabilities of bridging devices.

In [39,40], the authors use the Software-in-loop simulation technique to evaluate the software to be deployed in a WSN and to facilitate the deployment of a WSN. This is implemented by integrating a simulated environment and sensor device with the software that will run on deployed wireless sensor nodes.

The work in [41] presents how radio hardware can be integrated into a simulation. In this Radio-in-the-Loop (RIL) approach, the IEEE 802.15.4 Physical Layer and the Physical Medium are implemented by real hardware and physical phenomena, respectively. The communication layers above these, including the IEEE 802.15.4 MAC layer, are implemented by software. As such, the results of the experiments are more realistic than the results of pure simulations. The authors employ the OMNeT++ network simulator, and the interface toward the hardware is based on Linux/Unix inter-process communication and the PcapNG data format. The radio used is the IEEE 802.15.4 compliant ATmega128RFA1, running a special application on Contiki OS [42]. However, this work does not measure the latency introduced by the adaptation between the software and hardware parts of the simulation. For some IWSN protocols (e.g., ISA100.11a), the latency at the MAC layer is critical for the proper functioning of the slotted transmission scheme; excessive delays could make the communication impossible.

3.2. Limitations of Current IWSN Simulation Solutions

In our opinion, an IWSN simulator must be accurate (to correctly predict the behavior of the physically deployed IWSN system), fast (and therefore computationally efficient), and versatile (to accommodate different types of use-case scenarios). To this extent, the existing network simulators are not comprehensive, as they provide simplified models that lack accuracy or focus only on certain features of the network, leading to discrepancies between the behaviors of the simulated and physically deployed networks. Moreover, some models are suitable only for simulations, which makes them difficult to use in HIL scenarios. Table 1 synthesizes the main features of the existing IWSN simulation solutions in comparison with the model proposed in this paper.

Table 1. Features of IWSN simulation solutions.

IWSN Simulator Reference	Focus of Work	Applicable Standards	Communication Stack, Network Management, External Interface Implementation	Validation with Physical Devices, Mixed HW/SW (HIL) Scenario	Evaluated Metrics
[24]	Security	WirelessHART	Complete WirelessHART Stack, Network Manager	No	Effect of security attack on the success rate of collected data transmission
[25]	Effects of imperfect wireless links on industrial process	WirelessHART	Complete WirelessHART Stack, Network Manager with static resource allocation; Tennessee Eastman Process Control Challenge Problem system	No, but the process is widely studied	Process parameter variation on imperfect wireless links
[28]	Comparison of simulated IWSN protocols	ISA100.11a WirelessHART ZigBee	PHY and MAC layers, simplistic application layer	No	Communication statistics, energy consumption, RF signal level
[29]	Optimization of WSN energy consumption	ISA100.11a, WirelessHART	PHY, MAC, routing layers	Model based on physical devices	Network lifetime based on energy consumption
[32]	Mathematical model for energy consumption	IEEE 802.15.4	Not applicable	Model based on physical devices	Energy consumption
[33]	Effects of processing load on communication	IEEE 802.15.4	IEEE 802.15.4 stack and PAN coordinator	No	Application performance as a function of network delay and CPU load
[36]	Discussion on IWSN deployment, focus on signal quality	ISA100.11a, WirelessHART	Not applicable	Model based on a physical network	RF signal quality
[37]	Simulation of complete WirelessHART network	WirelessHART	Complete WirelessHART Stack, Network Manager	Model validated by comparison with a physical network	Management overhead, communication statistics, reliability, energy consumption
[38]	Hardware-in-the-loop testbed	IEEE 802.15.4	TinyOS stack and network management	Validated using HIL	Energy consumption

Table 1. *Cont.*

IWSN Simulator Reference	Focus of Work	Applicable Standards	Communication Stack, Network Management, External Interface Implementation	Validation with Physical Devices, Mixed HW/SW (HIL) Scenario	Evaluated Metrics
[41]	Integrating simulation with real HW at RF level	IEEE 802.15.4	PHY, MAC layers, and application model	Validated using HIL	Signal strength
The proposed ISA100.11a simulation model	Accurate implementation of ISA100.11a communication stack functionalities	ISA100.11a	Complete ISA100.11a Stack, System Manager; external interface available in Gateway module	Validation by comparison with the identical physical network, and two HIL scenarios	Start-up timings, communication statistics

As a result of the simplifications, the use of the existing models could lead to false predictions. In contrast, by integrating a field-tested communication stack (i.e., according to certified commercial devices) into the simulator, it should be possible to generate more reliable results. Having a realistic communication stack available in the simulation tools allows an earlier start of the wireless systems integration, by using the simulator interfaces toward external systems. In addition, an accurate simulation tool should help the industry by providing aid in the deployment and maintenance of IWSNs.

In the domain of generic WSNs, there are some cycle-accurate simulators (such as [21–23]) that simulate a software stack very similar to the software running on real devices, but these simulators do not implement standardized industrial communication protocols such as ISA100.11a or WirelessHART. To the best of our knowledge, the work in [37] presents the most thorough IWSN simulation model so far, but it is bound to the WirelessHART standard and does not consider a HIL validation method. NIVIS LLC [43] released an ISA100.11a open-source implementation but is not compatible with a simulator and runs on hardware that is not commercially available. An ISA100.11a model for ns-3 is available in [30] but cannot run on real hardware.

In this paper, we propose a comprehensive IWSN simulation model integrated with ns-3 that implements the ISA100.11a communication stack available on certified commercial devices. We make use of HIL scenarios to validate the implementation accuracy of the IWSN entities (i.e., ISA100.11a field and infrastructure devices).

To validate the overall IWSN model implementation, we compare the simulated IWSN behavior to that of the physical network replica.

3.3. *ns-3 Simulator Selection Criteria*

In our opinion, to facilitate the integration of a simulation model, the simulation environment must be open-source, actively maintained, mainstream, and general purpose. Moreover, to achieve high model accuracy, the environment should allow the integration of the same software stack running on real devices. As such, it should be compatible with C or C++ (which is used to implement the software running on the microprocessor of the ISA100.11a devices) and able to integrate proprietary C and C++ source code into the simulation.

Table 2 synthesizes the relevant criteria for selecting the simulation environment.

Given the above-mentioned requirements, we consider ns-3 [11] to be the best candidate for integration with the ISA100.11a IWSN simulation model. It is an open-source, actively maintained general-purpose simulator that integrates natively with C, C++ source code.

Table 2. Criteria used for selecting the IWSN simulator.

IWSN Simulator	License	Actively Maintained	General-Purpose Simulator	Programming Language
OMNeT++ [14]	Commercial/ Non-commercial	Yes	Yes	C++
QualNet simulator [20]	Commercial	Yes	Yes	C/C++
TOSSIM [21]	Open-source, BSD [44]	Yes	No, TinyOS specific	nesC
ns-2 [34]	Open-source, GNU GPLv2	No	Yes	C++, TCL
ns-3 [11]	Open-source, GNU GPLv2	Yes	Yes	C++, Python
Avrora [22]	Open-source, BSD [45]	No	No, WSN specific	Java
Cooja [23]	Open-source, BSD [46]	Yes [47]	No, WSN specific	Java
RTSim [35]	Open-source, GNU GPLv2	No [48]	Yes	C++
Riverbed Modeler [19]	Commercial/Academic	Yes	Yes	C/C++

4. ISA100.11a Wireless Communication Standard

4.1. ISA100.11a Devices

An IWSN defined by the ISA100.11a standard consists of two categories of devices: field devices and infrastructure devices. Each device can implement one or more roles specific to their category.

Field devices can implement the Input/Output role, denoting the capability of collecting sensor data or managing actuators and the router role, denoting that they are capable of forwarding traffic to other field devices.

Infrastructure devices can implement System Manager, Security Manager, Gateway, and Backbone Router roles. The System Manager is responsible for all the communication-related aspects of the network, including the configurations necessary for devices to join the network, the set-up of the radio network, the management of Quality of Service (QoS) requirements of devices, and ensuring redundancy inside the network. The Security Manager handles the security-related functions of the network, including the authentication of devices and management of cryptographic keys. The Gateway provides standardized methods to get data out of the IWSN and to externally access the devices inside the IWSN, typically in order to configure non-communication-related aspects of devices—for example, the sampling rate of the data collected by devices. The Backbone Router provides means of communication between the wireless network and the Infrastructure Devices.

Figure 1 depicts a simplified ISA100.11a IWSN, showing the roles and functionalities associated with each device.

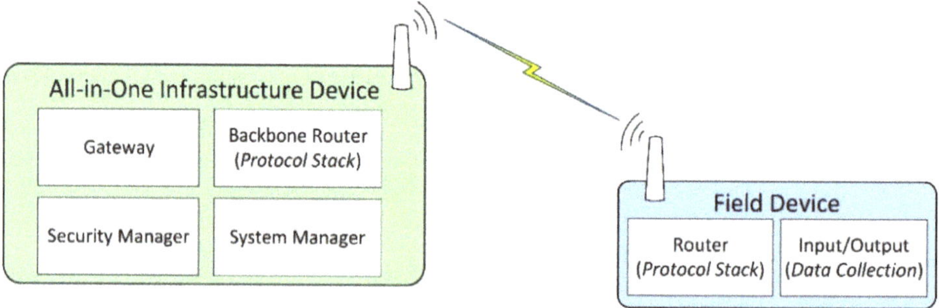

Figure 1. Roles and associated functionalities of ISA100.11a devices.

The IWSN is formed by an all-in-one infrastructure device, which in this case implements the ISA100.11a roles of System Manager, Security Manager, Gateway, and Backbone Router. An all-in-one infrastructure device might include functionalities not required by the ISA100.11a standard, for example, external protocol adapters.

Most field devices implement data collection functionalities. Data collection means that the devices include one or more transducers, and at some moments in time, the value is read from the transducer(s) and is inserted in the ISA100.11a protocol stack for transmission toward the infrastructure device. This data collection mechanism is typically integrated into the application layer of ISA100.11a by the means of User Application Processes (UAPs) and inside the UAPs by specialized process industries user application objects such as Analog Input Objects, Analog Output Objects, Binary Input Objects, and Binary Output Objects. The data from such objects are transmitted to the infrastructure device, where it reaches the Gateway module. External automation systems connect to the Gateway module through the External Protocol Adapter and retrieve the data.

4.2. Communication Stack

Table 3 summarizes the protocol layers defined in the ISA100.11a standard in the context of the OSI reference model.

Table 3. Protocol layers of ISA100.11a communications stack.

ISA100.11a Protocol Layers	OSI Model
User Application Processes (UAPs) Application sub-layer (ASL)/ISA native and legacy protocols (tunneling)	Application
Transport Layer (TL)/UDP (IETF RFC 768)	Transport
Network Layer (NL)/6LoWPAN (IETF RFC 4944)	Network
Upper data link layer/ISA100.11a upper Data Link Layer IEEE 802.15.4 MAC extension Media Access Control (MAC) sub-layer: IEEE 802.15.4 MAC	Data link
Physical layer: IEEE 802.15.4 PHY (2.4 GHz)	Physical

Each device in an ISA100.11a IWSN implements the protocol layers defined by the ISA100.11a standard. Field devices and Backbone Routers implement all the protocol layers while the System Manager, Security Manager, and Gateway do not implement the physical and data link layers. The PHY layer and MAC sub-layer implement the IEEE 802.15.4 standard [49] specifications.

On top of the MAC sub-layer, the ISA100.11a standard defines a MAC extension sub-layer and an upper data link layer. These two layers use some of the functionalities defined by IEEE 802.15.4 (e.g., frame transmission and reception) and add extra functionalities (e.g., periodic transmission and reception opportunities by introducing ISA100.11a super-frames and links, routing inside an ISA100.11a IWSN using source routing and graph routing, authentication of the received frame using AES cryptography, or performing time synchronization between wireless devices during acknowledged (ACK) data transmissions).

The network layer of ISA100.11a performs addressing and routing. It uses the IPv6 addressing scheme. To differentiate between the QoS classes, the network layer is aware of the ISA100.11a contracts and passes this information to the data link layer to enable traffic prioritization.

The transport layer multiplexes traffic between the network layer and multiple Transport Layer Service Access Points (TSAPs), and it is capable of handling the encryption, decryption, and authentication of traffic passing through it. With each TSAP, an Application Sub-Layer (ASL) entity exchanges protocol data units (PDUs) with the application layer. The ASL sub-layer provides access primitives to the object-oriented structure of User Application Processes (UAPs). These primitives include read attribute, write attribute, execute method, publish request, and publish notify. Each ASL entity is connected to one UAP. On each ISA100.11a device, at least one UAP must exist: the Device Management Application Process (DMAP). The DMAP exposes the configuration interface of the device in order to be integrated into the network. Typically, sensor or actuator devices have an additional UAP exposing the sensing or actuating capabilities of the device to the ISA100.11a network.

5. ISA100.11A Communication Stack Implementation in ns-3

Ns-3 is a discrete event-based simulator, written in C++, but simulations can also be defined in Python. It uses WAF [50] as a build system and offers models for various networking protocols and applications, including wireless technologies. Conceptually, a simulation scenario consists of several Nodes, which run Applications and communicate through NetDevices, which are attached to Channels. During simulations, various statistics, metrics, and raw communication data can be collected, and various simulation events can be logged. The collected data can be further analyzed with external tools.

5.1. Implementation of an ISA100.11a IWSN

In this work, we integrate a proprietary ISA100.11a software stack developed by the Control Data System (CDS) [51–53] with ns-3. This software stack runs on certified commercial devices and has been tested for interoperability with devices from different vendors, such as Yokogawa, Honeywell, or Draeger.

The internal architecture of physical ISA100.11a IWSN CDS devices employed by our hardware-based scenarios is illustrated in Figure 2.

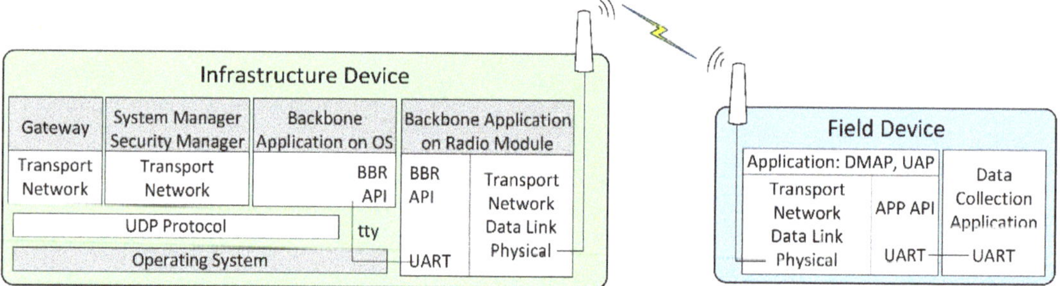

Figure 2. The internal architecture of physical ISA100 IWSN devices.

The implementation architecture is based on the CDS VR950 gateway [52] as the infrastructure device and the CDS VS210 development board [53] as the field device. These devices are commercially available, and they contain an ISA100.11a certified radio module with a software stack.

The VS210 development board contains two parts: (1) the radio modem running the ISA100.11a stack, including the application layer with DMAP and UAP, and an integration API denoted APP API, which is exposed through a Universal Asynchronous Receiver/Transmitter (UART) interface; (2) the data collection application, running on a microcontroller separate from the radio modem and communicating with the radio modem through a UART interface.

The VR950 gateway also consists of two major parts:

- The transceiver of the Backbone Router, which is connected to the antenna of the gateway. It runs a complete ISA100.11a protocol stack, and it includes an integration API with the rest of the gateway, which is denoted here as BBR API. Its application layer only has a DMAP. The communication with the other part of the gateway is performed through a UART.
- The operating system environment inside the Gateway, which is running on an embedded computer. The operating system is based on Linux. This part is connected to the transceiver over a UART (tty) interface, and it runs several network management applications that implement the gateway logical modules. These applications are the

Gateway application, the System Manager (SM) and Security Manager application, and the Backbone Application on OS (BBR). All of these applications communicate with each other through User Datagram Protocol (UDP). The Gateway and SM applications include the Network, Transport, and Application layers defined by ISA100.11a, while the BBR application performs the translation between the UART and UDP communication. The gateway has IP connectivity through an Ethernet interface.

This work considers the ISA100.11a software stack implementations of the field device and the all-in-one infrastructure device.

5.2. The ISA100.11a IWSN Simulation Model

The goal of the proposed simulation model is to achieve a simulation as accurate as possible by porting the ISA100.11a software stack of commercial devices. This unique approach should allow the simulation to replicate the exact behavior of a physical IWSN.

Figure 3 illustrates the architecture of the developed ISA100.11a simulation model, replicating the internal architecture of the physical ISA100.11a devices presented in Figure 2.

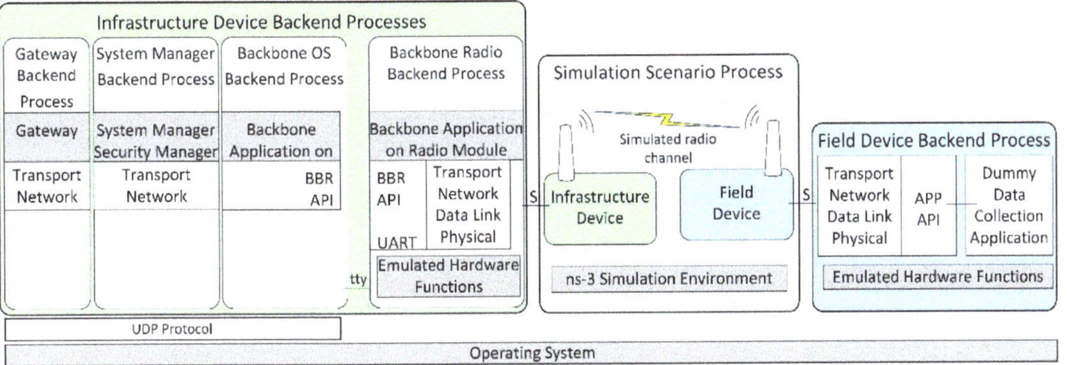

Figure 3. The architecture of the implemented ISA100.11a simulation model.

The simulated infrastructure and field device models are formed by a set of backend processes that run inside the operating system (Ubuntu 18.04 Linux distribution) instead of running on dedicated hardware.

The infrastructure device backend processes are as follows:

- Backbone Radio Backend Process, corresponding to the transceiver part of the real infrastructure device. It includes the emulated hardware functions submodule, which allows running the embedded software Backbone Application on a Radio Module without physical hardware.
- Backbone OS Backend Process, corresponding to the BBR application of the infrastructure device;
- System Manager Backend Process, corresponding to the SM application of the infrastructure device;
- Gateway Backend Process, corresponding to the Gateway application of the infrastructure device.

The Backbone OS, System Manager, and Gateway Backend Processes communicate with each other by means of the UDP protocol. A serial interface (tty) ensures communication between the Backbone OS Backend Process and the Backbone Radio Backend Process.

The Field Device Backend Process hosts several backend threads that communicate through sockets with the simulated field devices and run their functionalities, such as the ISA100.11a communication stack, the integration API, and a dummy data collection application. In the implemented basic scenarios, there is only one field device and thus only one backend thread, while in the scenarios evaluating the scalability of the simulator, there are multiple field devices and multiple backend threads.

The Simulation Scenario Process includes ns-3 and manages simulation time, simulated events, and network topology. It also implements the radio channel model, as well as simple models of the field and infrastructure devices that dynamically launch their backend processes/threads and forward simulation events to/from these backends (through Unix Domain Sockets, denoted with S in Figure 3).

During simulation startup, the Infrastructure Device and the Field Devices are created in ns-3 by instantiating ns-3 Nodes and specific ns-3 Applications. The Field Device Application is initialized, which configures the EUI64 address of its corresponding field device and loads provisioning data for it. A thread in the Field Device Backend Process is created, and communication with it is established by a Unix Domain Socket. In addition, at startup, the Infrastructure Device is initialized, and it first configures its EUI64 address and provisioning data; then, it launches its Backend Processes: (1) the Backbone Radio, (2) the Backbone OS, (3) the System Manager, and (4) the Gateway Backend Processes. Communication between the ns-3 Simulation Environment and the Backbone Radio Backend Process is established by using another Unix Domain Socket.

At the start of the simulation, an event is generated by each Field Device Application and Infrastructure Device Application. On the execution of these events, an Event Request simulation message is sent to each corresponding Field Device Backend Process thread and Backbone Radio Backend Process. In addition to their type, these simulation messages contain the current time in the simulation. When each Backend Process receives this event, it (1) updates timing-related simulated registers in its Emulated Hardware Functions module, (2) possibly executes interrupt handler routines, and (3) runs its internal main loop. While executing the program sequences from above (interrupt handlers and main loop), the ported software stack might (a) change the RF channel on which its virtual transceiver might receive RF frames, or (b) it might schedule an IEEE 802.15.4 compliant RF frame for transmission at a specific time moment—typically in less than 5 milliseconds time from the current simulation time. In the above two situations, specific simulation messages are transmitted back asynchronously to the ns-3 Simulation Environment using the Unix Domain Socket corresponding to the device. When a Backend Process' thread finishes processing the current event, it sends an event response message to the ns-3 Simulation Environment. This response message contains the latest time-moment when the specific Backend Thread expects to be run again. It might run earlier if some event has an effect on it, for example, when a Field Device receives an incoming RF frame.

In case of RF frame transmission, an RF frame transmission complete simulation message is sent to the Backend Thread initiating the transmission, at the simulation moment when this message is generated on physical hardware. This type of message contains the success/fail status of the transmission.

RF frames transmitted by Field Devices are processed in the ns-3 Simulation Environment by the model of physical topology and RF channel. The Carrier-Sense Multiple Access medium access scheme is modeled, and in case an RF frame is received by a device, a simulation message request is sent to its corresponding Backend Process thread, containing the timestamp, RF channel, and data payload. If the virtual transceiver's status in the Backend Process thread is in receive mode, then the RF frame is delivered to the ISA100.11a protocol stack. When the processing of the RF frame is complete, the Backend Process sends a simulation message response to the ns-3 Simulation Environment.

The Backbone Radio Backend process works similarly to a Field Device Backend Process, but it also communicates on its serial interface (tty) using BBR API messages. This interface is used to transfer data between the Backbone Radio Backend Process and other processes from the Infrastructure Device Backend Process group.

For achieving real-time functioning, the ns-3 Simulation Environment periodically waits for the wall-clock time to "catch up" with the simulated time. For this mechanism to work, the time in simulation must be at least as fast as the wall-clock time. In the current implementation, the synchronization period is 50 ms.

The simulation events include: passing of time, start/end of a radio frame transmission/reception, start/end of transmission/reception on emulated UART, and debug information transmission. The simulation scenario mirrors, as close as possible, the start-up sequence of an IWSN. As such, the devices startup, and the modules of the infrastructure device interconnect and begin the creation of the IWSN by sending advertisement radio frames. When the field devices receive the advertisements, they initiate the joining process by transmitting radio frames. The joining requests are processed by the infrastructure device, which sends responses back to the field devices. After several messages, the joining procedure is finalized, and the infrastructure device configures the field devices to start sending the collected environmental data. When this configuration is complete, the IWSN enters its steady state where most of the traffic consists of collected data and diagnostic data transmitted by the devices.

6. Evaluation of the ISA100.11a Model Implementation

To evaluate the accuracy of the implemented simulation model, we compare the behavior of the simulated network with the behaviors of one physical IWSN and two HIL IWSNs. To evaluate the scalability of the IWSN model, we compare metrics collected from two extended physical networks with the equivalent simulated IWSNs. The extended IWSNs contain one infrastructure device and five and ten field devices, respectively.

The *Physical network* scenario implements a real IWSN, employing a CDS VR950 gateway and a CDS VS210 node, both being commercial physical devices implementing the ISA100.11a IWSN standard. This scenario allows us to assess the performance of the real IWSN and consider the network behavior as a reference for the scenarios that employ simulation.

To validate the simulation model and make sure of its compatibility with the real ISA100.11a specifications, we make use of two HIL testing scenarios: one that tests the functionality of the infrastructure device, and another that tests the functionality of the field device. The *Gateway-in-the-loop* scenario employs a physical gateway to manage the simulated ISA100.11a network, while the *Node-in-the-loop* scenario uses a simulated gateway to manage a physical ISA100.11a network.

To avoid the potential issue of excessive delay between the hardware and software parts of the HIL system, we set the Hardware–Software (HW-SW) interface in a less time-critical part of the system. This HW-SW interface separates the physical hardware components of the testing scenario from the simulated components. In our case, this is the interface between the Backbone Router and the Gateway modules of the infrastructure device. In the HIL scenarios, the physical and simulated components can communicate only through the HW-SW interface of the infrastructure device, as illustrated in Figure 4. This way, the time-critical parts, including the Physical and Data-Link layers of the devices in the simulation, run in the same time-domain: either all of them are simulated in case of the *Gateway-in-the-loop* scenario or they use the real hardware implementation in case of the *Node-in-the-loop* scenario. As such, the testing results should showcase similar behaviors for the networks that have the same type of PHY and Data link implementations (i.e., Simulated vs. *Gateway-in-the-loop* and Physical vs. *Node-in-the-loop*).

A basic IWSN is implemented in each test scenario, consisting of one infrastructure device and one field device placed 1 m from one another. The network performance is evaluated in terms of specific network formation and operation metrics.

During a test scenario, the infrastructure and field devices are started at the same time. Next, the infrastructure device forms the IWSN, and the field device joins the network and sends data to the infrastructure device at 15-s intervals. The run-time of each test scenario is 40 min, and each test scenario has been repeated 40 times. The simulations run by wall-clock time, so that one simulated second corresponds to one real-time second.

The results of the simulations are logged in files containing the history of events from the simulator. For the hardware-based scenarios that employ a real wireless communication channel (i.e., *Physical Network* and *Node-in-the-loop*), the ISA100.11a messages transmitted over the radio interface are captured using a generic IEEE 802.15.4 sniffer that listens simultaneously to all 16 radio channels defined in the standard. The model of the sniffer is a WirelessHART Wi-Analys Network Analyzer [54].

For the simulated network, the architecture in Figure 3 is used. The hardware-based scenario architectures are described below.

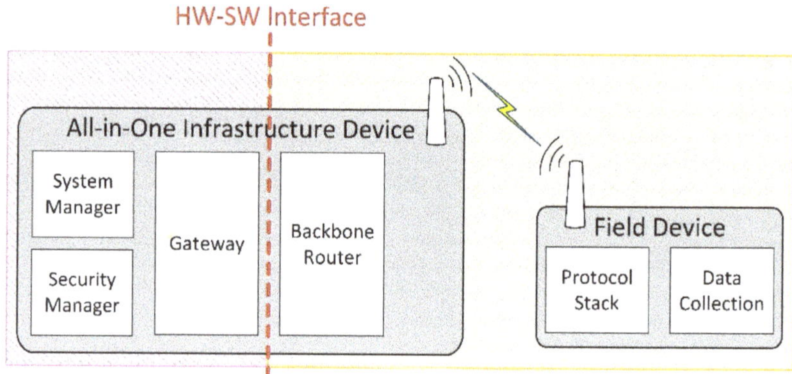

Figure 4. HW-SW interface in the HIL scenarios.

6.1. Physical Network Scenario

The *Physical network* scenario implements a real IWSN, employing a CDS VR950 gateway and a CDS VS210 field device. Figure 5 shows the hardware used in the experiment. On the left side is the infrastructure device. It uses its Ethernet connector for both communication and power (Power over Ethernet technology). On the right side is the physical field device. It is powered through the USB connection. Both devices have been previously configured for the same ISA100.11a subnet ID and join key. For the VR950, these configurations have been introduced by using its web interface, while for the VS210, the Field Tool FT210 [55] has been used.

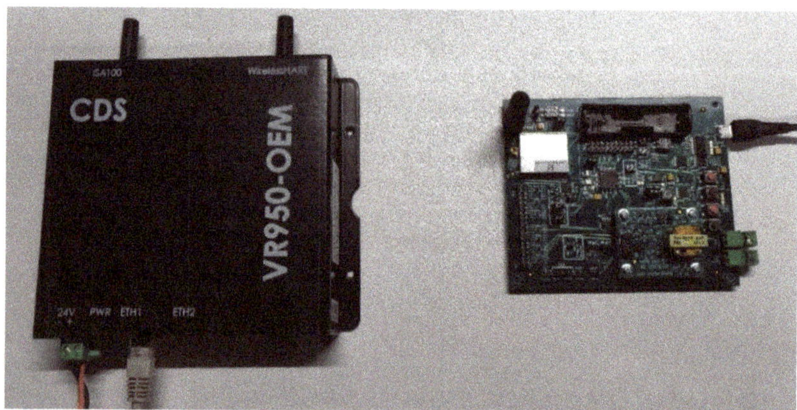

Figure 5. The ISA100.11a certified devices employed in the experimental setup: CDS VR950 gateway (**left**) and CDS VS210 field device (**right**).

To evaluate this scenario, the log files from VR950 and the logs of the RF sniffer are used.

6.2. Gateway-in-the-Loop to Manage a Simulated Node

This Gateway-in-the-loop scenario is presented in Figure 6. The System Manager, Security Manager, and Gateway applications are the ones implemented on the physical infrastructure device, while the Backbone modules (i.e., the Backbone OS Backend Process and Backbone Radio Backend Process) are simulated and together with the Simulation Scenario Process and Device Backend Process are running on a PC inside an operating system. The VR950 and the PC are connected through an Ethernet network, and their IP stack is properly configured for communication.

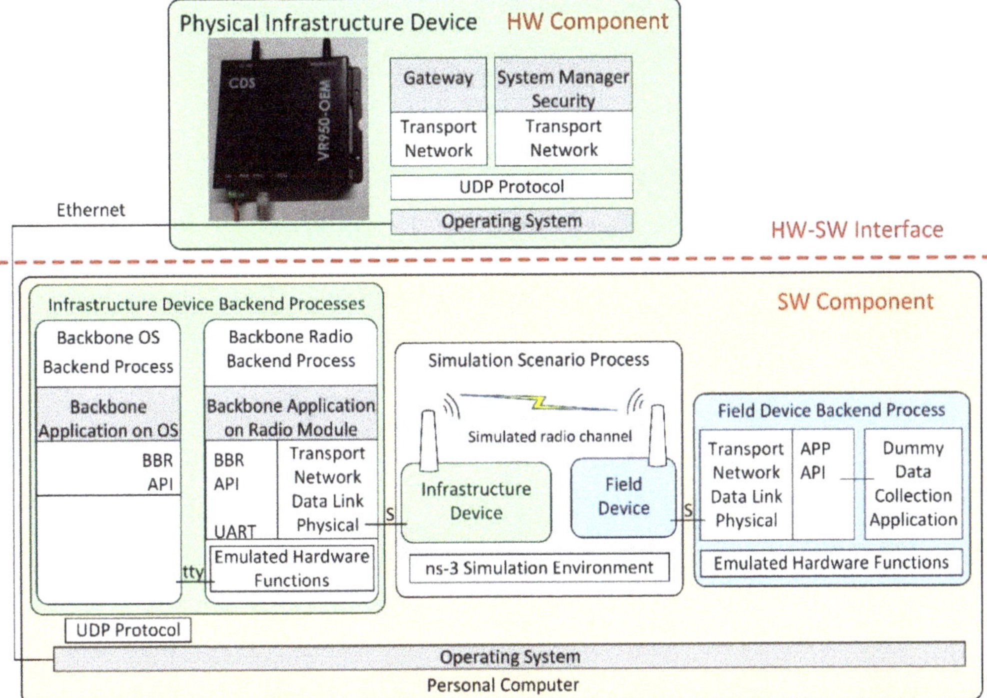

Figure 6. Gateway-in-the-loop scenario architecture.

The System and Security Manager application communicates with the Backbone OS Backend Process through the UDP protocol. It configures the Backbone Radio Backend Process as it would be running on a physical gateway, and thus, the physical gateway manages the simulated network.

For this scenario, the modified physical gateway has to be started up manually before the simulation can be started on the PC. An important aspect is that the time of the VR950 and that of the PC have to be well synchronized for this scenario to succeed. Time information is used for encryption and authentication in ISA100.11a, and if the time difference between modules is greater than 60 s, then the authentication of packets fails, and the applications or processes refuse to communicate with each other. Since no physical RF transmissions take place, the necessary experimental data have been extracted from the simulation log files and the log files of the VR950 gateway.

6.3. Node-in-the-Loop Managed by a Physical Gateway

In the *Node-in-the-loop* scenario, illustrated in Figure 7, the System and Security Manager and Gateway applications of the infrastructure device are running on a PC, while the Backbone application is running on the physical infrastructure device and communicates with the other modules through UDP protocol. The Backbone module creates a physical IWSN and allows the physical field device to communicate with the infrastructure device.

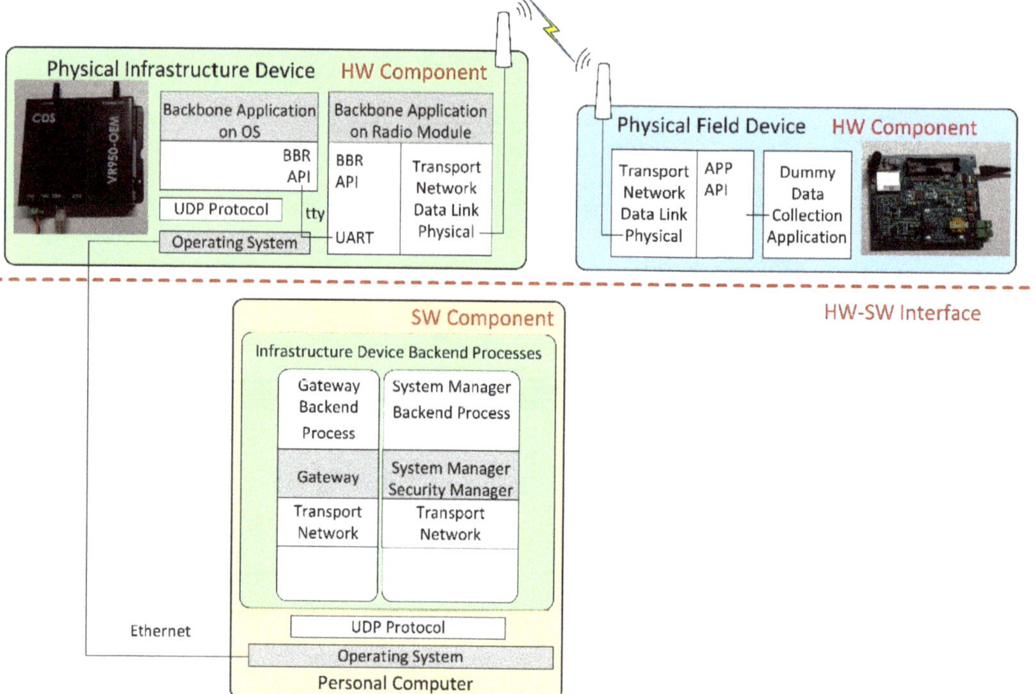

Figure 7. Node-in-the-loop scenario architecture.

Time synchronization between the hardware and simulated parts of the scenario is important to prevent the ISA100.11a applications or backend processes from discarding received packets due to decryption or authentication errors.

The logs from the simulated System and Security Manager and Gateway and the data collected by the RF sniffer are used to evaluate this experiment.

6.4. Extended Real and Simulated Network Scenario

For evaluating the scalability of the proposed simulator, two extended IWSNs have been implemented both in physical form and in the simulator. These IWSNs consist of a CDS VR950 gateway, as infrastructure device, and five or ten CDS VS210 nodes connected to the gateway. The extended IWSNs are deployed in the same laboratory as the basic networks, experiencing similar environmental conditions.

Figure 8 shows the testbed for the extended physical network. There is one infrastructure device and ten field devices. Depending on the test scenario, either five or 10 devices are powered up.

Figure 8. Extended real scenario implementation.

To successfully simulate IWSNs with multiple field devices, there are two key aspects that require special attention: (1) the implementation of the medium access channel model, and (2) the selection of randomness source. For (1), we modeled the RF channels such that if there is an ongoing transmission on a given RF channel, then that channel is considered busy, without taking into account the RF propagation delay between a transmission of one field device and carrier sense operation of another device. This is needed because ISA100.11a IWSNs work with synchronized timing, and at startup, each device will try to transmit at the exact same time. For (2), randomness is used in conjunction with CSMA in the exponential back-off algorithm and used for generating some cryptography-related messages. Not having random values with sufficient unpredictability can cause the IWSN to not function correctly, for example by generating too many collisions during RF medium access and practically jamming the starting-up IWSN. The pseudo-random number generator provided by the Linux OS (e.g., rand() function) is sufficiently unpredictable for making the simulated IWSN work correctly, although using it makes it harder to debug the simulator. Using a constant seed value for the pseudo-random number generator allows reproducibility of the simulation runs. Predictable and easier to debug approaches such as deriving values from current simulation time and in-simulation device identifiers by using bit-level operations make the simulated IWSN work incorrectly.

Figure 9 presents the architecture of the extended simulation scenario. The architecture follows the generic simulation architecture presented in Section 5.2. As illustrated in Figure 9, in the simulation scenario, there is one simulated infrastructure device and there are up to ten simulated field devices, each consisting of the field device model in ns-3 and a thread in the Field Device Backend Process. Each field device model communicates with its corresponding Field Device Backend Process by a dedicated Unix Domain Socket, which is denoted with S1 to S10. In the evaluation scenarios with five simulated field devices, there are five field device models and backend process threads, instead of 10.

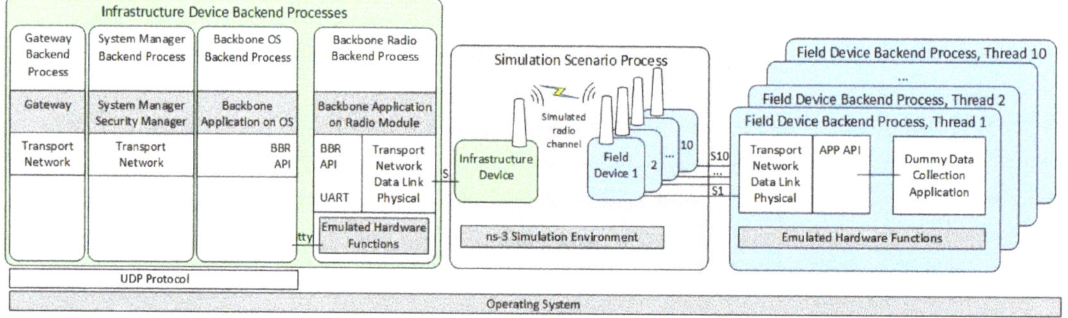

Figure 9. Extended simulated scenario architecture.

These evaluation scenarios have similar behavior to the other evaluation scenarios: an IWSN is started, and it is run for a total of 40 min. All field devices have been configured to transmit data at 15-s intervals. The evaluated metrics are collected for 40 runs of each scenario. The experimental runs have been executed automatically by implementing and using controller software that starts and stops the software modules on the infrastructure device and turns on and off the field devices by controlling a programmable power switch. In case of physical IWSNs, the metrics have been extracted from the log files of the VR950 gateway and from the logs of the RF sniffer. For simulated IWSNs, the metrics have been extracted from the log files of the simulator.

7. Results Analysis

The evaluation of the test scenarios is performed using two sets of metrics: one related to the timing of events in the IWSN and one related to communication statistics. Several key events are taken into account when analyzing the network behavior and defining these metrics. Figure 10 illustrates these events and the message exchanges between the infrastructure device and field device during IWSN formation (including initialization, advertising, and joining processes).

As depicted in Figure 10, first, the all-in-one infrastructure device and the field device perform their initialization. Next, the infrastructure device starts broadcasting the presence of an ISA100.11a network by transmitting advertisement RF frames while the field device listens for advertisements. When the field device receives a suitable advertisement (i.e., with the network identifier and security key matching the device provisioning configuration), it starts the network joining process by sending a specific join request to the infrastructure device and then waits for a response. After the join process messages exchange, the field device is admitted in the standard-compliant network, and a management communication channel is configured between the infrastructure and field devices. Next, the field device requests a communication channel (specifically an ISA100.11a Contract) to transmit collected data and the infrastructure device allocates communication resources. When the communication channel is set up, the field device starts transmitting collected data to the infrastructure device. At this point, the field device is "joined to application" [2].

Timing information is based on the identification of specific events in the log files and storing the associated timestamp of the events. The timing information metrics are the following:

- *Start time of the infrastructure device software modules*—time elapsed between the start of the test scenario and the moment the software modules of the infrastructure device become operational.
- *Time of first RF transmission of the infrastructure device*—time elapsed since the infrastructure device becomes operational until the first RF frame is transmitted by the infrastructure device; the first RF transmission is an advertisement frame signaling the presence of the ISA100.11a IWSN.

- *Time of first RF transmission of the field device*—time elapsed since the infrastructure device becomes operational until the first RF frame is transmitted by the field device.
- *Join time of the field device*—time elapsed since the infrastructure device becomes operational until the field device is admitted in the network.
- *Data collecting initialization time*—time elapsed since the completion of the joining process until the first collected data sample is received by the infrastructure device.

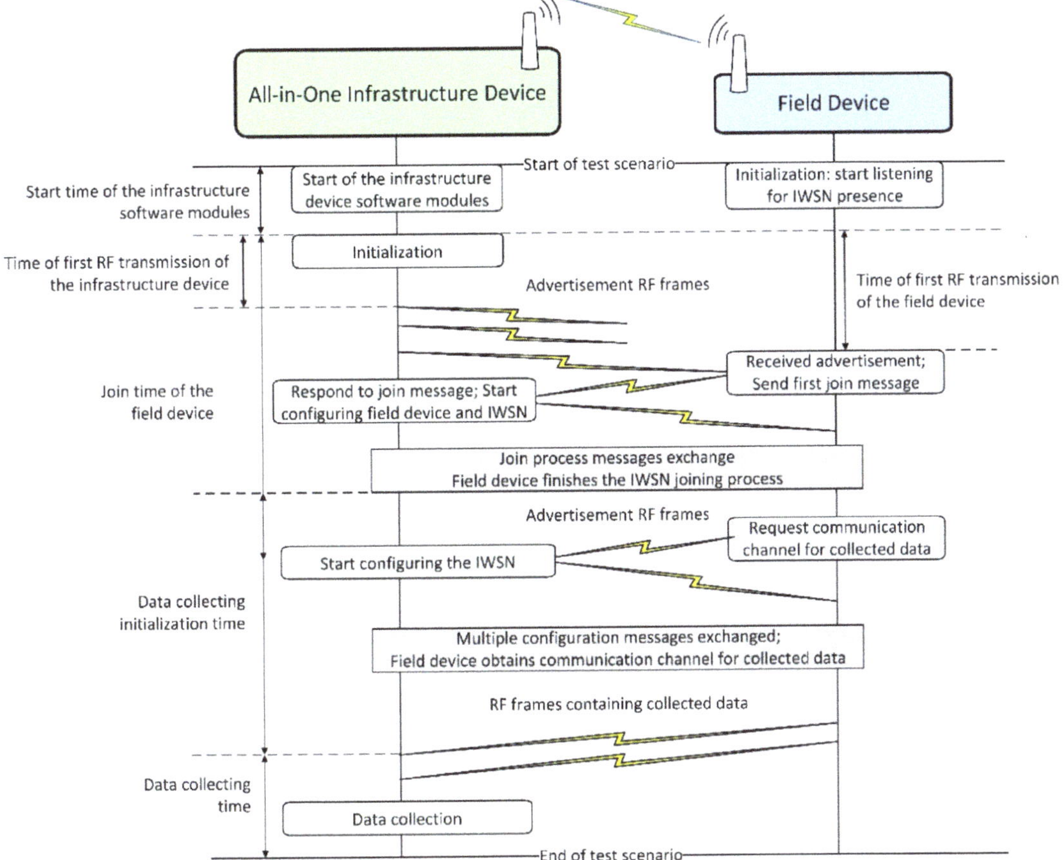

Figure 10. Summary of field device joining to application messages sequence in an ISA100.11a IWSN.

The mean values of the evaluated metrics related to the timing information for all scenarios are presented in Figure 11.

Communication statistics have been obtained by parsing the IEEE 802.15.4 compliant RF communication frames collected in the log files of the RF sniffer and of the simulator. The extracted communication metrics are the following:

- *Number of collected data samples received by the infrastructure device;*
- *Number of RF advertisement frames transmitted by the infrastructure device;*
- *Number of RF advertisement frames transmitted by the field device;*
- *Number of RF communication frames transmitted by the infrastructure device*—these are management messages transmitted during the *Data collecting initialization time;*
- *Number of RF communication frames transmitted by the field device*—these are transmitted during the *Data collecting initialization time* and during *Data collection* and include the

answers to the RF communication frames transmitted by the infrastructure device, collected data publishing, and device diagnostic messages;
- *Number of RF acknowledgment (ACK) frames transmitted by infrastructure device;*
- *Number of RF ACK frames transmitted by field device.*

The mean values of the evaluated metrics related to communication statistics for all scenarios are presented in Figure 12.

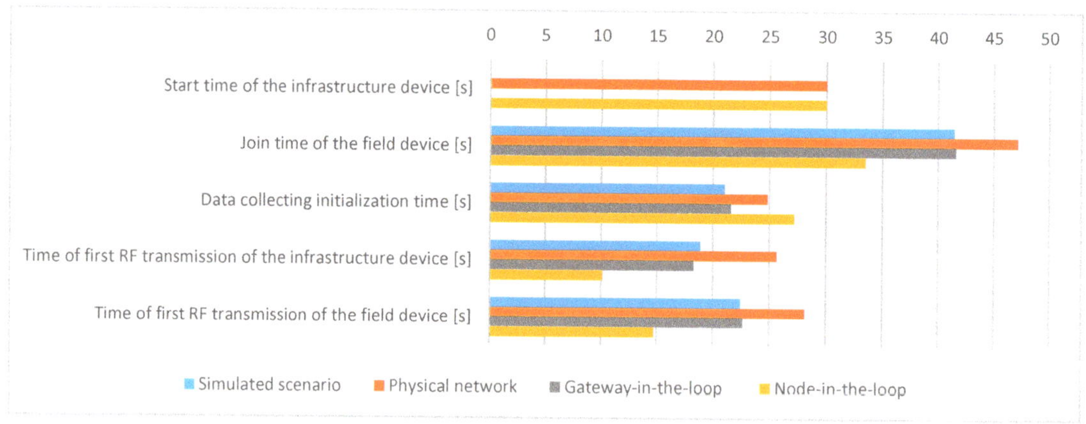

Figure 11. Mean values of the timing information metrics.

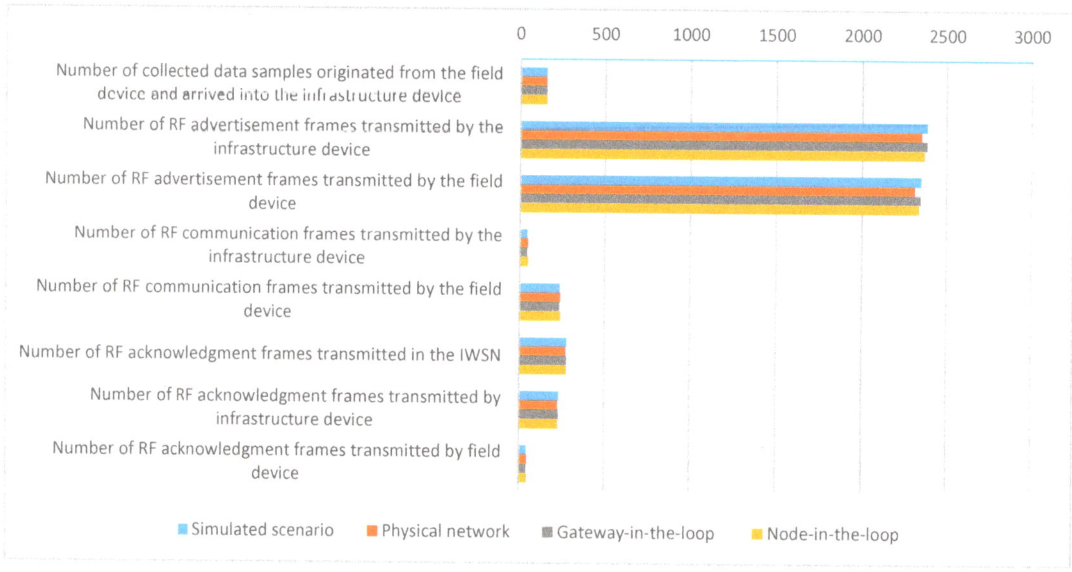

Figure 12. Mean values of the communication statistics metrics.

These results show that after power-on, the physical infrastructure device requires some time to initialize its hardware, load and start the OS, and start the management modules, resulting in a 30 s delay during IWSN formation. In the simulated scenario, the OS is already running, and the start-up time of the software modules on the infrastructure device is negligible.

The tests show that more time is needed in the experimental testbed for a device to join the network once the IWSN is formed. This may occur due to retransmissions or due to CPU load on the physical infrastructure device.

The time of first RF transmission of the infrastructure device shows the moment when the infrastructure device becomes fully operational. The results show that this latency is the biggest in the *Physical network* scenario, being caused by the limited processing capability (CPU performance) of the physical devices. As part of the modules are simulated in the HIL scenarios, the CPU available on the computer speeds up the process, resulting in a smaller value for the *Gateway-in-the-loop* scenario and reaching the shortest time of the first RF transmission for the *Node-in-the-loop* scenario. Moreover, an inspection of the System Manager log files shows that the frequency of exchanged messages is highest for the *Node-in-the-loop* scenario.

Thus, the *Physical* scenario has the biggest value because all processing is performed on the infrastructure device, the *Gateway-in-the-loop* scenario has an intermediate value as the BBR application has been moved into the simulator, and the *Node-in-the-loop* scenario has the smallest value as the whole SM has been moved into the simulator.

Regarding the *Simulated* scenario, the log files show that there is an implementation limitation on the communication path between the Backbone Radio Backend Process and Backbone OS Backend Process (serial interface and emulated hardware functions), limiting the frequency of message exchange between the simulated transceiver and SM and by extension increasing the value of this metric. This tells us that an improved version of the proposed simulator should allow the configuration of the processing capabilities of the hardware. As such, the software modules' initialization time would be more accurate.

The *Simulated* and *Gateway-in-the-loop* scenarios have very close values for this metric, showing that the simulated and physical network management modules behave very similarly when managing a simulated field device.

The *Time of first RF transmission of the field device* has to be correlated with the *Time of first RF transmission of the infrastructure device*. The difference between these two metrics shows the amount of time needed for the field device to find the IWSN formed by the infrastructure device. The values for these differences are similar in all scenarios. The absolute values show that the data is valid, as the field device starts the join procedure only after the infrastructure device has started signaling the presence of the IWSN.

Data collecting initialization time, which is defined as the time required for a device to start sending data once the IWSN is formed, has similar values in all scenarios. This metric depends on the completion of a sequence of messages; thus, it is very sensitive to delays in communication. Such delays might appear due to non-ideal RF communication conditions. Therefore, the *Physical network* and *Node-in-the-loop* scenarios are characterized by a slightly longer *Data collecting initialization time*.

The networks in the testing scenarios show comparable results for the IWSN communication metrics, the closest similarities being obtained for the networks with the same type of PHY and Data link implementations (i.e., *Simulated* vs. *Gateway-in-the-loop* and *Physical* vs. *Node-in-the-loop*).

The simulated infrastructure and field devices transmit more RF advertisement and communication frames than the physical ones because the simulated IWSN forms a few seconds earlier.

7.1. Accuracy of the Evaluated Metrics

For each metric and for each scenario type (i.e., Physical network, Simulated scenario, Gateway-in-the-loop, and Node-in-the-loop), we have obtained a set of results, each corresponding to the execution of a specific scenario. The values in these sets of results differ between scenario executions, so for the purpose of evaluating the simulator, we compute the mean value for each metric over multiple executions of each scenario type. There is an uncertainty attached to the mean value of each metric, because it is not known if the mean value of a finite number of scenario executions is equal to the mean value that would be obtained if the experimental scenarios would have been executed many more times.

The accuracy of the evaluated metric is defined as the closeness of the computed mean value of each metric in the simulated scenario or in the HIL experiment to the corresponding reference mean value collected in the physical deployed replica.

Moreover, we define statistical confidence intervals to quantify the precision of the computed mean value of each evaluated metric. The confidence intervals are computed from the set of the values obtained by multiple executions of a specific scenario/experiment. For each evaluated metric, its associated confidence interval is the range of values in which the computed mean value of that metric is located with the probability of the confidence level selected for the evaluation. We have selected the confidence level of 95%.

Table 4 summarizes the statistical confidence interval of the extracted metrics at a confidence level of 95%.

Table 4. The confidence intervals of the evaluated metrics.

Metric	Simulated Scenario	Physical Network	Gateway-in-the-Loop	Node-in-the-Loop
	IWSN timing information			
Start time of the infrastructure software modules	0.0 s	30.0 s	0.0 s	30.0 s
Time of first RF transmission of the infrastructure device	18.8 ± 0.75% s	25.6 ± 1.13% s	18.3 ± 6.73% s	10.1 ± 1.75% s
Time of first RF transmission of the field device	22.4 ± 2.36% s	28.1 ± 2.07% s	22.7 ± 4.48% s	14.8 ± 3.92% s
Join time of the field device	41.4 ± 1.29% s	47.2 ± 1.34% s	41.6 ± 2.43% s	33.5 ± 1.89% s
Data collecting initialization time	21.0 ± 6.52% s	24.8 ± 5.10% s	21.6 ± 6.09% s	27.2 ± 5.17% s
	IWSN communication statistics			
Number of collected data samples received by the infrastructure device	156.4 ± 0.10%	156.0 ± 0.05%	156.2 ± 0.09%	156.5 ± 0.10%
Number of RF advertisement frames transmitted by the infrastructure device	2382.2 ± 0.01%	2349.9 ± 0.58%	2382.3 ± 0.05%	2367.4 ± 0.35%
Number of RF advertisement frames transmitted by the field device	2345.1 ± 0.02%	2309.5 ± 0.77%	2344.6 ± 0.06%	2335.2 ± 0.17%
Number of RF communication frames transmitted by the infrastructure device	40.4 ± 0.55%	49.7 ± 1.27%	40.3 ± 0.55%	50.7 ± 1.06%
Number of RF communication frames transmitted by the field device	236.5 ± 0.08%	241.2 ± 0.30%	236.3 ± 0.08%	243.1 ± 0.25%
Number of RF ACK frames transmitted by infrastructure device	234.5 ± 0.08%	228.1 ± 0.77%	234.3 ± 0.08%	230.2 ± 1.41%
Number of RF ACK frames transmitted by field device	43.4 ± 0.51%	46.5 ± 2.28%	43.3 ± 0.51%	47.5 ± 2.73%

For the metric *Start time of the infrastructure software modules*, there is no confidence interval defined as it has no variation.

The confidence interval is presented as a percent of the mean value of its associated metric. The confidence interval for most metrics is under ±2.5%, showing that the metrics have a sufficiently precise value for our study. There are three metrics with confidence intervals greater than ±2.5%: *Data collecting initialization time, Time of first RF transmission of the infrastructure device*, and *Time of first RF transmission of the field device*. All these metrics are in the category of timing information, so any delay in processing or communication affects them, including retransmissions, packet loss, and interference.

Considering the order of the events from which the metrics are calculated, it can be observed that the confidence interval becomes larger with later events. This shows that randomness is added to the metrics as each test scenario is executed. Interestingly, in the case of the *Join time of the field device*, randomness seems to be decreasing, as it has a smaller confidence interval than the time of the *First RF transmission of the field device*. It is possible that this decrease in variation is related to the periodic nature of the ISA100.11a communication resources and that events tend to align with time moments when the periodic communication resources are active (i.e., ISA100.11a links, superframes, communication contracts).

7.2. Variations of the Evaluated Metrics

The variations of the collected metrics show how much uncertainty is associated with the metrics collected in the test scenarios. Firstly, this shows how easily repeatable the experiments are, and secondly, for the simulated scenario, it characterizes the robustness of the proposed simulator.

The variations of the evaluated metrics are quantified as the relative standard deviation of each metric, which are calculated on the 40 executions of the scenario where the metric belongs. The relative standard deviation is the statistical standard deviation of the values of its associated metric in one type of scenario, converted to a percent of the metric's mean value. Table 5 presents an overview of the relative standard deviations of each evaluated metric.

Table 5. Relative standard deviations of the evaluated metrics.

Metric	Simulated Scenario	Physical Network	Gateway-in-the-Loop	Node-in-the-Loop
	IWSN timing information			
Start time of the infrastructure software modules	0.0 s	30.0 s	0.0 s	30.0 s
Time of first RF transmission of the infrastructure device	18.8 ± 2.42% s	25.6 ± 3.65% s	18.3 ± 21.71% s	10.1 ± 5.64% s
Time of first RF transmission of the field device	22.4 ± 7.63% s	28.1 ± 6.67% s	22.7 ± 14.47% s	14.8 ± 12.64% s
Join time of the field device	41.4 ± 4.16% s	47.2 ± 4.31% s	41.6 ± 7.83% s	33.5 ± 6.11% s
Data collecting initialization time	21.0 ± 21.03% s	24.8 ± 16.46% s	21.6 ± 19.64% s	27.2 ± 16.69% s
	IWSN communication statistics			
Number of collected data samples received by the infrastructure device	156.4 ± 0.31%	156.0 ± 0.18%	156.2 ± 0.30%	156.5 ± 0.32%
Number of RF advertisement frames transmitted by the infrastructure device	2382.2 ± 0.02%	2349.9 ± 1.87%	2382.3 ± 0.17%	2367.4 ± 1.12%
Number of RF advertisement frames transmitted by the field device	2345.1 ± 0.06%	2309.5 ± 2.49%	2344.6 ± 0.19%	2335.2 ± 0.54%
Number of RF communication frames transmitted by the infrastructure device	40.4 ± 1.76%	49.7 ± 4.09%	40.3 ± 1.78%	50.7 ± 3.42%
Number of RF communication frames transmitted by the field device	236.5 ± 0.27%	241.2 ± 0.95%	236.3 ± 0.25%	243.1 ± 0.82%
Number of RF ACK frames transmitted by infrastructure device	234.5 ± 0.27%	228.1 ± 2.47%	234.3 ± 0.26%	230.2 ± 4.53%
Number of RF ACK frames transmitted by field device	43.4 ± 1.64%	46.50 ± 7.35%	43.3 ± 0.00%	47.5 ± 8.81%

In case of the *Start time of the infrastructure software modules* metric, there is no standard deviation defined, as it has constant values and no variation

The metrics related to timing information have generally a bigger standard deviation as a percent of their mean value than the metrics related to communication statistics. This can be explained by the fact that getting to a specific moment in the evolution of a test scenario is more sensitive to any delay than the total number of occurrences of a specific event. In the case of timing-related metrics, each random delay increases the variation of the metric.

The IWSN communication statistics indicate that both the reference scenario and the simulated and HIL scenarios show similar behaviors across multiple executions, having their relative standard deviation under ±5%.

The evaluated metrics show that the simulated IWSN behavior closely resembles the one of the IWSN deployed on the experimental testbed, demonstrating that the ISA100.11a model is accurately integrated with ns-3.

7.3. Scalability of the Proposed IWSN Model

The scalability of the proposed simulator has been evaluated in terms of metrics related to IWSN functionality and metrics related to computer resource usage. The metrics related to IWSN functionality are the following:

- *Data collecting initialization time for the first device to start transmitting data*, which is measured from the start of the scenario until the first collected data sample arrives to the infrastructure device. This metric is highly dependent on the behavior of medium access mechanism, as during start-up of the IWSNs RF, collisions are expected.
- *Data collecting initialization time for the last device to start transmitting data*, from the start of the scenario. This metric marks the end of the initialization of an IWSN, and it is very sensitive to the behavior of the medium access mechanism, as any failed transmission can cause significant increase in this metric.
- *Number of collected data samples received by the infrastructure device*. This metric characterizes the correct functioning and stability of the IWSN.
- *Total number of RF advertisement frames transmitted in the IWSN*. Since RF advertisement frames are transmitted periodically, their number is correlated with the level of RF interference detected by the IWSN.
- *Total number of RF communication frames transmitted in the IWSN* by the infrastructure device or by any of the field devices. This number characterizes the stability of the IWSN, as the System Manager located in the infrastructure device reacts to communication diagnostics sent by the field devices and transmits configuration messages into the network for avoiding potential issues and to optimize the network. These configuration messages are sent in RF communication frames; thus, a greater number of RF communication frames indicates that the System Manager has sent more configuration change messages to the field devices.

The relative standard deviation values of the scalability metrics are summarized in Table 6. Each scenario, simulated and physical, has been executed 40 times.

Table 6. Relative standard deviations of scalability metrics related to IWSN functionality.

Metric	Simulated Scenario, 5 Devices	Physical Network, 5 Devices	Simulated Scenario, 10 Devices	Physical Network, 10 Devices
Data collecting initialization time for the first device	57.0 ± 10.27% s	66.7 ± 10.27% s	57.8 ± 9.44% s	70.3 ± 9.47 s
Data collecting initialization time for the last device	527.1 ± 31.95% s	235.4 ± 12.94 s	1284.1 ± 40.19% s	339.1 ± 6.49% s
Number of collected data samples received by the infrastructure device	704.3 ± 2.33%	748.8 ± 0.71%	949.1 ± 6.53%	1460.6 ± 0.85%
Total number of RF advertisement frames transmitted in the IWSN	9045.2 ± 1.46%	9055.8 ± 1.30%	8653.4 ± 3.22%	8502.3 ± 2.93%
Total number of RF communication frames transmitted in the IWSN	7309.9 ± 11.62%	5248.2 ± 9.37%	16436.2 ± 10.57%	11839.8 ± 8.97%

The data collecting initialization time of the simulated and physical networks is quite similar, both as mean value and as relative standard deviation of the experimental runs. The simulated networks have started data collecting some seconds earlier than the physical networks—this is probably related to the implementation of the carrier sense multiple access (CSMA) mechanism in the simulator. There might be a difference between the CSMA behavior of a physical and simulated medium, such that the simulation allows more aggressive access to the RF channel, thus allowing a simulated field device to finish the configuration message exchanges earlier and to start sending data to the infrastructure device earlier than the first device in the physical network scenario.

The fact that all five and respectively 10 field devices started collecting and sending data to the infrastructure device show the viability of the proposed simulator architecture. The mean value of data collection initialization time for the last device is much larger in the case of simulated scenarios compared to the physical networks, for both scenario types. In addition to the large difference between simulated and physical IWSNs, the relative standard deviation of the two simulated scenario types shows a large variation of this metric across multiple runs of the scenarios. Apart from the probable difference in CSMA behavior between physical and simulated scenarios mentioned above, the large variation between simulation runs shows that there is a higher amount of randomness in the simulation than in the physical network. Randomness is required for simulating an ISA100.11a communication stack, for example, in the implementation of the exponential back-off algorithm used in conjunction with CSMA; however, such large differences show that either (a) the expected distribution of the used random values is not the expected one or (b) there is some software limitation that generates this behavior.

The collected data samples in simulated and physical IWSN scenarios have similar mean and standard deviation values. The simulated scenarios have smaller values for this metric than their physical network equivalents, which is consistent with the fact that during simulations, the last device has finished data collection initialization considerably later than the last device in a physical network. These similar numbers for this metric show that after initialization, the IWSNs have been performing their task, and they have been collecting data until the end of the scenario.

The number of RF advertisement frames is very similar for the simulated and physical scenarios. Increasing the number of devices from five to 10 in the IWSN decreases a little the total number of RF advertisement frames transmitted in the IWSN, showing increased interference. The used implementation of the ISA100.11a System Manager has an upper limit on the total rate at which RF advertisement frames are transmitted by the IWSN, and it splits the maximum rate of advertisement transmission between the available field devices with routing role. So, having a slight decrease in the total number of advertisements is expected.

The total number of RF communication frames has been larger in simulated scenarios than in the physical network. This shows that in simulated scenarios, the ISA100.11a System Manager has sent more notification messages to field devices than in the physical scenarios. This observation indicates that field devices in simulated scenarios have reported less-ideal communication diagnostics than the physical field devices. The cause might be related to the implementation of CSMA or to some other software limitation.

The metrics used to evaluate the computer resource usage of the simulator are the following:

- Maximum CPU usage of the simulator while running a specific scenario.
- Average CPU usage of the simulator while running a specific scenario.

All scenarios were run on a computer equipped with an Intel Core i5-4300M CPU, having a maximum clock speed of 2.60 GHz. The CPU has been configured to use the performance mode CPU frequency governor, so variations in its clock speed have been minimized. The values of the resource usage metrics are summarized in Table 7. The CPU usage is important for this real-time simulator because the simulation needs to be at least as fast as the wall-clock, as otherwise, timing-related issues might appear in the simulation.

Table 7. Values of scalability metrics related to computer resource usage.

Metric	Simulated Scenario, 1 Device	Simulated Scenario, 5 Devices	Simulated Scenario, 10 Devices
Average CPU usage of the simulator	4%	10%	21%
Maximum CPU usage of the simulator	5%	13%	24%

The results show that the simulated scenario with 10 devices has not been limited by the available CPU capacity. Probably, a scenario with more devices should work well on the same computer. It can be observed that the CPU usage increases more than linearly with the number of devices in a scenario. This might be related to the nature of the RF propagation medium, where each transmitted frame is received by every other radio transceiver in its range. Probably, the simulator can be optimized for reducing CPU usage, for example by not delivering RF frames to transceivers, which are not in receiver mode. As another direction for scaling up the size of the simulated scenarios, a more powerful computer can be used, as the one used in the evaluation is a relatively old CPU equipped in a laptop computer.

8. Conclusions

This paper proposes a highly accurate ISA100.11a simulation model that integrates the communication stack available on commercial ISA100.11a certified devices with the ns-3 simulator.

Using the complete implementation of the communication stack and of the network management applications allows detailed system-level modeling of the IWSN. On the one hand, the observable interactions in this model span across all layers of the communication stack, from the physical layer to the application layer. On the other hand, the model allows observing all interactions between the communication stack of the field devices, the communication stack of the infrastructure device, and the network management applications (e.g., Gateway, System Manager). This wide-ranging model is expected to closely match the behavior of a physical IWSN, helping to solve several problems related to deploying IWSNs.

The accuracy and scalability of the proposed ISA100.11a simulation model are evaluated in a number of test scenarios.

Accuracy is analyzed in four different test scenarios: (1) through simulation using the ISA100.11a model implementation integrated with ns-3, (2) on an experimental testbed employing ISA100.11a certified devices, (3) by implementing a *Gateway-in-the-loop* HIL scenario, and (4) by implementing a *Node-in-the-loop* HIL scenario.

The behavior and performances of the tested IWSNs are evaluated in terms of two sets of metrics: one related to the timing of events in the IWSN and one related to communication statistics. Moreover, the confidence interval and relative standard deviation of each metric are computed to assess the precision of the proposed IWSN model.

The results of the HIL scenarios show that from a functional point of view, the simulation model is indeed equivalent to its physical counterpart. A physical infrastructure device successfully managed a simulated field device and a physical field device successfully joined a network managed by a simulated infrastructure device. The difference between the mean values of timing-related metrics of simulated and physical IWSN is under 7 s, while the difference between HIL scenarios and physical scenarios is under 16 s. For the mean values of communication statistics, the difference between the simulated and physical IWSN for each metric is under 15%. Functional equivalence allows easy exportation of provisioning data from simulation to deployable devices. Moreover, integration testing can be first performed with the simulator, which saves time and allows greater flexibility during testing and integration.

Scalability is analyzed by implementing an extended set of simulated and real scenarios with five or 10 field devices. The performance metrics for characterizing the scalability of the ISA100.11a simulation model are related to the timing of the events, communication statistics, and computer resource usage of the simulator. For the first two types of metrics, the mean value and relative standard deviation are calculated, while for the third type, the mean and maximum values are recorded. The difference between the mean value of the number of collected data samples between simulated and physical IWSNs is under 45 data samples or 7% for an IWSN consisting of five devices, and under 512 or 54% for IWSN consisting of 10 devices. The greatest difference between mean values of metrics is in the case of *Data collecting initialization time* for the last device, having a difference of 292 s or 55.34% for an IWSN of five devices and 945 s or 73.59% for an IWSN consisting of 10 devices.

The evaluation results show that the simulated IWSN behavior closely resembles the one of the deployed networks, highlighting the fact that the proposed ISA100.11a model integrated with ns-3 accurately implements the ISA100.11a communication stack functionalities.

For future work, the authors aim to extend the research and evaluate and improve the performance of the proposed simulation model for larger IWSNs. To this extent, radio channel and network topology modeling for complex scenarios will be considered. Moreover, energy consumption models will be included to enable the proposed simulation model to provide useful information regarding the trade-off between improving the QoS of the system and reducing the overall energy consumption.

Author Contributions: Conceptualization, Z.P. and E.P.; methodology, E.P.; software, Z.P.; validation, A.P.; formal analysis, Z.P., and E.P.; investigation, Z.P.; resources, O.R., and T.P.; data curation, Z.P., E.P., and A.P.; writing—original draft preparation, Z.P.; writing—review and editing, A.P., T.P., O.R., and E.P.; visualization, Z.P., and E.P.; supervision, E.P.; project administration, E.P.; funding acquisition, E.P. All authors have read and agreed to the published version of the manuscript.

Funding: This research was funded by the Romanian Ministry of Education and Research, CCCDI–UEFISCDI, grant number PN-III-P2-2.1-PED-2019-3427. The APC was funded by the Technical University of Cluj-Napoca.

Acknowledgments: This work was supported by a grant of the Romanian Ministry of Education and Research, CCCDI–UEFISCDI, project number PN-III-P2-2.1-PED-2019-3427, within PNCDI III. The ISA100.11a certified devices employed in the experimental setup were provided by Control Data Systems SRL.

Conflicts of Interest: The authors declare no conflict of interest.

References

1. Candell, R.; Kashef, M.; Liu, Y.; Lee, K.; Foufou, S. Industrial wireless systems guidelines: Practical considerations and deployment life cycle. *IEEE Ind. Electron. Mag.* **2018**, *12*, 6–17. [CrossRef]
2. *Wireless Systems for Industry Automation: Process Control and Related Applications*; ISA-100.11a-2011 Standard; International Society of Automation (ISA): Research Triangle Park, NC, USA, 2011.
3. IEC. *Industrial Networks—Wireless Communication Network and Communication Profiles*; ISA 100.11a. IEC-62734 Standard; International Electrotechnical Commission (IEC): Geneva, Switzerland, 2014.
4. IEC. *Industrial Communication Networks—Wireless Communication Network and Communication Profiles–WirelessHART*; IEC-62591 Standard; International Electrotechnical Commission (IEC): Geneva, Switzerland, 2010.
5. *ZigBee Specification*; ZigBee Document 05-3474-21; ZigBee Alliance: Davis, CA, USA, 2015.
6. Bluetooth SIG. *Mesh Profile Specification 1.0.1*; Bluetooth Special Interest Group: Kirkland, WA, USA, 2019.
7. IEC. *Industrial Networks—Wireless Communication Network and Communication Profiles—WIA-FA (Wireless Networks for Industrial Automation—Factory Automation)*; IEC/PAS 62948 Standard; International Electrotechnical Commission (IEC): Geneva, Switzerland, 2015.
8. Liang, W.; Zheng, M.; Zhang, J.; Shi, H.; Yu, H.; Yang, Y.; Liu, S.; Yang, W.; Zhao, X. WIA-FA and its applications to digital factory: A wireless network solution for factory automation. *Proc. IEEE* **2019**, *107*, 1053–1073. [CrossRef]
9. IEC. *Industrial Networks—Wireless Communication Network and Communication Profiles—WIA-PA (Wireless Networks for Industrial Automation—Process Automation)*; IEC/PAS 62601 Standard; International Electrotechnical Commission (IEC): Geneva, Switzerland, 2015.
10. Florencio, H.; Dória Neto, A.; Martins, D. ISA 100.11a networked control system based on link stability. *Sensors* **2020**, *20*, 5417. [CrossRef] [PubMed]
11. ns-3, A Discrete-Event Network Simulator for Internet Systems. Available online: https://www.nsnam.org/ (accessed on 26 June 2020).
12. Padrah, Z.; Pop, C.; Jecan, E.; Pastrav, A.; Palade, T.; Ratiu, O.; Puschita, E. An ISA100.11a model implementation for accurate industrial WSN simulation in ns-3. In Proceedings of the 2020 International Workshop on Antenna Technology (iWAT), Bucharest, Romania, 25–28 February 2020; Institute of Electrical and Electronics Engineers (IEEE): Piscataway Township, NJ, USA, 2020; pp. 1–4. [CrossRef]

13. Chéour, R.; Jmal, M.W.; Kanoun, O.; Abid, M. Evaluation of simulator tools and power-aware scheduling model for wireless sensor networks. *IET Comput. Digit. Tech.* **2017**, *11*, 5. [CrossRef]
14. OMNeT++ Discrete Event Simulator. Available online: https://omnetpp.org/ (accessed on 26 June 2020).
15. Boulis, A. Castalia: Revealing pitfalls in designing distributed algorithms in WSN. In Proceedings of the 5th International Conference on Embedded Networked Sensor Systems (SenSys '07), Sydney, Australia, 6–9 November 2007; Association for Computing Machinery: New York, NY, USA, 2007; pp. 407–408. [CrossRef]
16. Varga, A.; Hornig, R. An overview of the OMNeT++ simulation environment. In Proceedings of the 1st International Conference On Simulation Tools and Techniques for Communications, Networks and Systems & Workshops (Simutools '08), Marseille, France, 3–7 March 2008; ICST (Institute for Computer Sciences, Social-Informatics and Telecommunications Engineering): Brussels, Belgium, 2008. Article 60. pp. 1–10.
17. Köpke, A.; Swigulski, M.; Wessel, K.; Willkomm, D.; Klein Haneveld, P.T.; Parker, T.E.V.; Visser, O.W.; Lichte, H.S.; Valentin, S. Simulating wireless and mobile networks in OMNeT++ the MiXiM vision. In Proceedings of the 1st International Conference On Simulation Tools and Techniques for Communications, Networks and Systems & Workshops (Simutools '08), Marseille, France, 3–7 March 2008; ICST (Institute for Computer Sciences, Social-Informatics and Telecommunications Engineering): Brussels, Belgium, 2008. Article 71. pp. 1–8.
18. INETMANET Framework for OMNEST/OMNeT++ 4.0. Available online: https://github.com/inetmanet/inetmanet/ (accessed on 26 June 2020).
19. Exceptional Network Performance and Visibility for Any Application. Available online: https://www.riverbed.com/dk/ (accessed on 26 June 2020).
20. QualNet–Network Simulation Software. SCALABLE Networks. Available online: https://www.scalable-networks.com/products/qualnet-network-simulation-software-tool/ (accessed on 26 June 2020).
21. Levis, P.; Lee, N.; Welsh, M.; Culler, D. TOSSIM: Accurate and scalable simulation of entire TinyOS applications. In Proceedings of the 1st International Conference on Embedded Networked Sensor Systems (SenSys '03), Los Angeles, CA, USA, 5–7 November 2003; Association for Computing Machinery: New York, NY, USA, 2003; pp. 126–137. [CrossRef]
22. Avrora–The AVR Simulation and Analysis Framework. Available online: http://compilers.cs.ucla.edu/avrora/ (accessed on 26 June 2020).
23. Osterlind, F.; Dunkels, A.; Eriksson, J.; Finne, N.; Voigt, T. Cross-level sensor network simulation with cooja. In Proceedings of the 31st IEEE Conference on Local Computer Networks, Tampa, FL, USA, 14–16 November 2006; pp. 641–648. [CrossRef]
24. Bayou, L.; Espes, D.; Cuppens-Boulahia, N.; Cuppens, F. WirelessHART NetSIM: A wirelesshart SCADA-based wireless sensor networks simulator. In *Security of Industrial Control Systems and Cyber Physical Systems. CyberICS 2015, WOS-CPS 2015. Lecture Notes in Computer Science*; Bécue, A., Cuppens-Boulahia, N., Cuppens, F., Katsikas, S., Lambrinoudakis, C., Eds.; Springer: Cham, Switzerland, 2016; Volume 9588. [CrossRef]
25. Liu, Y.; Candell, R.; Lee, K.; Moayeri, N. A simulation framework for industrial wireless networks and process control systems. In Proceedings of the 2016 IEEE World Conference on Factory Communication Systems (WFCS), Aveiro, Portugal, 3–6 May 2016; pp. 1–11. [CrossRef]
26. Downs, J.; Vogel, E. A plant-wide industrial process control problem. *Comput. Chem. Eng.* **1993**, *17*, 245–255. [CrossRef]
27. Liu, Y.; Candell, R.; Lee, K.; Moayeri, N. Tennessee Simulator Federated with OMNET++ Networking Model. 2016. Available online: https://github.com/usnistgov/tesim_omnetpp (accessed on 26 June 2020).
28. Al-Yami, A.; Abu-Al-Saud, W.; Shahzad, F. Simulation of Industrial Wireless Sensor Network (IWSN) protocols. In Proceedings of the 2016 IEEE Conference on Computer Communications Workshops (INFOCOM WKSHPS), San Francisco, CA, USA, 10–15 April 2016; pp. 527–533. [CrossRef]
29. Herrmann, M.J.; Messier, G. Cross-layer lifetime optimization for practical industrial wireless networks: A petroleum refinery case study. *IEEE Trans. Ind. Inform.* **2018**, *14*, 3559–3566. [CrossRef]
30. Messier, G. ISA100.11a ns3 Simulation Code. Available online: https://github.com/ggmessier/ns3-isa100.11a (accessed on 26 June 2020).
31. Polley, J.; Blazakis, D.; McGee, J.; Rusk, D.; Baras, J.S. ATEMU: A fine-grained sensor network simulator. In Proceedings of the 2004 1st Annual IEEE Communications Society Conference on Sensor and Ad Hoc Communications and Networks (SECON 2004), Santa Clara, CA, USA, 4–7 October 2004; pp. 145–152. [CrossRef]
32. Martinez, B.; Montón, M.; Vilajosana, I.; Prades, J.D. The Power of Models: Modeling Power Consumption for IoT Devices. *IEEE Sens. J.* **2015**, *15*, 5777–5789. [CrossRef]
33. Pagano, P.; Chitnis, M.; Lipari, G.; Nastasi, C.; Liang, Y. Simulating Real-Time Aspects of Wireless Sensor Networks. *EURASIP J. Wirel. Commun. Netw.* **2010**, *2010*, 107946. [CrossRef]
34. The Network Simulator–ns-2. Available online: https://www.isi.edu/nsnam/ns/ (accessed on 26 June 2020).
35. Palopoli, L.; Lipari, G.; Abeni, L.; Di Natale, M.; Ancilotti, P.; Conticelli, F. A tool for simulation and fast prototyping of embedded control systems. In Proceedings of the 2001 ACM SIGPLAN Workshop on Optimization of Middleware and Distributed Systems (OM '01), Snowbird, UT, USA, 19 June 2001; Association for Computing Machinery: New York, NY, USA, 2001; pp. 73–81. [CrossRef]

36. Savazzi, S.; de Souza, R.H.; Becker, L.B. Wireless Network Planning and Optimization in Oil and Gas Refineries. In Proceedings of the 2013 III Brazilian Symposium on Computing Systems Engineering, Niteroi, Rio De Janeiro, Brazil, 4–8 December 2013; pp. 29–34. [CrossRef]
37. Zand, P.; Mathews, E.; Havinga, P.; Stojanovski, S.; Sisinni, E.; Ferrari, P. Implementation of WirelessHART in the NS-2 Simulator and Validation of Its Correctness. *Sensors* **2014**, *14*, 8633–8668. [CrossRef] [PubMed]
38. Li, W.; Zhang, X.; Tan, W.; Zhou, X. H-tossim: Extending tossim with physical nodes. *Wirel. Sens. Netw.* **2009**, *1*, 324–333. [CrossRef]
39. Alonso-Eugenio, V.; Guerra, V.; Zazo, S.; Perez-Alvarez, I. Software-in-loop simulation environment for electromagnetic underwater wireless sensor networks over STANAG 5066 protocol. *Electronics* **2020**, *9*, 1611. [CrossRef]
40. Clavijo-Rodriguez, A.; Alonso-Eugenio, V.; Zazo, S.; Perez-Alvarez, I. Software-in-loop simulation of an underwater wireless sensor network for monitoring seawater quality: Parameter selection and performance validation. *Sensors* **2021**, *21*, 966. [CrossRef]
41. Boehm, S.; Koenig, H. SEmulate: Seamless network protocol simulation and radio channel emulation for wireless sensor networks. In Proceedings of the 2019 15th Annual Conference on Wireless On-demand Network Systems and Services (WONS), Wengen, Switzerland, 22–24 January 2019; pp. 111–118. [CrossRef]
42. Dunkels, A.; Gronvall, B.; Voigt, T. Contiki–A lightweight and flexible operating system for tiny networked sensors. In Proceedings of the 29th Annual IEEE International Conference on Local Computer Networks, Tampa, FL, USA, 16–18 November 2004; pp. 455–462. [CrossRef]
43. ISA100.11a—Field-Device Communication Stack. NIVIS LLC. Available online: https://github.com/irares/ISA100.11a-Field-Device (accessed on 26 June 2020).
44. Main Development Repository for TinyOS. Available online: https://github.com/tinyos/tinyos-main (accessed on 29 November 2020).
45. Avrora Project on SourceForge. Available online: https://sourceforge.net/projects/avrora/ (accessed on 29 November 2020).
46. The Official Git Repository for Contiki. Available online: https://github.com/contiki-os/contiki (accessed on 29 November 2020).
47. A Fork of the Cooja Network Simulator from Contiki-os/Contiki. Available online: https://github.com/contiki-ng/cooja (accessed on 29 November 2020).
48. RTSim Project on SourceForge. Available online: https://sourceforge.net/projects/rtsim/ (accessed on 29 November 2020).
49. IEEE. *IEEE Standard for Low-Rate Wireless Networks (Revision of IEEE Std 802.15.4-2011), IEEE Std 802.15.4-2015*; IEEE: New York, NY, USA, 2016. [CrossRef]
50. Nagy, T. Waf: The Meta Build System. Available online: https://waf.io/ (accessed on 26 June 2020).
51. Control Data Systems. CDS Wireless Website. Available online: http://www.cds.ro/ (accessed on 26 June 2020).
52. CDS VersaRouter 950—Control Data Systems. Available online: http://www.cds.ro/products/versarouter-950/ (accessed on 26 June 2020).
53. Development Board VS210—Control Data Systems. Available online: http://www.cds.ro/products/development-board-vs210/ (accessed on 26 June 2020).
54. HART Wi-Analys—FieldComm Group. HART Wi-Analys Sniffer. Available online: https://fieldcomm-group.myshopify.com/products/hrt-wa (accessed on 7 June 2020).
55. Field Tool FT210—Control Data Systems. Available online: http://www.cds.ro/products/field-tool-ft210/ (accessed on 26 June 2020).

Communication

The Design and Manufacturing Process of an Electrolyte-Free Liquid Metal Frequency-Reconfigurable Antenna

Peng Qin [1,2], Lei Wang [1], Tian-Ying Liu [1,2], Qian-Yu Wang [1,2], Jun-Heng Fu [1,2], Guan-Long Huang [3], Lin Gui [1,2], Jing Liu [1,2] and Zhong-Shan Deng [1,2,*]

1 CAS Key Laboratory of Cryogenics, Technical Institute of Physics and Chemistry, Chinese Academy of Sciences, Beijing 100190, China; qinpeng17@mails.ucas.edu.cn (P.Q.); wanglei_820425@163.com (L.W.); liutianying17@mails.ucas.edu.cn (T.-Y.L.); wangqianyu19@mails.ucas.edu.cn (Q.-Y.W.); fujunheng17@mails.ucas.edu.cn (J.-H.F.); lingui@mail.ipc.ac.cn (L.G.); jliu@mail.ipc.ac.cn (J.L.)
2 School of Future Technology, University of Chinese Academy of Sciences, Beijing 100049, China
3 The College of Electronics and Information Engineering, Shenzhen University, Shenzhen 518060, China; guanlong.huang@ieee.org
* Correspondence: zsdeng@mail.ipc.ac.cn; Tel.: +86-010-82543483

Abstract: This communication provides an integrated process route of smelting gallium-based liquid metal (GBLM) in a high vacuum, and injecting GBLM into the antenna channel in high-pressure protective gas, which avoids the oxidation of GBLM during smelting and filling. Then, a frequency-reconfigurable antenna, utilizing the thermal expansion characteristic of GBLM, is proposed. To drive GBLM into an air-proof space, the thermal expansion characteristics of GBLM are required. The dimensions of the radiating element of the liquid metal antenna can be adjusted at different temperatures, resulting in the reconfigurability of the operating frequency. To validate the proposed concept, an *L*-band antenna prototype was fabricated and measured. Experimental results demonstrate that the GBLM in the antenna was well filled, and the GBLM was not oxidized. Due to the GBLM being in an air-proof channel, the designed liquid metal antenna without electrolytes could be used in an air environment for a long time. The antenna is able to achieve an effective bandwidth of over 1.25–2.00 GHz between 25 °C and 100 °C. The maximum radiation efficiency and gain in the tunable range are 94% and 2.9 dBi, respectively. The designed antenna also provides a new approach to the fabrication of a temperature sensor that detects temperature in some situations that are challenging for conventional temperature sensing technology.

Keywords: antioxidation; frequency-reconfiguration; liquid metal; temperature sensor; antenna sensor; thermal expansion; electrolyte-free

Citation: Qin, P.; Wang, L.; Liu, T.-Y.; Wang, Q.-Y.; Fu, J.-H.; Huang, G.-L.; Gui, L.; Liu, J.; Deng, Z.-S. The Design and Manufacturing Process of an Electrolyte-Free Liquid Metal Frequency-Reconfigurable Antenna. *Sensors* **2021**, *21*, 1793. https://doi.org/10.3390/s21051793

Academic Editor: Luca Catarinucci

Received: 11 January 2021
Accepted: 25 February 2021
Published: 5 March 2021

1. Introduction

In recent years, the reconfigurable antenna has emerged as a promising candidate to face the challenges and requirements of high gain, broadband and multifunction in advanced communication systems. Generally, switching components such as radio frequency microelectromechanical systems (RF MEMS) [1], varactor diodes [2] and p-type intrinsic n-type (PIN) diodes [3] are frequently applied to ensure sensitive control of antennas' reconfigurable performance. More and more radio frequency (RF) switches are used in antennas to seek better reconfigurable effects, while the complex auxiliary circuits and nonlinear effects remain inevitable. Moreover, some possible challenges, like constrained tuning and poor harmonics, require more strategies and solutions. Currently, ionic solutions [4], liquid crystals [5] and liquid metals, for example, provide new methods for reconfiguration. Among them, the outstanding characteristics of fluidity, electrical conductivity and deformability of liquid metal (LM) promise to become especially important in the field of reconfigurable electronic devices. Up to now, the potential of LM-based antennas has been demonstrated via pattern-, frequency- and polarization-reconfigurable prototypes [6–23], in which the LM mainly performs as a deformable structure for shape-patterning or switching in the radiating elements.

As a conventional LM, mercury has been applied to antennas since 1939 [6–10]. Although mercury is not easily oxidized at room temperature, it is volatile and highly toxic. The effects of mercury exposure can be very severe and subtle. At present, the mercury ban is a global trend. Gallium-based LM (GBLM) is a good substitute for mercury due to its non-volatile and non-toxic properties [11–23]. However, the oxidation of gallium greatly limits the antenna application of GBLM, since gallium oxide is facile to stick to the surface of the cavity [9,24,25]. Under the effects of surface tension and oxidation, mercury and GBLM fill, and withdraw from, microchannels, showing different behaviors [24]. GBLM requires more pressure to move into the microchannel than mercury. When the pressure is removed, GBLM cannot withdraw from the microchannel like mercury does, but requires HCl solution to eliminate gallium oxide. Compared with mercury, it is more difficult to control the flow of GBLM in the microchannel. Only in a glove box, where the oxygen concentration is less than 1 ppm, can GBLM maintain its unoxidized morphology [25]. Due to the low oxygen concentration threshold, the reaction between GBLM and oxygen occurs very easily. In [9], scientists tried to make Galinstan antennas in a glove box with an oxygen concentration of less than 1 ppm, but the antenna could only be used in the glove box and, once removed, it was ineffective. Additionally, the radiation efficiency of the antenna may be reduced since the electrical conductivity of gallium oxide (5×10^{-4} S/m [26]) is much lower than that of GBLM. Almost all existing research works on GBLM antennas either ignore the trouble of oxidation or add acid/alkaline electrolytes [12–23]. The electrolytes used in the deoxidation process produce three serious problems. Firstly, GBLM and acid/alkaline solutions react with and consume one another, so the system cannot coexist. Secondly, through the chemical reaction, the system (especially after the electrolysis reaction) generates bubbles, and multiple fluids can easily disconnect the GBLM. Thirdly, the presence of electrolytes will reduce antenna efficiency [27]. Therefore, avoiding oxidation or deoxidation with an electrolyte-free method is crucial for a GBLM reconfigurable antenna, both in terms of physical fabrication and practical application.

This paper presents a novel idea to drive GBLM into a confined space, and proposes a new type of GBLM-based reconfigurable antenna. Herein, by adopting the integrated manufacturing process of metal smelting under high vacuum, and GBLM infusion under a high-pressure shielding-gas atmosphere, an electrolyte-free antenna radiating element nicely filled with non-oxidized GBLM is obtained. In addition, in order to drive GBLM into enclosed spaces, a thermally controlled strategy is applied to generate the volumetric expansion of GBLM. Thus, non-contact and accurate control for the height of the metallic cylinder in the capillary of the antenna can be realized. In short, this work proposes a new idea of a thermally reconfigurable control method, and a new electrolyte-free manufacturing process for liquid metal antennas, which may pave a possible way toward the simplification of reconfigurable antennas in the future.

2. Antenna Design

2.1. Material and Structure

Figure 1 shows the proposed monopole antenna structure, which consists of five main parts: a SubMiniature version A (SMA) connector, a heating film, a ground plane, an antenna radiating element and a ceramic tube. Enlarged views of the top and bottom parts of the radiating element are shown in Figure 1b,c. The shell of the antenna radiating element is made of quartz glass (relative permittivity: $\varepsilon_r = 3.7$, the tangent of dielectric loss angle: $\tan\delta = 0.00011$, at 1 GHz [28]). The bottom part of the radiating element is a temperature-sensitive bulb filled with EGaIn (a kind of GBLM with the composition of 75.5% gallium and 24.5% indium, electrical conductivity $\sigma = 3.46 \times 10^6$ S/m) to make a reconfigurable wire. The platinum wire ($\delta = 9.43 \times 10^7$ S/m) at the bottom of the radiating element is welded to the inner conductor of the SMA connector, which could be used to connect the feeder.

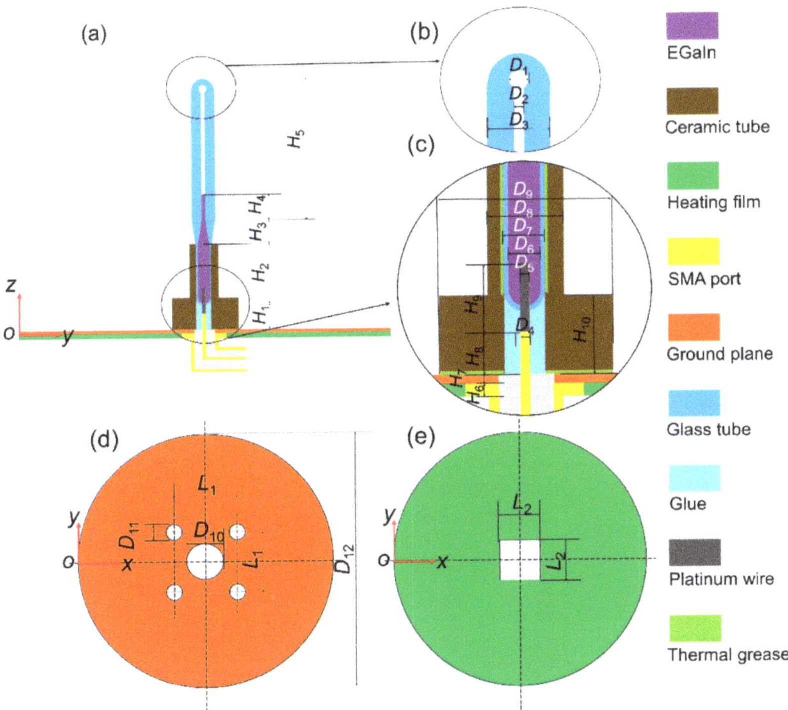

Figure 1. Configuration of the proposed monopole antenna. (**a**) Front view. (**b**) Top zoom. (**c**) Bottom zoom. (**d**) Ground plane. (**e**) Heating film.

The bottom views of the ground plane and heating film are shown in Figure 1d,e, respectively. The heating wire of the heating film is wrapped in rubber, and its temperature is adjusted through a temperature control system. The heating film and the ground plane form a heating platform. The heating film, placed under the ground plane, is used to avoid the influence of the metal heating wire on the antenna radiating element. Ceramic tubes and thermal grease are applied to ensure that the heat on the heating platform can be effectively transferred to the EGaIn. The temperature change of the liquid metal is obtained by adjusting the electric current loads of the heating film. A calibrated thermocouple with ± 0.5 °C accuracy was applied to detect the temperature of the liquid metal.

2.2. Principles of Reconstruction

The degree of liquid expansion is usually quantified by the cubic expansion coefficient γ, which is defined by:

$$\gamma = \frac{1}{V_0}\left(\frac{\partial V}{\partial T}\right)_P \tag{1}$$

where V is the volume, T is the temperature, P is the pressure and V_0 is the initial volume at the initial temperature T_0.

EGaIn usually remains in liquid state from 15.5 °C to 2000 °C [29] and has the characteristics of cold shrinkage and thermal expansion. Its liquid-phase temperature range is much larger than that of mercury, from −38.8 °C to 356.7 °C [30]. The thermal expansion coefficient of quartz glass is 5.5×10^{-7} 1/K [31], which can be ignored compared to the thermal expansion coefficient of GBLM [32].

Hence, V, the volume of liquid metal, can be expressed by the dimensions marked in Figure 1, i.e., D_2, D_6, H_2, H_3 and H_4:

$$V = \frac{\pi}{12}\left(D_6^3 + 3D_6^2H_2 + D_2^2H_3 + D_6^2H_3 + D_2D_6H_3 + 3D_2^2H_4\right) = \frac{\pi}{4}C + \frac{\pi}{4}D_2^2H_4 \quad (2)$$

where C is a constant value during temperature variation, and it can be expressed as:

$$C = \frac{1}{3}\left(D_6^3 + 3D_6^2H_2 + D_2^2H_3 + D_6^2H_3 + D_2D_6H_3\right) \quad (3)$$

According to (1), the volume of liquid metal can be rewritten as (4) when the temperature increases by ΔT:

$$V = V_0 + \Delta V = V_0(1 + \gamma\Delta T) \quad (4)$$

Additionally, the relationship between H_4 and ΔT can be derived from (2) and (4):

$$H_4 = H_{40} + \frac{C + D_2^2}{D_2^2}\gamma\Delta T \quad (5)$$

where H_{40} is the initial length of H_4 at the initial temperature T_0. After the geometric structure is determined, T is the only variable of H_4. As the length L ($L = H_1 + H_2 + H_3 + H_4$) of the radiating element of the monopole antenna generally should satisfy the condition of $L = \lambda/4$, the resonant frequency of the antenna can be changed by controlling the temperature.

2.3. Parametric Analysis

The working frequency band of the antenna proposed in this work is designed in the *L*-band (1–2 GHz). As mentioned above, the lengths of EGaIn mainly determine the working frequencies of the proposed antenna. All the initial geometric dimensions of the antenna structure (for example, H_2, H_3, H_{40}) need to be uniquely confirmed. A full-wave electromagnetic high-frequency structure simulator (HFSS) was used to simulate the proposed antenna in this communication. The entire antenna, except the SMA connector, was considered as the simulation model. Figure 2a shows a clear change in reflection coefficient values (represented by S_{11}), as H_{40} and H_2 changes. It can be seen that the smaller H_{40} is, the higher the operating frequency of the antenna and the lower the reflection coefficient would be. However, the initial value of H_{40} should be large enough so that EGaIn would not flow back into the temperature-sensitive bulb when the temperature drops. Therefore, trade-off values should be chosen, while taking the highest working frequency of 2 GHz into consideration. In this work, the initial lengths of H_{40} and H_2 were chosen to be 17.5 mm and 6 mm, respectively. Figure 2b shows a little change in reflection coefficient values as H_3 changes. The length of H_3 is finally assigned to be 7.0 mm with a lower reflection coefficient value. By using this method of controlling variables, other dimensions can also be optimized similarly. The optimized dimensions of the proposed antenna shown in Figure 1 are given in Table 1.

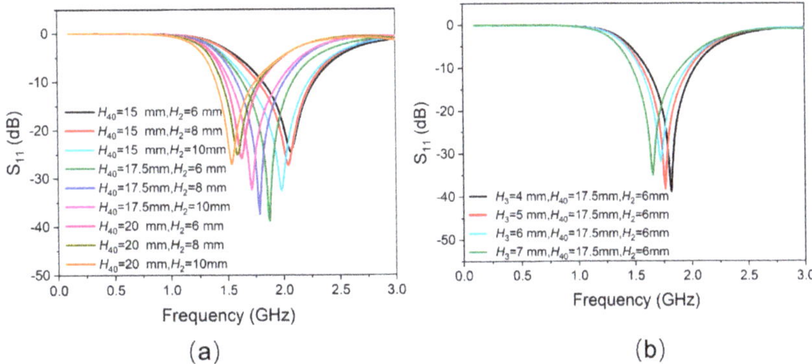

Figure 2. (**a**) Simulated results of $|S_{11}|$ versus the variation of EGaIn column H_{40} and the height of the cylinder of temperature-sensitive bulb H_2. (**b**) Simulated results of $|S_{11}|$ versus the variation of the height of the cone of temperature-sensitive bulb H_3.

Table 1. Dimensions of the Optimized Antenna (Unit: mm).

H_1	H_2	H_3	H_{40}	H_5	H_6
6.0	6.0	7.0	17.5	37.0	2.0
H_7	H_8	H_9	H_{10}	D_1	D_2
2.0	4.0	4.2	6.0	2.0	0.32
D_3	D_4	D_5	D_6	D_7	D_8
6.3	0.5	0.5	3.0	4.0	8.6
D_9	D_{10}	D_{11}	D_{12}	L_1	L_2
14.8	4.0	2.0	100.0	5.0	14.0

3. Fabrication and Measurement

3.1. Fabrication Process

Initially, in a glove box with nitrogen protection, an attempt was made to inject GBLM through a syringe into an open glass tube. However, the antenna failed after being removed from the glove box because the GBLM was still oxidized. Later, a mechanical pump was used to extract air from the semi-closed glass tube. The prepared liquid metal was filled into the glass tube by the pressure difference between the inside and outside of the glass tube. The fabricated antenna still had the problem that GBLM had been oxidated and incompletely filled. Through many previous failed experiments, two important pieces of information were confirmed. Firstly, the GBLM would be partially oxidized rapidly (when exposed to air) before it was applied in antenna manufacturing. Secondly, it was difficult to completely fill the channel using both manual perfusion and pressure perfusion with a mechanical pump. Therefore, an integrated solution for preparing GBLM under high-vacuum conditions and filling GBLM under high-pressure conditions was proposed.

The final schematic manufacturing process of the proposed liquid metal antenna is shown in Figure 3a. Firstly, the glass tube used as the temperature-sensitive bulb, and the capillary tube, were sintered and connected. Secondly, the platinum feeding electrode and the glass tube were sintered together. The feeding electrode was made of platinum wire because platinum is not easily oxidized and has high conductivity. Thirdly, EGaIn was prepared and filled into the glass tube under a vacuum environment. The relationship between temperatures and heights was properly marked. The temperature-sensitive bulb was heated so that the excess EGaIn was discharged. Meanwhile, the top of the glass tube was sintered and sealed. Steps 1, 2 and 4 are common measures for the manufacture of thermometers, which were entrusted to Wuqiang Huayang Instrument Co., Ltd., China. Step 3 will be described in more detail below.

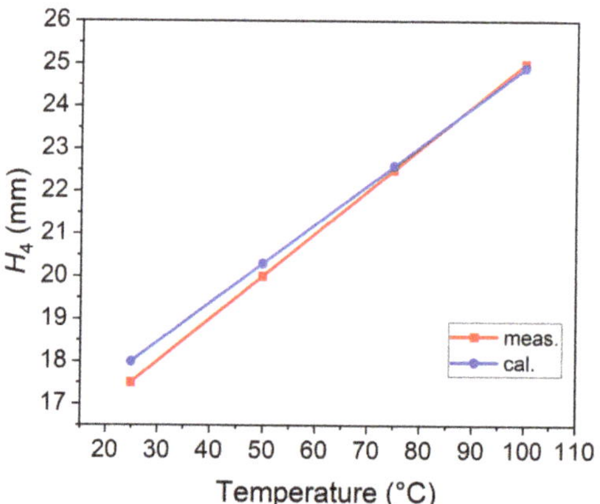

Figure 5. The lengths of H_4 at different temperatures.

The reflection coefficients of the proposed antenna at different temperature states are shown in Figure 6a, and the relationship between impedance bandwidth and temperature is shown in Figure 6b. Under ambient temperature (25 °C), simulated and measured bandwidths with a reflection coefficient below −10 dB are 24.4% (1.55–1.97 GHz, resonated at 1.72 GHz) and 29.4% (1.50–2.00 GHz, resonated at 1.70 GHz), respectively. As the temperature gradually rises from 25 °C to 100 °C, EGaIn in the glass tube expands. The measured length of H_4 increases from 17.5 mm to 25 mm linearly. Hence, the resonant frequency of the antenna decreases. The simulated frequency range with $|S_{11}| < -10$ dB reduces to 1.24–1.43 GHz (resonated at 1.32 GHz), while the corresponding measured results decrease to 1.25–1.44 GHz (resonated at 1.32 GHz) by degrees.

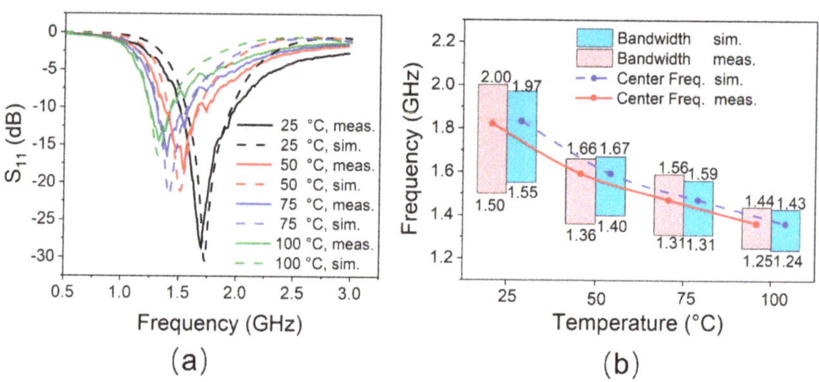

Figure 6. (**a**) Reflection coefficients of the antenna at different temperatures. (**b**) Impedance bandwidths performance of the antenna at different temperatures.

This frequency reconstruction process would be expected to be continuously changing since the temperature changes are successive. The working frequency band basically covers the *L*-band, and the minimum reflection coefficient is better than −15 dB for all the cases. The simulated results are in good agreement with the measured ones. The slight difference is mainly observed in the lower reflection coefficient obtained by the measurement, which may be due to manufacturing errors and reflections from the environment.

The normalized radiation patterns of the E-plane (X-Z plane) and H-plane (Y-Z plane) at four different temperatures and resonant frequencies of the antenna are shown in Figure 7. Similarly to simulated results, this antenna shows typical radiation patterns for a classical monopole antenna with a stable omnidirectional radiation on the H-plane and a bidirectional radiation on the E-plane. Furthermore, radiation efficiencies in different cases were calculated using the maximum gain and directivity. As shown in Figure 8, the maximum radiation efficiency and gain are 94% and 2.9 dBi, respectively, in the tunable range. There are differences in gain and efficiency trends between the simulations and measurements, because the simulations represent ideal conditions, which are not completely consistent with the actual conditions. The measurement results show that the loss of the antenna increases as the frequency increases, resulting in a decrease in efficiency and gain. The loss comes from dimensional errors, dielectrics, cables and interfaces.

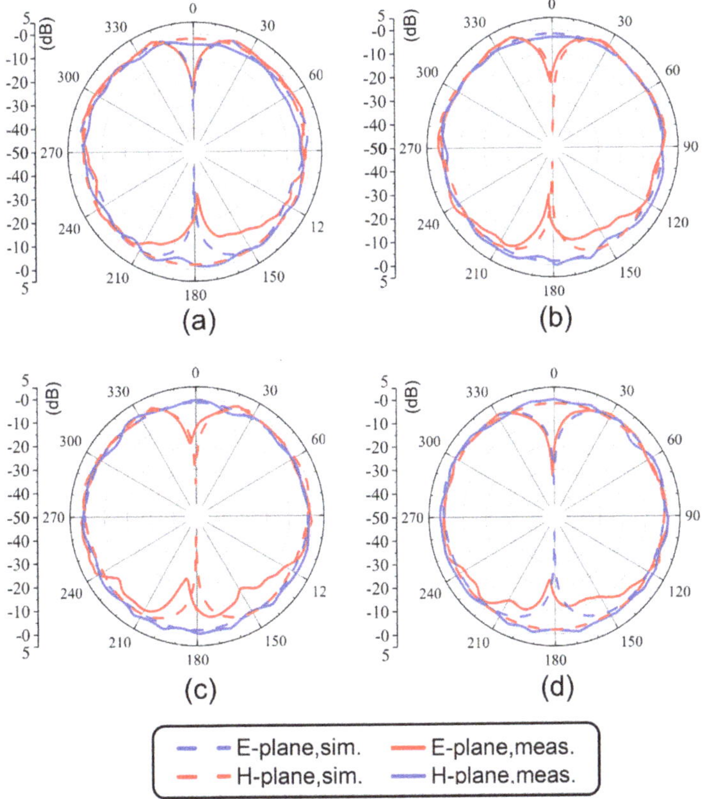

Figure 7. Radiation patterns at (**a**) 1.7 GHz (25 °C), (**b**) 1.51 GHz (50 °C), (**c**) 1.41 GHz (75 °C) and (**d**) 1.32 GHz (100 °C).

The prototype antenna was fabricated to verify the feasibility of the thermal expansion-based strategy, currently with a tuning range of 1.25–2.00 GHz, but this is not the limit for the thermal-expansion-based antenna. For example, higher operating frequencies can be obtained by reducing the geometric size, especially by tuning H_1, H_2, H_3, H_4 and D_2, or lowering the operating temperature, unless the liquid–solid phase transition occurs. When the geometry is determined, the lowest working frequency is determined by the maximum operating temperature. Considering the heat resistance of common materials and the safety of measurements, as a prototyped antenna, its operating temperature range in the aforementioned measurements is only 25–100 °C. Under the geometric dimensions shown in Table 1, the trend in Figure 6 is that when the length of H_4 increases by 10 mm every time, the temperature rises by 100 °C. The increased operating temperature can further reduce the resonance frequency. In short, the smaller geometric size and wider operating temperature range can greatly broaden the working bandwidth of the antennas, based on the proposed control method. Moreover, the antenna proposed in this study can also serve as a temperature sensor for high-temperature measurement, since the boiling point of liquid metal is extremely high. When working as a temperature sensor, the heat source to be measured replaces the heating platform. The designed antenna will have both temperature sensing and signal transmission functions. In measuring high temperatures, the major concern is the heat resistance of the antenna materials. It is necessary to replace the organic glue in this study with high-temperature-resistant inorganic glue (heat resistance temperature can exceed 1300 °C [33]) and replace the Polytetrafluoroethylene radio frequency connector with the ceramic radio frequency connector (heat resistance temperature can exceed 600 °C [34]).

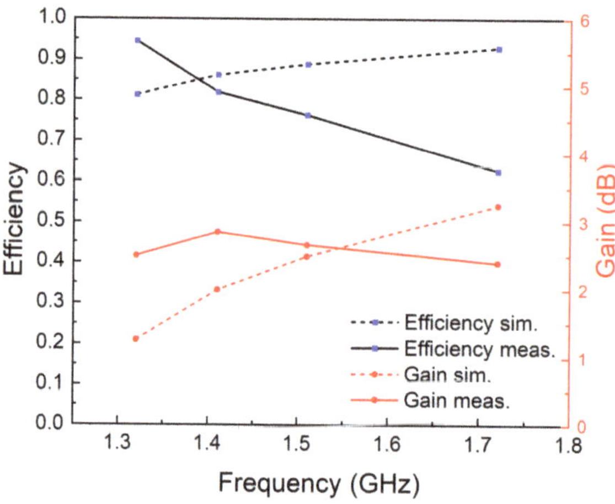

Figure 8. The radiation efficiencies and gains at different resonant frequencies.

Several LM frequency-reconfigurable antennas, and traditional frequency-reconfigurable antennas, are listed in Table 3 for comparison. In [1], only two discrete frequencies can be reconfigured. Gains are not given in [3,11]; furthermore, efficiencies are not given in [2,3,9,11]. The tuning range is relatively small in [11] and the antenna needs an external force to achieve reconfigurability. The tuning range is broad in [9]. Since the oxidation problem of Galinstan was not solved well, highly toxic mercury was used instead of Galinstan in [9]. The principles of continuous electrowetting (CEW) and electrically controlled capillarity (ECC) are used to drive LM in [12,13], respectively. In [13], the peak efficiency is 70%, which is lower than the levels seen in this work, because the NaOH solution is used as a conductive liquid in the bias circuit and to eliminate gallium oxide in [12,13]. Compared with the above work, the antenna proposed in this communication is safer and more convenient to apply, since the oxidation problem is avoided without use of another protective device.

Table 3. Performance comparison between different antennas.

Ref.	Radiator	Methods of Reconstruction	Liquid Metal	Solution Environment	Reconstruction Speed	Control Circuit Power	Reconfigurable Characteristics (GHz)	Peak Gain (Mea.)	Peak Efficiency (Mea.)
[1]	Planar inverted F	RF-MEMS	None	None	Fast	Small	0.718 and 4.96	3.3 dBi	85%
[2]	Quasi-Yagi	Varactor diodes	None	None	Fast	Small	6.0–6.6	6.35 dBi	N/A
[3]	Planar inverted U	PIN diodes	None	None	Fast	Small	2.63–3.7	N/A	N/A
[9]	Monopole	Micro pump	Mercury	Teflon solution	16 s	Medium	1.29–5.17	2.5 dBi	N/A
[11]	Dipole	External force	EGaIn	None	N/A	None	1.91–1.99	N/A	≈90%
[12]	Slot	Continuous electrowetting	Galinstan	NaOH solution	0.9 mm/s	Medium	2.2–2.6	2 dBi	N/A
[13]	Monopole	Electrically controlled capillarity	EGaIn	NaOH solution	3.6 mm/s	Medium	0.66–3.4	3.4 dBi	70%
This work	Monopole	Thermal expansion	EGaIn	None	≈1 s/°C	High	1.25–2.00	2.9 dBi	94%

Nevertheless, the speed of the antenna's response to temperature is expected to be improved. During the tests, it took 20–30 s to obtain a temperature difference of 25 °C. In other words, the temperature change rate is about 1 °C/s. In fact, a long reconfiguration time is a common problem for antennas using liquid metal through fluid flow, and the thermal-expansion-based actuation is no exception. Whether using external force methods [11], pumping methods [9], CEW [12] or ECC [13], the speed of liquid metal antennas is currently not comparable to control methods using radio frequency switches [1–3]. On the other hand, heating the plane requires more power. In the future, heating methods with a high density such as laser heating may be used to accelerate the reconfiguration procedure of thermal-expansion-based liquid metal antennas.

4. Conclusions

A frequency-reconfigurable antenna based on the thermal expansion of GBLM has been investigated and verified by measurement. A continuous operating frequency range of 1.25–2.00 GHz is acquired via manipulating the temperature between 25 °C and 100 °C using EGaIn. The tunable working bandwidths cover the majority of the *L*-band. Under the conditions of four representative temperatures, the maximum radiation efficiency and gain in the tunable range are 94% and 2.9 dBi, respectively. Moreover, during the fabrication process in this work, the GBLM did not become oxidized when it was smelted and filled into the antenna because an integrated process route was proposed, in which GBLM was smelted in a high vacuum and injected into the antenna in a high-pressure protective gas. In brief, a thermal-control strategy is proposed in order to realize, for the first time, frequency reconfigurability of a monopole antenna via the thermal expansion of GBLM. This provides a promising solution for reconfigurable antennas based on liquid metal techniques. The designed antenna can be used as a temperature sensor in some situations that are challenging for conventional temperature sensing methods.

Author Contributions: Conceptualization, P.Q. and Z.-S.D.; validation, P.Q. and Q.-Y.W.; formal analysis, P.Q. and J.-H.F.; investigation, P.Q., L.W. and T.-Y.L.; resources, P.Q. and L.W.; methodology, P.Q., G.-L.H. and T.-Y.L.; software, P.Q.; data curation, P.Q., and L.W.; writing—original draft preparation, P.Q.; writing—review and editing, P.Q., L.W., G.-L.H., Z.-S.D. and T.-Y.L.; visualization, P.Q., J.-H.F. and G.-L.H.; supervision, L.W., L.G., J.L. and Z.-S.D.; project administration, L.G., J.L. and Z.-S.D.; funding acquisition, L.G., J.L. and Z.-S.D. All authors have read and agreed to the published version of the manuscript.

Funding: This work is supported by the National Key Research and Development Program of China (2019YFB2204903) and the National Science Foundation for Young Scientists of China (Grant No. 51706234).

Institutional Review Board Statement: Not applicable.

Informed Consent Statement: Not applicable.

Data Availability Statement: Not applicable.

Acknowledgments: The authors thanks anonymous reviewers for providing valuable advices.

Conflicts of Interest: The authors declare no conflict of interest.

References

1. Zouhur, A.; Mopidevi, H.; Rodrigo, D.; Unlu, M.; Jofre, L.; Cetiner, B.A. RF MEMS reconfigurable two-band antenna. *IEEE Antennas Wirel. Propag. Lett.* **2013**, *12*, 72–75. [CrossRef]
2. Qin, P.-Y.; Weily, A.R.; Guo, Y.J.; Bird, T.S.; Liang, C.-H. Frequency reconfigurable quasi-yagi folded dipole antenna. *IEEE Trans. Antennas Propag.* **2010**, *58*, 2742–2747.
3. Mansoul, A.; Ghanem, F.; Hamid, M.R.; Trabelsi, M. A selective frequency-reconfigurable antenna for cognitive radio applications. *IEEE Trans. Antennas Propag. Lett.* **2014**, *13*, 515–518. [CrossRef]
4. Cristina, B.-F.; Tong, K.-F.; Allann, A.-A.; Wong, K.-K. A low-cost fluid switch for frequency-reconfigurable Vivaldi antenna. *IEEE Trans. Antennas Propag. Lett.* **2017**, *16*, 3151–3154.
5. Li, J.; Chu, D. Liquid crystal-based enclosed coplanar waveguide phase shifter for 54–66 GHz applications. *Crystals* **2019**, *9*, 650. [CrossRef]
6. Werndl, E. Antenna Tunable in Its Length. U.S. Patent 2,278,601, 7 April 1942.
7. Kosta, Y.; Kosta, S. Realization of a microstrip-aperture-coupled-passive-liquid patch antenna. In Proceedings of the IEEE International RF and Microwave Conference, Kuala Lumpur, Malaysia, 2–4 December 2008; pp. 135–138.
8. Yee, S.; Weinstein, D.; Fiering, J.; White, D.; Duwel, A. A Miniature Reconfigurable Circularly Polarized Antenna Using Liquid Microswitches. In Proceedings of the IEEE 16th Annual Wireless and Microwave Technology Conference, WAMICON, Cocoa Beach, FL, USA, 13–15 April 2015; pp. 1–5.
9. Dey, A.; Guldiken, R.; Mumcu, G. Microfluidically reconfigured wideband frequency tunable liquid metal monopole antenna. *IEEE Trans. Antennas Propag.* **2016**, *64*, 2572–2576. [CrossRef]
10. Huang, T.; Zeng, L.; Liu, G.B.; Zhang, H.F. A novel tailored coplanar waveguide circularly polarized antenna controlled by the gravity field. *Int. J. RF Microw. Comput. Aided Eng.* **2019**, *29*, 21823–21832. [CrossRef]
11. So, J.H.; Thelen, J.; Qusba, A.; Hayes, G.J.; Lazzi, G.; Dickey, M.D. Reversibly deformable and mechanically tunable fluidic antenna. *Adv. Funct. Mater.* **2009**, *19*, 3632–3637. [CrossRef]
12. Gough, R.C.; Morishita, A.M.; Dang, J.H.; Hu, W.; Shiroma, W.A.; Ohta, A.T. Continuous electrowetting of non-toxic liquid metal for RF applications. *IEEE Access* **2014**, *2*, 874–882. [CrossRef]
13. Wang, M.; Trlica, C.; Khan, M.R.; Dickey, M.D.; Adams, J.J. A reconfigurable liquid metal antenna driven by electrochemically controlled capillarity. *J. Appl. Phys.* **2015**, *117*, 194901–194905. [CrossRef]
14. Morishita, A.M.; Kitamura, C.K.Y.; Ohta, A.T.; Shiroma, W.A. Two-octave tunable liquid-metal monopole antenna. *Electron. Lett.* **2014**, *50*, 19–20. [CrossRef]
15. Dang, J.H.; Gough, R.C.; Morishita, A.M.; Ohta, A.T.; Shiroma, W.A. Liquid-metal frequency-reconfigurable slot antenna using air-bubble actuation. *Electron. Lett.* **2015**, *51*, 1630–1632. [CrossRef]
16. Sarabia, K.J.; Ohta, A.T.; Shiroma, W.A. Pixelated dual-dipole antenna using electrically actuated liquid metal. *Electron. Lett.* **2019**, *55*, 1032–1034. [CrossRef]
17. Liu, Y.; Wang, Q.; Jia, Y.; Zhu, P. A Frequency- and Polarization- Reconfigurable Slot Antenna Using Liquid Metal. *IEEE Trans. Antennas Propag.* **2020**, *68*, 7630–7635. [CrossRef]
18. Yang, X.; Liu, Y.; Lei, H.; Jia, Y.; Zhu, P.; Zhou, Z. A Radiation Pattern Reconfigurable Fabry Pérot Antenna Based on Liquid Metal. *IEEE Trans. Antennas Propag.* **2020**, *68*, 7658–7663. [CrossRef]
19. Arbelaez, A.; Goode, I.; Gomez-Cruz, J.; Escobedo, C.; Saavedra, C.E. Liquid Metal Reconfigurable Patch Antenna for Linear, RH, and LH Circular Polarization with Frequency Tuning. *Can. J. Electr. Comput. Eng.* **2020**, *43*, 218–223. [CrossRef]
20. Huang, G.L.; Liang, J.J.; Zhao, L.; He, D. Package-in-Dielectric Liquid Patch Antenna Based on Liquid Metal Alloy. *IEEE Antennas Wirel. Propag. Lett.* **2019**, *18*, 2360–2364. [CrossRef]
21. Wang, M.; Khan, M.R.; Dickey, M.D.; Adams, J.J. A compound frequency- and polarization- reconfigurable crossed dipole using multidirectional spreading of liquid metal. *IEEE Antennas Wirel. Propag. Lett.* **2016**, *16*, 79–82. [CrossRef]
22. Zhou, Y.; Fang, S.; Liu, H.; Wang, Z.; Shao, T. A function reconfigurable antenna based on liquid metal. *Electronics* **2020**, *9*, 873. [CrossRef]
23. Mazlouman, S.J.; Jiang, X.J.; Mahanfar, A.N.; Menon, C.; Vaughan, R.G. A reconfigurable patch antenna using liquid metal embedded in a silicone substrate. *IEEE Trans. Antennas Propag.* **2011**, *59*, 4406–4412. [CrossRef]
24. Dickey, M.D.; Chiechi, R.C.; Larsen, R.J.; Weiss, E.A.; Weitz, D.A.; Whitesides, G.M. Eutectic aallium-indium (EGaIn): A liquid metal alloy for the formation of stable structures in microchannels at room temperature. *Adv. Funct. Mater.* **2008**, *18*, 1097–1104. [CrossRef]

25. Liu, T.; Sen, P.; Kim, C.J. Characterization of liquid-metal galinstan for droplet applications. In Proceedings of the 2010 IEEE 23rd International Conference on Micro Electromechanical Systems (MEMS), Wanchai, Hong Kong, China, 24–28 January 2010; pp. 560–563.
26. Ramana, C.V.; Rubio, E.J.; Barraza, C.D.; Miranda Gallardo, A.; McPeak, S.; Kotru, S.; Grant, J.T. Chemical bonding, optical constants, and electrical resistivity of sputter-deposited gallium oxide thin films. *J. Appl. Phys.* **2014**, *115*, 043508–043517. [CrossRef]
27. Wang, M.; Lei, W.T.; Dong, J.; Dickey, M.D.; Adams, J.J. Investigation of biasing conditions and energy dissipation in electrochemically controlled capillarity liquid metal electronics. *Electron. Lett.* **2020**, *56*, 323–325. [CrossRef]
28. Sergolle, M.; Castel, X.; Himdi, M.; Besnier, P.; Parneix, P. Structural composite laminate materials with low dielectric loss: Theoretical model towards dielectric characterization. *Compos. Part C-Open AC* **2020**, *3*, 100050–100058. [CrossRef]
29. Li, X.; Li, M.; Zong, L.; Wu, X.; You, J.; Du, P.; Li, C. Liquid metal droplets wrapped with polysaccharide microgel as bio-compatible aqueous ink for flexible conductive devices. *Adv. Funct. Mater.* **2018**, *28*, 1804197–1804204. [CrossRef]
30. Lide, D.R. *CRC Handbook of Chemistry and Physics*; CRC Press: Boca Raton, FL, USA, 2005; pp. 4–22.
31. Mackenzie, J.D. Glasses from melts and glasses from gels, a comparison. *J. Non-Cryst. Solids* **1982**, *48*, 1–10. [CrossRef]
32. Li, K.; He, H.Y.; Xu, B.; Pan, B.C. The stabilities of gallium nanowires with different phases encapsulated in a carbon nanotube. *J. Appl. Phys.* **2009**, *105*, 54308–54312. [CrossRef]
33. Chen, Y.; Wang, X.; Yu, C.; Ding, J.; Deng, C.; Zhu, H. Properties of inorganic high-temperature adhesive for high-temperature furnace connection. *Ceram. Int.* **2019**, *45*, 8684–8689. [CrossRef]
34. Zhou, X.; Guo, Y. High Temperature Resistant Unsteady Welding Type Radio Frequency Coaxial-Cable Connector. Chinese Patent CN203967313U, 26 November 2014.

 sensors 

Article

Multi-Layer and Conformally Integrated Structurally Embedded Vascular Antenna (SEVA) Arrays

Amrita Bal [1], Jeffery W. Baur [2], Darren J. Hartl [3,4], Geoffrey J. Frank [2,5], Thao Gibson [5], Hong Pan [1] and Gregory H. Huff [6,*]

1 Department of Electrical and Computer Engineering, Texas A&M University, College Station, TX 77843, USA; abal@tamu.edu (A.B.); hongpan0507@gmail.com (H.P.)
2 Materials and Manufacturing Directorate, U.S. Air Force Research Laboratory, WBAFB, Dayton, OH 45433, USA; jeffery.baur@us.af.mil (J.W.B.); geoffrey.frank.ctr@us.af.mil (G.J.F.)
3 Department of Aerospace Engineering, Texas A&M University, College Station, TX 77843, USA; darren.hartl@tamu.edu
4 Department of Material Science and Engineering, Texas A&M University, College Station, TX 77840, USA
5 University of Dayton Research Institute, 300 College Park, Dayton, OH 45469, USA; thao.gibson@us.af.mil
6 Department of Electrical Engineering, Pennsylvania State University, State College, PA 16801, USA
* Correspondence: ghuff@psu.edu

Abstract: This work presents the design and fabrication of two multi-element structurally embedded vascular antennas (SEVAs). These are achieved through advances in additively manufactured sacrificial materials and demonstrate the ability to embed vascular microchannels in both planar and complex-curved epoxy-filled quartz fiber structural composite panels. Frequency-reconfigurable antennas are formed by these structures through the pressure-driven transport of liquid metal through the embedded microchannels. The planar multi-layer topology examines the ability to fabricate two co-located radiating structures separated by a single ply of quartz fabric within the composite layup. The multi-element linear array topology composed of microchannels embedded on to a single-layer are used to demonstrate the ability to conformally-integrate these channels into a complex curved surface that mimics an array of antennas on the leading edge of an Unmanned Aerial Vehicle (UAV). A parallel-strip antipodal dipole feed structure provides excitation and serves as the interface for fluid displacement within the microchannels to facilitate reconfiguration. The nominal design of the SEVAs achieve over a decade of frequency reconfiguration with respect to the fundamental dipole mode of the antenna. Experimental and predicted results demonstrate the operation for canonical states of the antennas. Additional results for the array topology demonstrate beam steering and contiguous operation of interconnected elements in the multi-element structure.

Keywords: unmanned aerial vehicle; phased array; frequency reconfiguration; beam steering

Citation: Bal, A.; Baur, J.W.; Hartl, D.J.; Frank, G.J.; Gibson, T.; Pan, H.; Huff, G.H. Multi-Layer and Conformally Integrated Structurally Embedded Vascular Antenna (SEVA) Arrays. *Sensors* **2021**, *21*, 1764. https://doi.org/10.3390/s21051764

Academic Editor: Antonio Lázaro

Received: 27 January 2021
Accepted: 25 February 2021
Published: 4 March 2021

1. Introduction

Unmanned Aerial Vehicles (UAV) are deployed to serve a wide variety of commercial, municipal, and defense applications including health care, public safety, mobile network coverage diagnostics, etc. [1–3]. The growing utilization of UAV-based application spaces and the growing complexity of their missions have increased the need for more robust communication and sensing systems that support these wireless links. This has facilitated a need for robust, shape-conformal, and frequency reconfigurable antennas and phased arrays that can adapt to dynamic operational scenarios; these vehicles (controlled remotely or autonomously) are also required in many scenarios to maintain reliable and long-range communication with a ground station. Reconfigurable, multi-band, and/or broadband omnidirectional antennas used in UAV applications have been proposed in response to this (e.g., [4–9]). Antenna arrays have also been used in-place of single antennas in these applications to provide enhanced gain and scanning capabilities (e.g., [10–12]).

Antennas considered for use in UAVs are often derived from canonical planar topologies. The integration of these antennas on/in to the UAV presented challenges and numerous efforts have been carried out to embed these antennas into load bearing structures (even conformal to their surfaces). These efforts mainly focused on improving structural efficiency and antenna performance. Various efforts in this direction resulted in development of vertically grown carbon nanotubes and carbon composite fibers used for fabrication of load bearing antennas having desirable RF and mechanical performance [13,14]. Similar load bearing antennas, like Conformal Load bearing Antenna Structures (CLAS) and Composite Smart Structures (CSS), have been proposed to overcome geometrical limitations and eases integration with a non-planar surface [15–19]. Examples of CLAS include MEMS reconfigurable pixelated patch antennas and bi-layer log periodic antenna arrays [20,21]. Antennas and antenna arrays designed using high strength metal coated fibers suited for stress, weight and shape critical applications provide synergistic advances [22–24].

The development of load bearing antenna technologies has been critical in the deployment of radiating systems that can withstand the extreme environments and physical conditions encountered by both traditional aircraft and UAVs [25–28]. Advances in additive manufacturing using liquid metal as a reconfiguration mechanism [29,30] have been demonstrated recently to impart a greater degree of electromagnetic agility into the design space of conformal load bearing antenna systems. Previous work on the load-bearing, structurally embedded vascular antenna (SEVA) [30] demonstrated the frequency reconfigurability of a dipole antenna meandered in exponentially increasing sinusoidal pattern. The effective (resonant) length of the embedded dipole was varied using pressure driven transport of liquid metal (eGaIn) [31] to operate the antenna over a wide frequency range. The prior work [32–35] on embedded structures emphasize on the mechanics, materials, and fundamental concepts used to emphasize the functionalization of antennas in structurally embedded environments. The repeatability and stability of antenna characteristics make this structure an extensible demonstration vehicle to use in future and ongoing efforts. In this work the design is used to demonstrate the ability to engineer multi-layer and conformally curved reconfigurable antennas enabled by transport of liquid metal through the embedded microvascular channels. This is illustrated through two unique designs. First, it is used to demonstrate an antenna with vascular networks fabricated on two closely spaced and consecutive layer of fabric in the composite lay-up. In the second design it is shown that vascular networks with distinct pattern can fabricated in a complexly curved substrate mimicking the shape of a UAV wing. These structurally embedded vascular antennas operate over a decade of frequency in a single footprint. The organization of this work on multi-element frequency reconfigurable SEVAs follows. Full-wave electromagnetic solver [36] is used for simulation of these structures. First the design and fabrication of a multi-layer SEVA (ML-SEVA) is presented. The multi-element SEVA conformal linear array (SEVA-CLA) integrated onto a complex-curved surface is discussed next. The last section includes measured and simulated results from these structures. A brief discussion concludes the work.

2. Design and Fabrication

2.1. Multi-Layer SEVA (ML-SEVA)

Figure 1 shows the CAD model and notional composite lay-up of the multi-layer SEVA (ML-SEVA) presented in this work. The design and operation of this two-layer ML-SEVA leverages advances in additive manufacturing techniques to synthesize complex co-located vascular networks that provide additional operational degrees of freedom with respect to the antenna performance metrics demonstrated by the original SEVA in [30]. The fabric lay-up for the ML-SEVA in this work features two unique vascular networks separated by a single ply low-dielectric epoxy/quartz prepreg fabric RM-2014/4581 Astroquartz® III with a 0.25 mm layer thickness. Each pair of the meandered lines that form the microchannels through which liquid metal is flowed are first extrusion-printed using the sacrificial PLA formulation in a single layer thickness of 0.6 mm onto Polyimide sheets. These two pairs of meandered lines create two distinct dipole-like configurations that are then thermally transferred to a fabric layer below the previous one and arranged according to the lay-up shown in Figure 1.

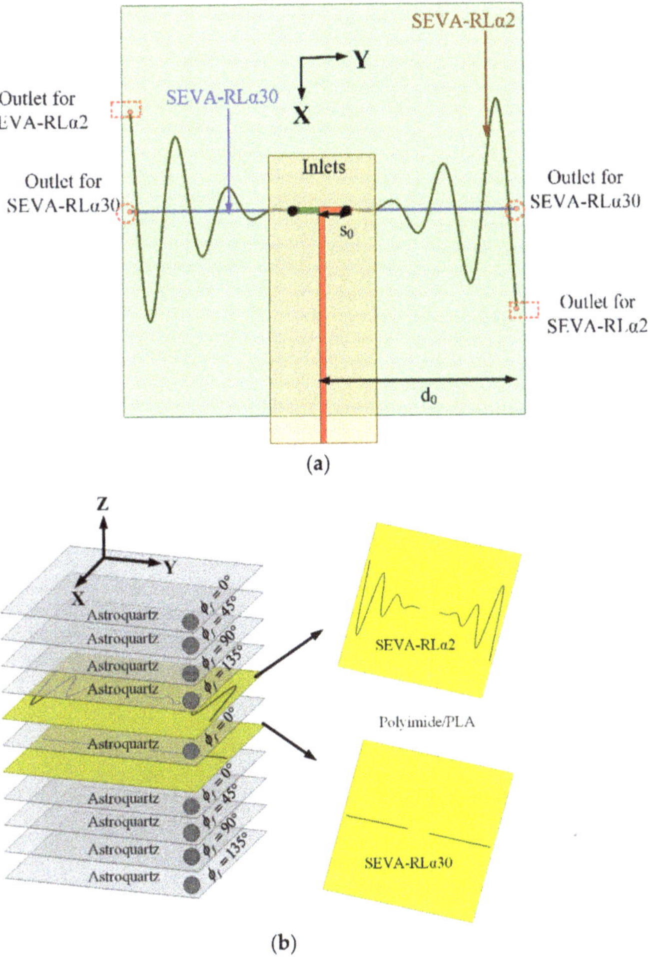

Figure 1. Detailed representation of multi-layer structurally embedded vascular antennas (ML-SEVA) (**a**) CAD model used for simulating the structure (**b**) Fabrication lay-up.

The feed structure consists of parallel strips connected to the meandering microvascular elements of the antenna through antipodal dipoles; this also forms a balun-like transition [30]. Equations (1) and (2) provide the parameterization of the sacrificial PLA used to create microvascular channels for the ML-SEVA in Figure 1. These meander outward in the y-dir. starting from the feed location $y = \pm s_0 = \pm 10.0$ mm and to $y = \pm d_0 = \pm 72.2$ mm and are modulated sinusoidally in the x-dir. with a period $p_0 = 2$ within a power series envelope with a growth parameter α. In this work, the first vascular network in the lay-up uses $\alpha = 2$ to provide a sinusoidally-meandering dipole antenna topology referred to as the SEVA-RLα2. The vascular network on the layer below uses $\alpha = 30$ to form a quasi-linear dipole-like antenna referred to as the SEVA-RLα30. The parameter t in these expressions denotes the location of liquid metal, or length of meandering along y-axis in SEVA, and the terms t_1 and t_2 (used later) denote the channel filling parameter in SEVA-RLα2 and SEVA-RLα30, respectively.

$$x(t) = \frac{t^\alpha}{d_0^{\alpha-1}} \sin(\pi p_0 t) \quad (1)$$

$$y(t) = t + s_0 \quad (2)$$

The processing steps for this structure follow [30] by first curing the composite with embedded microchannels in an autoclave. Vertical cylindrical channels are then machined down from the top of the composite to the lowest of the two sacrificial layers (SEVA-RLα30) to provide release holes for the ablation of sacrificial materials. This is followed by a post-cure in a vacuum oven to vaporize the sacrificial material and create microvascular voids in the composite. These are inspected visually and then tested with compressed air to ensure all channels cleared of any blockage. After this the vertical channel at each feed location $(x, y) = (0, \pm s_0)$, with $s_0 = 3.94$ in. (10 mm), connects the two arms of the SEVA-RLα2 and SEVA-RLα30; the remaining four only contact a single arm where they intersect $y = \pm d_0 = 2.6$ in.

The resulting composite with embedded microvasculature is electromagnetically functionalized by first adhering hollow copper cylindrical vias 2 mm in diameter at the two feed locations. These are extended just below the surface for mechanical stability and provide access for the pressure-driven displacement of fluids into each of the four channels. These metallic vias extend upwards through drilled holes in a 31 mil (0.7874 mm) FR4 substrate ($\varepsilon_r = 4.4$ and $tan\ \delta = 0.02$) where they are electrically connected (soldered) to the arms of an antipodal dipole having a length $l = 2y_0$. This antipodal dipole is fed by a parallel strip feed line with a 50 Ω characteristic impedance that is terminated in a Sub-Miniature Version A (SMA) connector for measurements. In this configuration, the transition from parallel strip line to antipodal dipole operates as a Dyson-style balun for the SEVA over the frequency ranges considered in this work. The ML-SEVA operates as a combination of two thin wire dipoles growing outward from the antipodal dipole; the selective filling of channels is achieved by leaving the vertical channel at its opposing end unobstructed or sealing the individual vertical channels with a thermal adhesive where it intersects $y = \pm d_0$.

2.2. Multi-Element SEVA Conformal Linear Array (SEVA-CLA)

Figure 2 shows the notional composite lay-up and CAD model of the mandrel used to form the composite into a complex-curved surface for the SEVA conformal linear array (SEVA-CLA). The complex-curved shape of the mandrel is not necessarily representative of a leading-edge design for a specific UAV and/or other aircraft, but it demonstrates the extensibility of the SEVA processing steps. This advancement in the additive manufacturing is leveraged in this work to fabricate a large contiguous channel embedded within a complex-curved composite with multiple access points. The structurally-embedded microvascular structure can operate in a number of different radiating manifold configurations; the focus here is on its operation as a three-element linear array as well as a single contiguous antenna that extends the bandwidth of the manifold. The SEVA-CLA generally follows the processing steps for the original SEVA [30] but expands on them in terms of the dimension and geometry of the complexity of the vascular network. As with other SEVA designs, debulking of the prepreg to remove volatiles from the resin and avoid the formation of porosity is an important step before curing in an autoclave. Once cured and released from the mandrel it is drilled at the prescribed locations (at the ends of the curvilinear channels) to access the sacrificial layer. The structure is then placed in a vacuum oven to post-cure and vaporize the sacrificial material and create the embedded microvascular network.

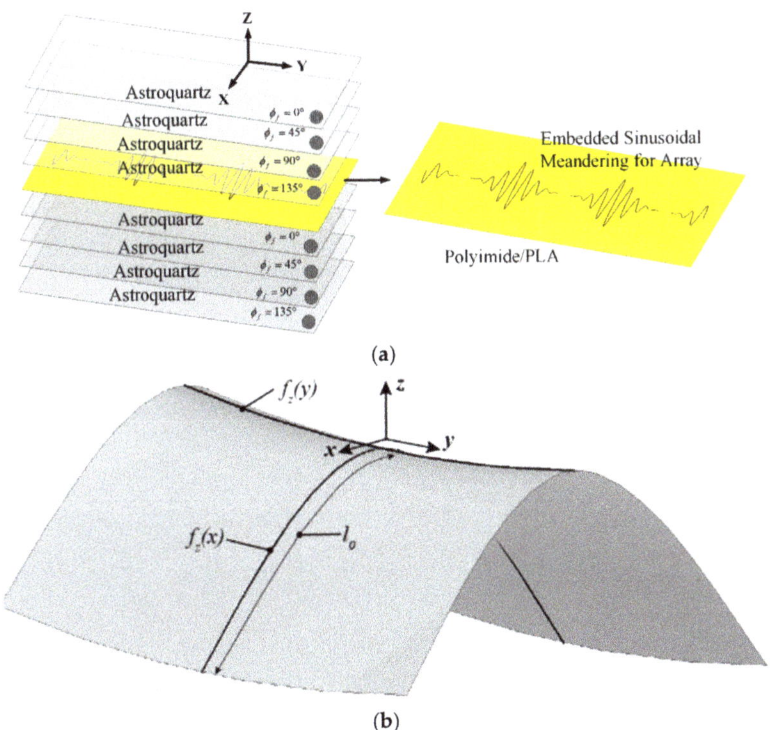

Figure 2. Detailed representation of multi-element SEVA conformal linear array (SEVA-CLA) (**a**) Fabrication lay-up (**b**) CAD model of complex curved mandrel used for shaping the composite.

The meandering vasculature of the three-element SEVA-CLA uses the parameterization of the SEVA-RLα2 for each of the three driven antenna elements. The curvature of the mandrel used to shape the composite is approximated for the high frequency modeling of antennas by sweeping the polynomial approximations of the "leading edge" curvature $f_z(x) \simeq 12.0 \times 10^{-3}x^2$ in the xz-plane along the contour "saddle path" $f_z(y) \approx 275.5 \times 10^{-6}y^2$ in the yz-plane. Three of the feed structures are first positioned then the array elements are first mapped conformally onto the complex-curved surface of the mandrel using [37]. The central element has symmetric arm lengths $d_0 = d_2 = 3.59$ in. (91.44 mm) while the arm lengths for the edge elements are asymmetric with the length of one arm fixed to $d_0 = d_1 = 2.92$ in. (74.12 mm) and the other one fixed to $d_0 = d_2 = 3.59$ in. (91.44 mm). The zero crossing points (denoted by variable t) on y-axis extend from 0 to 2.4 in. for sinusoidal arms of length d_1 and it extends from 0 to 3.2 in. for sinusoidal arms of length d_2 repeated at regular intervals of 0.4 in. along the y-axis.

The topology and operation/role of the feed network for this structure is similar to the structure presented in Figure 1. Its 50 Ω impedance parallel strip feed lines have a length $l_0 = 5.78$ in. (147 mm) and is fabricated on 31 mil (0.7874 mm) thick RT/Duroid 5880 ($\varepsilon_r = 2.2$ and $tan\ \delta = 0.0004$); this is used in place of the FR4 substrate to provide a feed network that can conform to the curvature of the composite. Figure 3 (top view) shows three feed elements connected to the sinusoidally meandered dipoles. Figure 3 (side view) shows the detailed connection between feed and embedded microvascular channels.

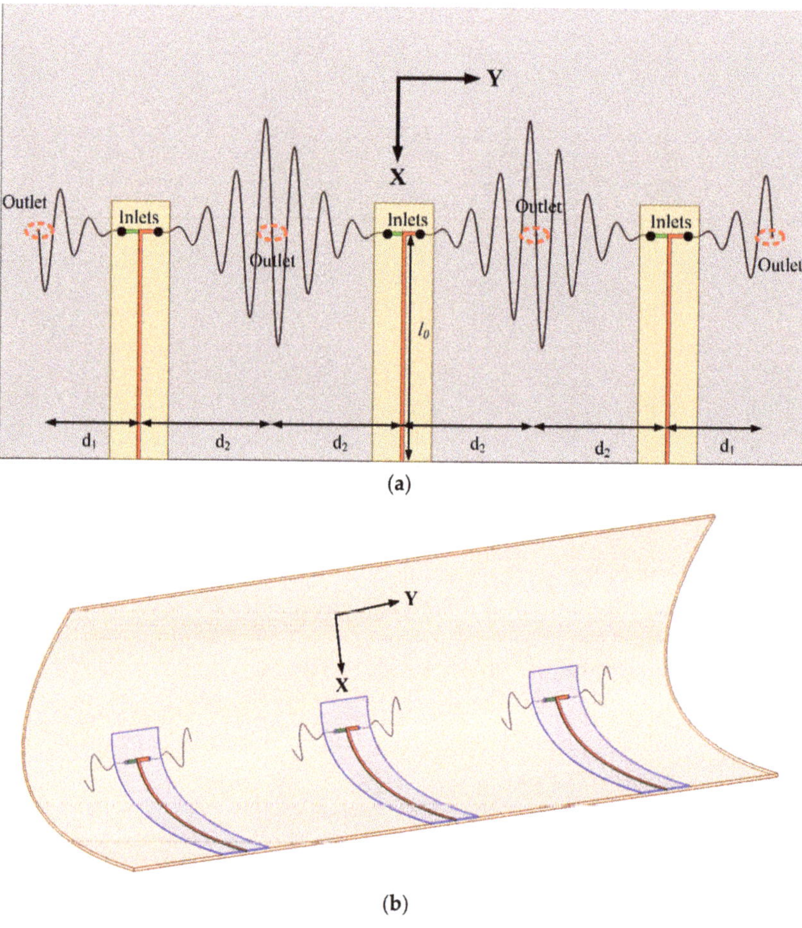

Figure 3. CAD Model of multi-element SEVA-CLA (**a**) Top view: Connection of feed elements with sinusoidally meandered dipoles (**b**) Side view: Connection of feed elements with sinusoidally meandered dipoles.

3. Measurement and Analysis

Discussion on the transport of liquid metal in the embedded vascular channels of ML-SEVA and SEVA_CLA is presented in this section. Comparison between simulated and measured results of these structures draw attention to the frequency reconfigurability of ML-SEVA and beam scanning ability of SEVA-CLA. Operation of SEVA-CLA as a contiguous antenna element operating in VHF is also presented in this section. Demonstration of fundamental array principles are presented by studying contour plots of isotropic antenna array and SEVA-CLA.

3.1. Multi-Layer SEVA (ML-SEVA)

Figure 4 shows the final fabricated model of ML-SEVA. The embedded microvascular channels are visible in the semi-translucent ML-SEVA. This offers a means to visibility track and measure the position of liquid metal being transported through the microchannels as well as its proximity to the zero-crossing points of the exponentially increasing sinusoidal structure (SEVA-RLα2) at t_1 = 0, 0.4, 0.8, 1.2, 1.6, 2, and 2.4 in. Corresponding values are chosen for t_2 (SEVA-RLα30) in this experiment. Selected combinations of t_1 and t_2 are presented in this article to demonstrate the principle of frequency reconfiguration in

ML-SEVA. Radiation patterns observed have remain consistent with simulation across different configurations and the selected combinations are shown for illustration.

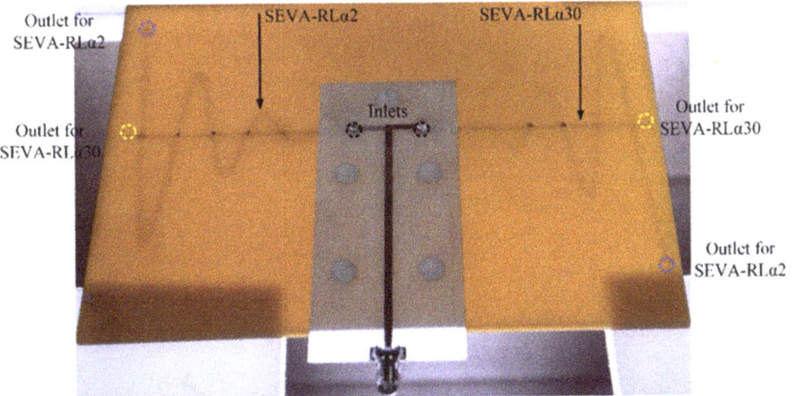

Figure 4. Fabricated ML-SEVA and feed network aligned and attached with nylon threads.

Cylindrical metallic tubes with a diameter of 2 mm are soldered to the feed and visually aligned with the opening on the embedded channels in the substrate to ensure continuous and leakage-proof flow of liquid metal from inlets towards the outlets. This also establishes electrical connection between liquid metal in the embedded channels and the feed. Additionally, for mechanical stability, the feed is affixed to the substrate by nylon screws ensuring a tight fitted connection between the feed and the substrate. Filling the embedded channels with non-conductive, low-loss, low dielectric, and heat transfer fluid Fluorinert FC-70 [38] electronic liquid before flowing liquid metal ensures smooth flow of liquid metal in the channels. When liquid metal is transported from the inlet of SEVA-RLα2 towards its outlet the outlet of SEVA-RLα30 is sealed with hot melt adhesive (HMA) to avoid leakage of liquid metal into the undesired channel. Similarly, the outlet of SEVA-RLα2 is sealed with HMA when liquid metal is transported through SEVA-RLα30.

Figure 5 shows the measured and simulated input reflection coefficient (in dB) for the chosen combinations of t_1 and t_2. The plots clearly demonstrate down shift of fundamental mode resonant frequency with increasing liquid metal flow in the embedded channels. These plots illustrate that the radiation pattern of the antenna remains quasi-omnidirectional for the different configurations. The close agreement observed between the measured and simulated results indicates that the operation of frequency reconfigurable ML-SEVA is also stable across the range of frequencies considered in this work. Figure 6 shows the measured and simulated E_θ and E_φ in xy- and xz- planes at the lowest value of magnitude (in dB) for the fundamental mode. When the antenna resonates at a frequency closer to the antipodal dipole (around 4 GHz) used to feed the microvascular component the antenna, the radiated pattern deviates from expectations due to increase interactions with the parallel strip feed network and other artifacts of experimental set-up. Thus, higher volume of metal in the embedded channels offers stable radiation pattern compared to lower metal volumes in the channels as with higher metal volumes in channel the antenna resonates at frequencies much lower than 4 GHz. Simulated maximum gains for configurations in Figure 6a–d is 2.3, −0.01, −3, 1.7, and −5 dBi, respectively.

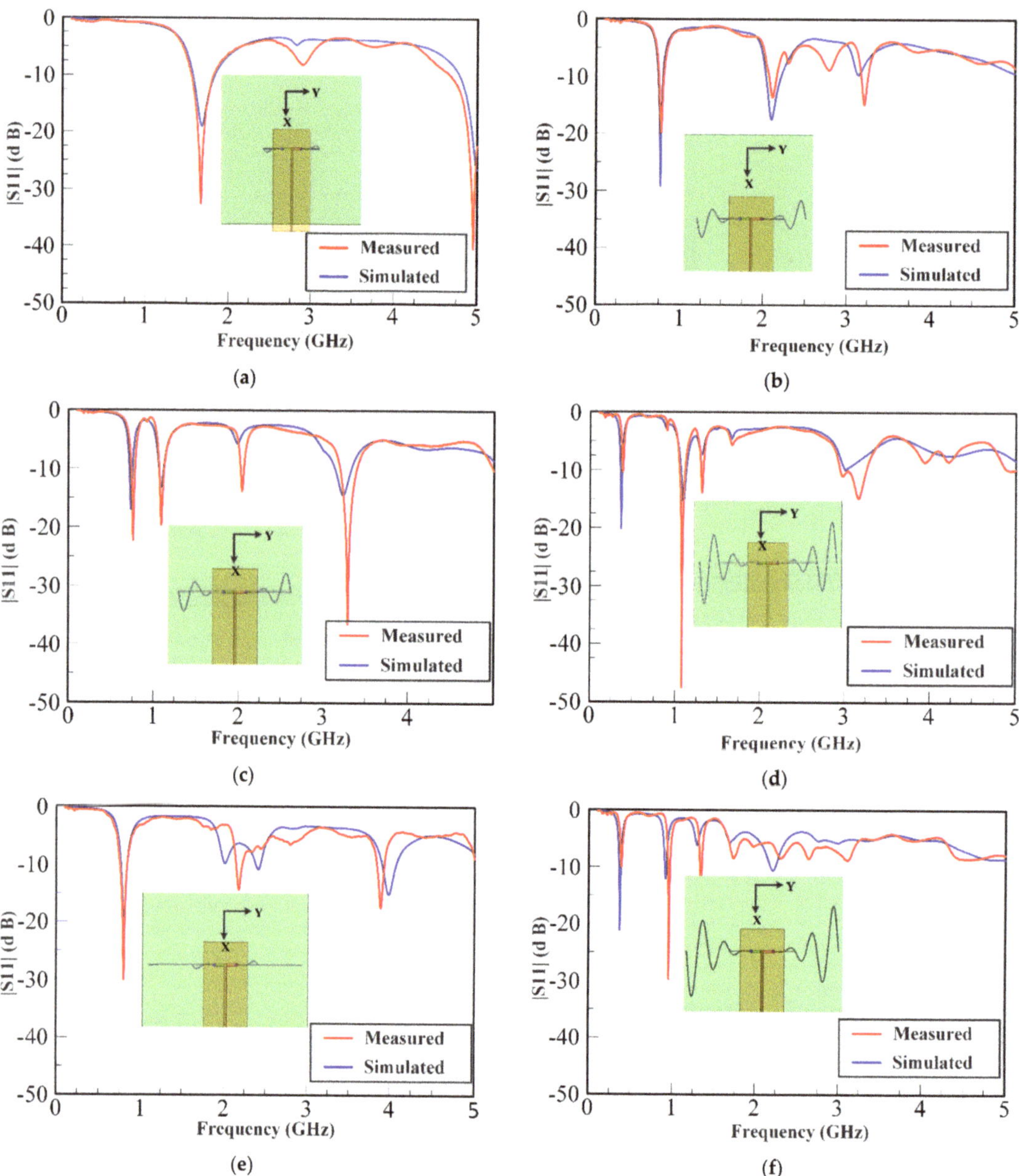

Figure 5. *Cont.*

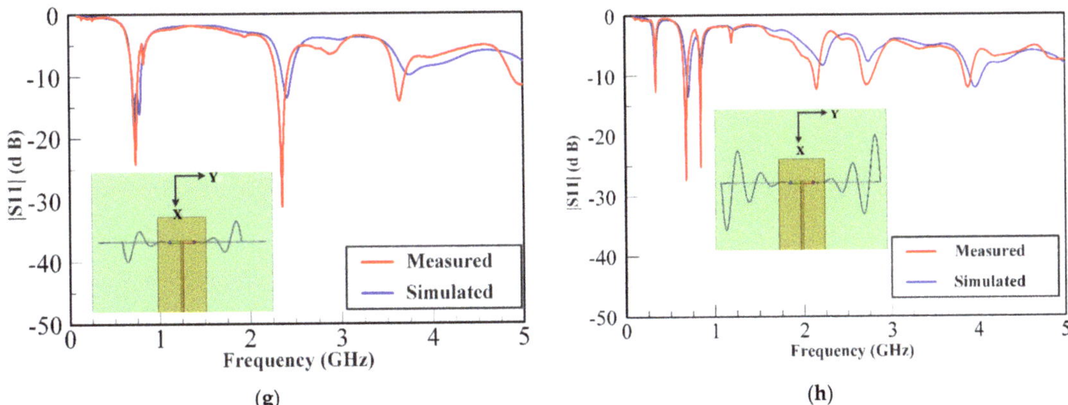

Figure 5. Measured and simulated S-parameter values (|S_{11}| (dB)) for different scenarios. (**a**) t_1 = 0.8 in. and t_2 = 0.8 in. (**b**) t_1 = 1.6 in. and t_2 = 0.8 in. (**c**) t_1 = 1.6 in. and t_2 = 1.6 in. (**d**) t_1 = 2.4 in. and t_2 = 1.6 in. (**e**) t_1 = 1.6 in. and t_2 = 2.4 in. (**f**) t_1 = 2.4 in. and t_2 = 0.8 in. (**g**) t_1 = 1.6 in. and t_2 = 2.4 in. (**h**) t_1 = 2.4 in. and t_2 = 2.4 in.

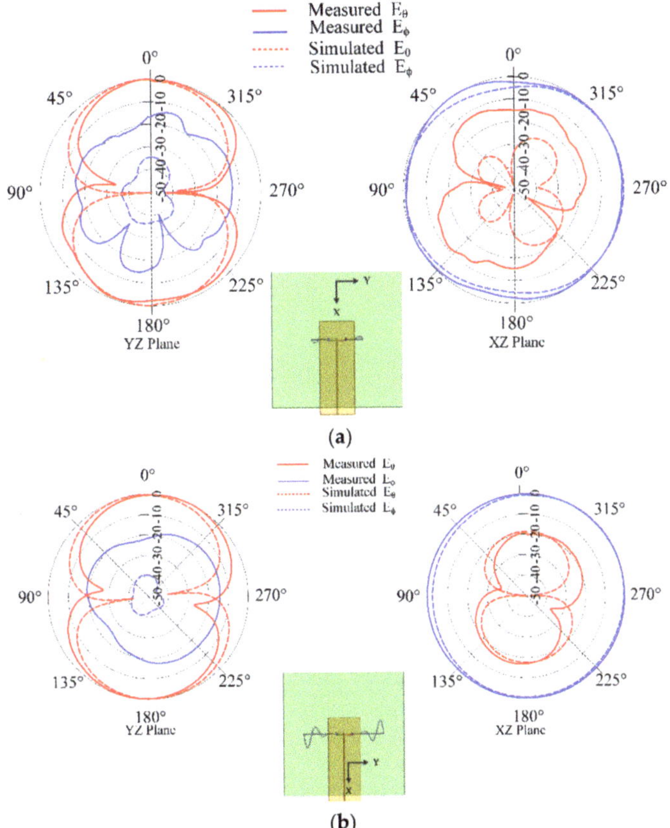

Figure 6. *Cont.*

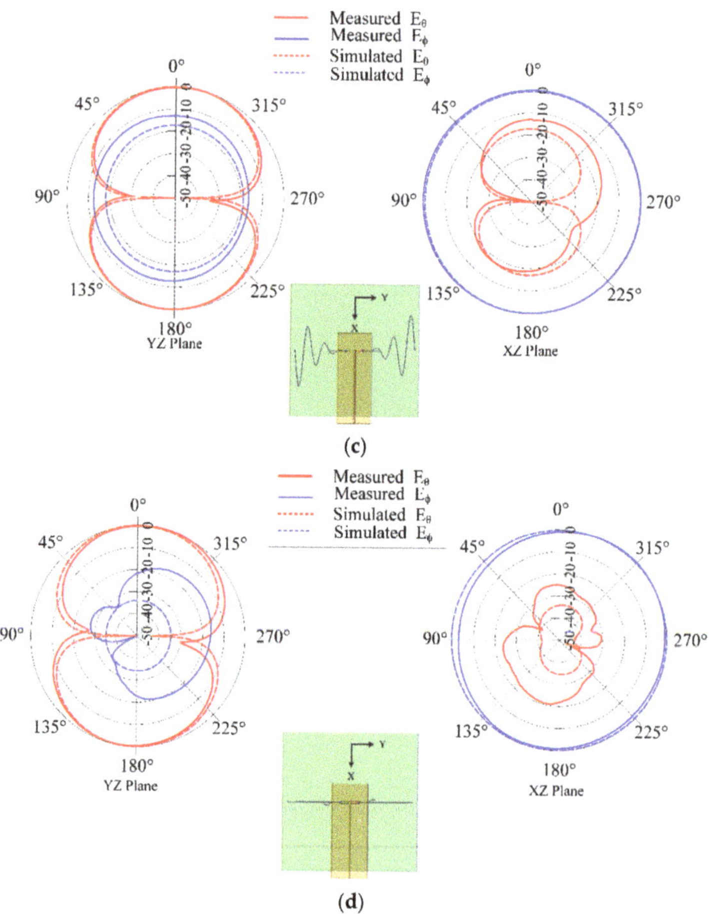

Figure 6. Measured and Simulated radiation pattern in yz- (left) and xz-planes (right) for different scenarios (**a**) t_1 = 0.8 in. and t_2 = 0.8 in. at f =1.681 GHz; (**b**) t_1 = 1.6 in. and t_2 = 1.6 in. at f = 0.74 GHz; (**c**) t_1 = 2.4 in. and t_2 = 0.8 in. at f = 0.372 GHz; and (**d**) t_1 = 0.8 in. and t_2 = 2.4 in. at f = 0.8 GHz.

3.2. Multi-Element SEVA Conformal Linear Array (SEVA-CLA)

Figure 7 shows the fabricated model of multi-element Conformal Linear Array (SEVA-CLA). Oxidation during the fabrication of the composite resulted in an opaquer panel compared to the ML-SEVA. This limits the visibility of the embedded channels in the composite thus, the embedded channels of multi-element SEVA-CLA are filled using a different approach than ML-SEVA. Additional modifications included the use of a torus-shaped gasket formed from an HMA that was used to affix the feed structure on the curved composite. This prevents leakage of liquid metal and ensures smooth flow of liquid metal in the embedded channels.

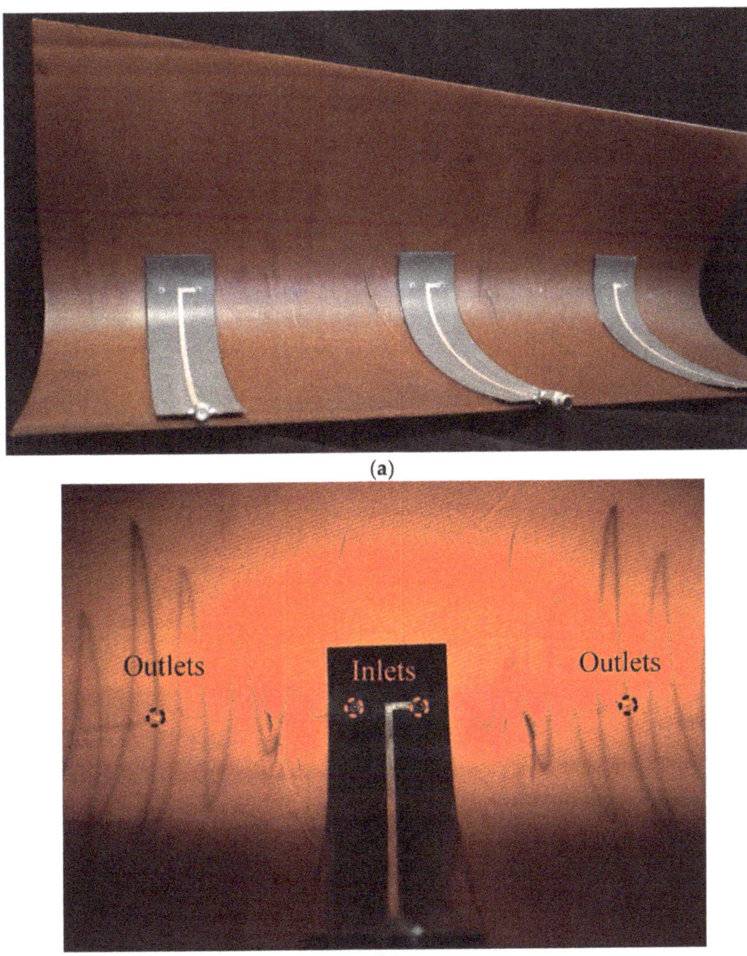

Figure 7. Fabricated model of SEVA-CLA (**a**) Top view showing the complete array with the three elements (**b**) Single element of the array with outlets and inlets indicated.

During the measurement campaign for this antenna the embedded microchannels were first filled with FC-70 in preparation for the transport of liquid metal (to regulate back-pressure, etc.). For each of the three antennas, the antipodal dipole feed structure was first measured while the other two feeds are terminated in matched loads. Pre-measured liquid metal was then injected into the channels through the hollow via in the feed section using an appropriately tipped syringe. The amount of FC-70 flowing out from the outlets along with input reflection coefficient reading from the network analyzer was used to provide a rough estimate of the position of liquid metal in the microchannel. This was necessary as the darker color of the panel prevented visual inspection of the exact position of the fluid. This process was repeated for each element, for each of the measured configurations.

Figure 8 shows the simulated and measured input reflection coefficient of the multi-element SEVA-CLA with a filling of t = 1.55 in. in each arm of sinusoidal meandering, operating at 824 MHz as shown in Figure 3b. A subset of the mutual impedance measurements of multi-element SEVA-CLA are compared. This configuration is chosen such that the center of the matched impedance bandwidth of the antennas (when filled with liquid metal) would provide a half-wavelength element spacing in the array. Operationally, at frequencies higher than this (when the dipole arms contain less liquid metal) the spacing become much greater than a half-wavelength and the pattern behaves as expected with the generation of grading lobes and other features. The radiation pattern of SEVA-CLA with zero-degree relative phasing between the elements is shown in Figure 9. Beam scanning by the array in the yz-plane is shown in Figure 10. Maximum beam scanning range of the array is within ±45 degrees. The principal beam of the array can be scanned by varying the progressive phase shift between the elements of the array. The phase difference between the antenna elements is varied by connecting microstrip transmission lines with pre-determined phases at 824 MHz.

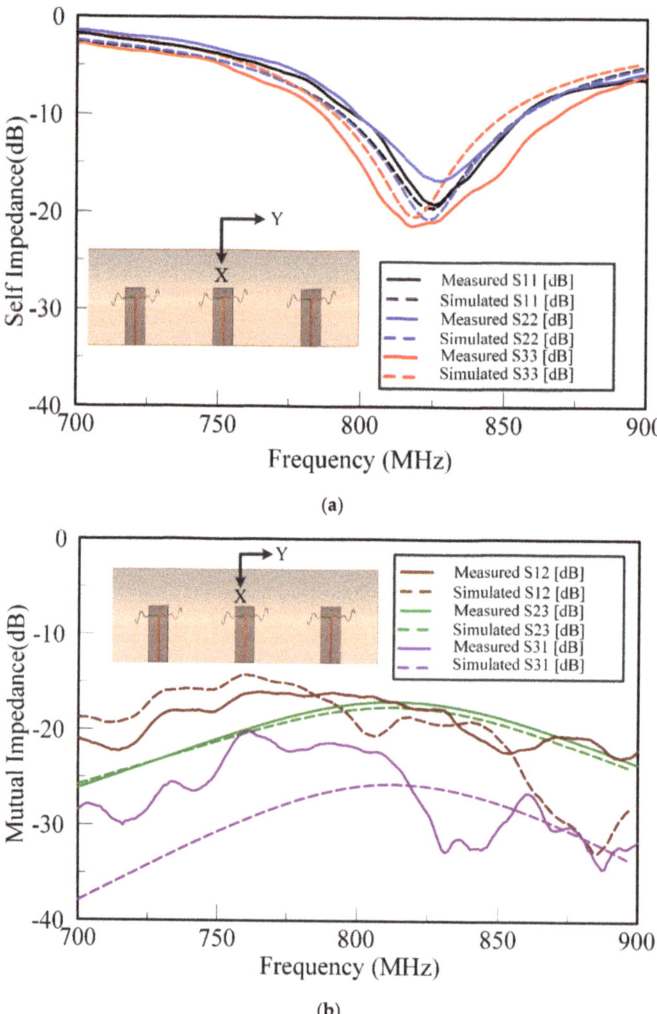

Figure 8. Measured and simulated impedance results of multi-element SEVACLA at 824 MHz for channel filling variable, t = 1.55 in (**a**) Self-impedance of the array elements (**b**) Mutual-impedance of array elements.

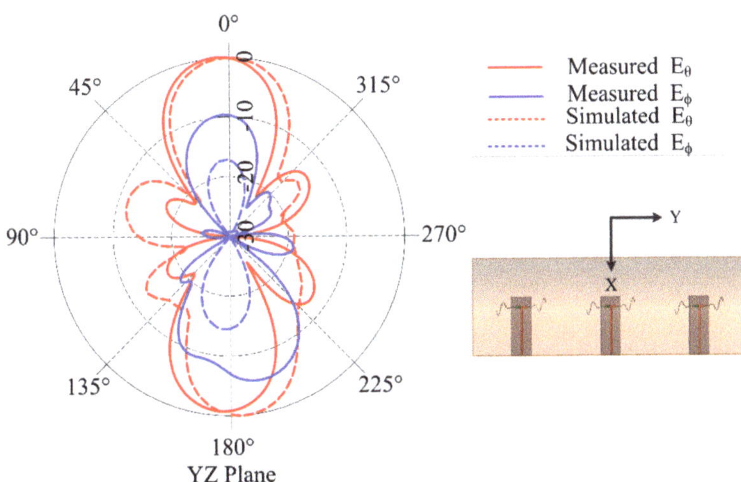

Figure 9. Radiation pattern of multi-element SEVA-CLA for an antenna array with 0° relative phasing at 824 MHz.

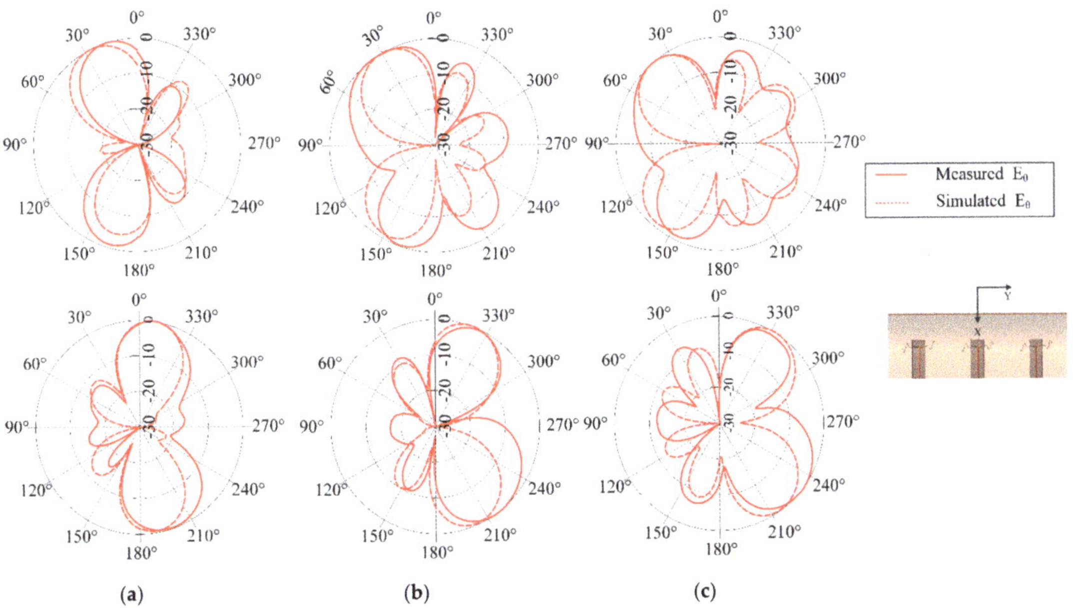

Figure 10. Beam scanning of multi-element SEVA-CLA for different scan angles at with channel filling factor t = 1.55 in. operating at 824 MHz (**a**) Radiated beam at ± 15° (**b**) Radiated beam at ± 30° (**c**) Radiated beam at ± 45°.

The contour plot in Figure 11 shows the array pattern for three isotropic elements spaced 183 mm apart and the three element SEVA-CLA with uniform excitation over a nominal operational bandwidth from 0.45 GHz to 4.05 GHz. The overall array pattern for multi-element SEVA-CLA is obtained by multiplying the single elemental pattern of multi-element SEVA-CLA with the array pattern of similarly spaced isotropic source. Careful examination of the two normalized gain plots testifies that the gain pattern of SEVA-CLA is obtained by multiplying the gain pattern for three-element isotropic array with the unit element pattern shown in Figure 11. Additional nulls at ±90 degrees in the SEVA-CLA contour plot are introduced by the individual elemental pattern of multi-element SEVA-CLA. The position and distribution of maxima at 0 degrees and grating lobes are similar between the two plots. Increase in the number of grating lobes in the visible region at higher frequencies is due to the widening gap between individual elements in terms of wavelength. Figure 12 shows the gain distribution of SEVA-CLA over similar frequency range (0.45–4.05 GHz). The variation in the gain pattern is attributed to the coupling between the elements with different filling factor, t, over the frequency range. The gain of SEVA_CLA observes a dip around 2 GHz as the grating lobes start to emerge in the visible range. The efficiency of the SEVA-CLA could not be measured experimentally in the facility due to lack of available tools or in simulation but a decrease in efficiency is expected to be seen with increasing grating lobes [39].

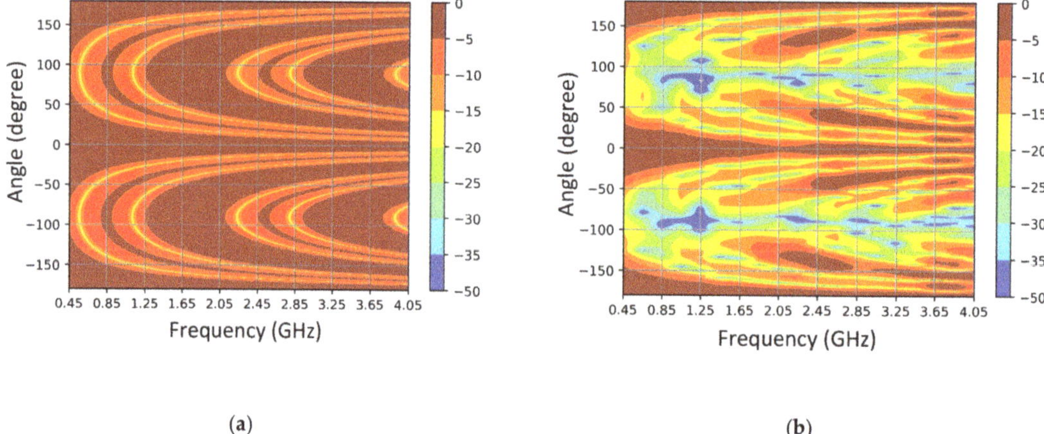

Figure 11. Simulated variation of radiation pattern with frequency (**a**) 3-element isotropic elements spaced 183 mm apart and (**b**) multi-element SEVA-CLA spaced 183 mm apart.

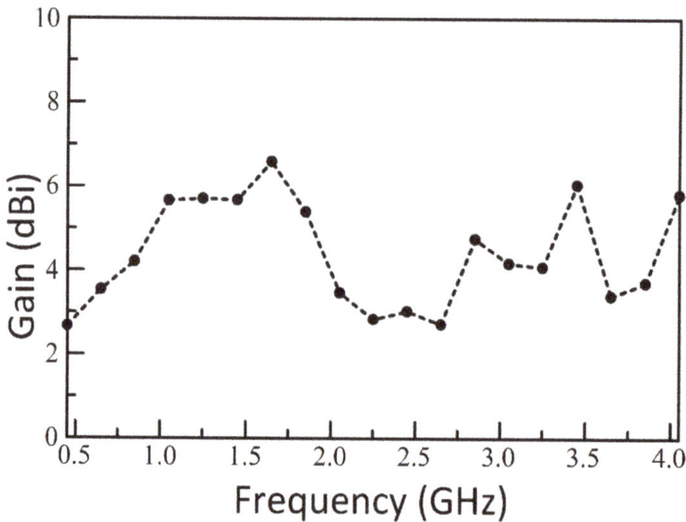

Figure 12. Simulated variation of gain with frequency multi-element SEVA-CLA.

Figure 13 shows the measured input reflection coefficient for another unique configuration of the SEVA-CLA. This corresponds to the complete filling of embedded microchannels with liquid metal. The central element is driven in the configuration shown with the other two ports terminated in open circuits—this is one of many unique configurations that can potentially be achieved using different reactive loads and feed locations. In the form shown, the contiguous antenna structure operates as single antenna element with a matched impedance bandwidth centered at 95 MHz. The shift between measured and simulated input reflection coefficient is attributed to differences arising from fabrication and processing as well as other variations between modeled and fabricated designs.

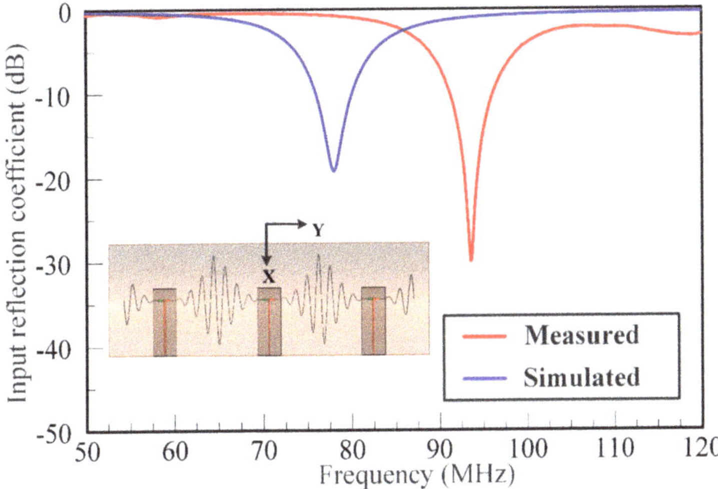

Figure 13. Measured and simulated results of multi-element SEVA-CLA operated as a single antenna element.

4. Discussion

Current advances in the design and implementation of antennas engineered to reconfigure using the controlled transport of liquid metal through microchannel has been used to enable frequency reconfiguration, polarization reconfiguration, and beam-steering [40–42]. These designs show the wide range of applications of antenna sensors using transport of liquid metal through microchannels suitable for varied applications over wide frequency range. These and other structures are typically single-layer design with microfluidic channels traversing a planar geometry. They open up questions on the integration of these techniques for most complex host. This study demonstrates the ability to engineer multilayer and conformally curved reconfigurable antennas enabled by the transport of liquid metal in structurally embedded vascular networks. This study provides insights into the fabrication of vascular networks in two consecutive layers of fabric in the composite lay-up. It also demonstrates the ability to embed distinct patterns in a complexly curved composite. The design, fabrication and electromagnetic functionalization of these structures leads to better understanding of the scope they provide and challenges they present. These fabrication techniques have been leveraged to create multilayer SEVA in which two co-located reconfigurable antennas have been utilized synergistically for impedance and radiation tuning. In SEVA-CLA this technique has been utilized to demonstrate beam scanning of three element array on a complex curved composite which can be reconfigured for a decade of frequency. The agreement between measured and simulated data testifies the concept and ideas presented in this paper. These design techniques can be successfully utilized in designing and fabricating embedded antennas to operate as sensors in UAVs or airplanes. This research also opens avenues in designing shape-conformal electromagnetic sensors using the concepts of microfluidics and structurally embedded vascular electromagnetic devices.

5. Conclusions

ML-SEVA and multi-element SEVA-CLA portrays advances in design and fabrication of SEVA. Pressure driven transport of liquid metal through these structures establishes frequency-reconfiguration over a wide frequency range. Multi-element SEVA-CLA conformal to the shape of UAV wing operates as phased array establishing its beam scanning ability. These structures open avenues towards design of load bearing frequency reconfigurable antennas and antenna arrays integrable on UAVs and/or aircrafts.

Author Contributions: Antenna design and simulation, A.B., H.P., and G.H.H.; composite and additive manufacturing, J.W.B., G.J.F., and T.G.; experimental campaign, A.B. and H.P.; project management and supervision, G.H.H., J.W.B., G.J.F., and D.J.H.; writing—original draft preparation, A.B.; writing—review and editing, A.B., J.W.B., and G.H.H.; All authors have read and agreed to the published version of the manuscript.

Funding: This work was supported by the AFRL Commander's Research and Development Fund.

Institutional Review Board Statement: Not applicable.

Informed Consent Statement: Not applicable.

Data Availability Statement: All the results and references are included here.

Acknowledgments: Pennsylvania State University and Texas A&M University gratefully acknowledge the support of the Air Force Research Laboratory (AFRL).

Conflicts of Interest: Any opinions, findings, and conclusion or recommendations expressed in this material are those of the authors and do not necessarily reflect the view of the AFRL or the US government. Approved for public release (88ABW-2017-5151).

References

1. Inata, H.; Say, S.; Ando, T.; Liu, J.; Shimamoto, S. Unmanned aerial vehicle based missing people detection system employing phased array antenna. In Proceedings of the 2016 IEEE Wireless Communications and Networking Conference, Doha, Qatar, 3–6 April 2016.
2. Fernandez, M.G.; Lopez, Y.A.; Andres, F.L.-H. On the Use of Unmanned Aerial Vehicles for Antenna and Coverage Diagnostics in Mobile Networks. *IEEE Commun. Mag.* **2018**, *56*, 72–78. [CrossRef]
3. Kim, S.J.; Lim, G.J.; Cho, J.; Côté, M.J. Drone-Aided Healthcare Services for Patients with Chronic Diseases in Rural Areas. *J. Intell. Robot. Syst.* **2017**, *88*, 163–180. [CrossRef]
4. Tran, H.H.; Nguyen-Trong, N.; Le, T.T.; Park, H.C. Wideband and Multipolarization Reconfigurable Crossed Bowtie Dipole Antenna. *IEEE Trans. Antennas Propag.* **2017**, *65*, 6968–6975. [CrossRef]
5. Yoon, S.; Tak, J.; Choi, J.; Park, Y.-M. Conformal monopolar antenna for UAV applications. In Proceedings of the IEEE International Symposium on Antennas and Propagation & USNC/URSI National Radio Science Meeting, San Diego, CA, USA, 9–14 July 2017.
6. Nosrati, M.; Jafargholi, A.; Pazoki, R.; Tavassolian, N. Broadband Slotted Blade Dipole Antenna for Airborne UAV Applications. *IEEE Trans. Antennas Propag.* **2018**, *66*, 3857–3864. [CrossRef]
7. Wu, D.; Chen, X.; Yang, L.; Fu, G.; Shi, X. Compact and Low-Profile Omnidirectional Circularly Polarized Antenna with Four Coupling Arcs for UAV Applications. *IEEE Antennas Wirel. Propag. Lett.* **2017**, *16*, 2919–2922. [CrossRef]
8. Hu, J.; Luo, G.Q.; Hao, Z.-C. A Wideband Quad-Polarization Reconfigurable Metasurface Antenna. *IEEE Access* **2018**, *6*, 6130–6137. [CrossRef]
9. Balderas, L.I.; Reyna, A.; Panduro, M.A.; Del Rio, C.; Gutierrez, A.R. Low-Profile Conformal UWB Antenna for UAV Applications. *IEEE Access* **2019**, *7*, 127486–127494. [CrossRef]
10. Seo, D.-G.; Jeong, C.-H.; Choi, Y.-S.; Park, J.-S.; Jeong, Y.-Y.; Lee, W.-S. Wide Beam Coverage Dipole Antenna Array with Parasitic Elements for UAV Communication. In Proceedings of the IEEE International Symposium on Antennas and Propagation and USNC-URSI Radio Science Meeting, Atlanta, GA, USA, 7–12 July 2019.
11. Costa, A.; Goncalves, R.; Pinho, P.; Carvalho, N.B. Design of UAV and ground station antennas for communications link budget improvement. In Proceedings of the IEEE International Symposium on Antennas and Propagation & USNC/URSI National Radio Science Meeting, San Diego, CA, USA, 9–14 July 2017.
12. Lee, C.U.; Noh, G.; Ahn, B.; Yu, J.-W.; Lee, H.L. Tilted-Beam Switched Array Antenna for UAV Mounted Radar Applications with 360° Coverage. *Electronics* **2019**, *8*, 1240. [CrossRef]

13. Zhou, Y.; Bayram, Y.; Du, F.; Dai, L.; Volakis, J.L. Polymer-Carbon Nanotube Sheets for Conformal Load Bearing Antennas. *IEEE Trans. Antennas Propag.* **2010**, *58*, 2169–2175. [CrossRef]
14. Mehdipour, A.; Sebak, A.-R.; Trueman, C.W.; Rosca, I.D.; Hoa, S.V. Reinforced Continuous Carbon-Fiber Composites Using Multi-Wall Carbon Nanotubes for Wideband Antenna Applications. *IEEE Trans. Antennas Propag.* **2010**, *58*, 2451–2456. [CrossRef]
15. Biswas, S. Fabrication of Conformal Load Bearing Antenna using 3D Printing. In Proceedings of the IEEE International Symposium on Antennas and Propagation & USNC/URSI National Radio Science Meeting, Boston, MA, USA, 8–13 July 2018.
16. Chen, C.; Zheng, H. Design of a dual-band conformai antenna on a cone surface for missle-borne. In Proceedings of the Sixth Asia-Pacific Conference on Antennas and Propagation (APCAP), Xi'an, China, 16–19 October 2017.
17. Kim, J.; Jang, J.-Y.; Ryu, G.-H.; Choi, J.-H.; Kim, M.-S. Structural design and development of multiband aero-vehicle smart skin antenna. *J. Intell. Mater. Syst. Struct.* **2014**, *25*, 631–639. [CrossRef]
18. Xu, F.; Wei, B.; Li, W.; Liu, J.; Qiu, Y.; Liu, W. Cylindrical conformal single-patch microstrip antennas based on three dimensional woven glass fiber/epoxy resin composites. *Compos. Part B Eng.* **2015**, *78*, 331–337. [CrossRef]
19. Liu, Y.; Du, H.; Liu, L.; Leng, J. Shape memory polymers and their composites in aerospace applications: A review. *Smart Mater. Struct.* **2014**, *23*. [CrossRef]
20. Bishop, N.A.; Miller, J.; Zeppettella, D.; Baron, W.; Tuss, J.; Ali, M. A Broadband High-Gain Bi-Layer LPDA for UHF Conformal Load-Bearing Antenna Structures (CLASs) Applications. *IEEE Trans. Antennas Propag.* **2015**, *63*, 2359–2364. [CrossRef]
21. Bishop, N.; Ali, M.; Baron, W.; Miller, J.; Tuss, J.; Zeppettella, D.; Ali, M. Aperture coupled MEMS reconfigurable pixel patch antenna for conformal load bearing antenna structures (CLAS). In Proceedings of the 2014 IEEE Antennas and Propagation Society International Symposium (APSURSI), Memphis, TN, USA, 6–11 July 2014; pp. 1091–1092.
22. Kiourti, A.; Volakis, J.L. Stretchable and Flexible E-Fiber Wire Antennas Embedded in Polymer. *IEEE Antennas Wirel. Propag. Lett.* **2014**, *13*, 1381–1384. [CrossRef]
23. Yao, L.; Jiang, M.; Zhou, D.; Xu, F.; Zhao, D.; Zhang, W.; Zhou, N.; Jiang, Q.; Qiu, Y. Fabrication and characterization of microstrip array antennas integrated in the three dimensional orthogonal woven composite. *Compos. Part B Eng.* **2011**, *42*, 885–890. [CrossRef]
24. Zhong, J.; Kiourti, A.; Sebastian, T.; Bayram, Y.; Volakis, J.L. Conformal Load-Bearing Spiral Antenna on Conductive Textile Threads. *IEEE Antennas Wirel. Propag. Lett.* **2016**, *16*, 230–233. [CrossRef]
25. Wang, Z.; Zhang, L.; Bayram, Y.; Volakis, J.L. Embroidered Conductive Fibers on Polymer Composite for Conformal Antennas. *IEEE Trans. Antennas Propag.* **2012**, *60*, 4141–4147. [CrossRef]
26. Du, C.-Z.; Zhong, S.-S.; Yao, L.; Qiu, Y.-P. Textile microstrip two-element array antenna on 3D orthogonal woven composite. In Proceedings of the 2010 International Symposium on Signals, Systems and Electronics, Nanjing, China, 17–20 September 2010; Volume 2, pp. 1–2. [CrossRef]
27. Ghorbani, K. Conformal load bearing antenna structure using Carbon Fibre Reinforced Polymer (CFRP). In Proceedings of the International Workshop on Antenna Technology: Small Antennas, Novel EM Structures and Materials, and Applications (iWAT), Sydney, Australia, 4–6 March 2014.
28. Kim, M.-S.; Park, C.-Y.; Cho, C.-M.; Jun, S.-M. A multi-band smart skin antenna design for flight demonstration. In Proceedings of the 8th European Conference on Antennas and Propagation (EuCAP 2014), The Hague, The Netherlands, 6–11 April 2014.
29. Morishita, A.; Kitamura, C.; Ohta, A.; Shiroma, W. Two-octave tunable liquid-metal monopole antenna. *Electron. Lett.* **2014**, *50*, 19–20. [CrossRef]
30. Huff, G.H.; Pan, H.; Hartl, D.J.; Frank, G.J.; Bradford, R.L.; Baur, J.W. A Physically Reconfigurable Structurally Embedded Vascular Antenna. *IEEE Trans. Antennas Propag.* **2017**, *65*, 2282–2288. [CrossRef]
31. Morales, D.; Stoute, N.A.; Yu, Z.; Aspnes, D.E.; Dickey, M.D. Liquid gallium and the eutectic gallium indium (EGaIn) alloy: Dielectric functions from 1.24 to 3.1 eV by electrochemical reduction of surface oxides. *Appl. Phys. Lett.* **2016**, *109*, 091905. [CrossRef]
32. Hartl, D.J.; Frank, G.J.; Baur, J.W. Effects of microchannels on the mechanical performance of multifunctional composite laminates with unidirectional laminae. *Compos. Struct.* **2016**, *143*, 242–254. [CrossRef]
33. Hartl, D.J.; Frank, G.J.; Malak, R.J.; Baur, J.W. A liquid metal-based structurally embedded vascular antenna: II. Multiobjective and parameterized design exploration. *Smart Mater. Struct.* **2016**, *26*, 025002. [CrossRef]
34. Hartl, D.J.; Huff, G.H.; Pan, H.; Smith, L.; Bradford, R.L.; Frank, G.J.; Baur, J.W. Analysis and characterization of structurally embedded vascular antennas using liquid metals. In Proceedings of the Sensors and Smart Structures Technologies for Civil, Mechanical, and Aerospace Systems, Las Vegas, NV, USA, 20 April 2016.
35. Hartl, D.J.; Frank, G.J.; Huff, G.H.; Baur, J.W. A liquid metal-based structurally embedded vascular antenna: I. Concept and Multiphysics Modelling. *Smart Mater. Struct.* **2017**, *26*, 1–15. [CrossRef]
36. Ansys HFSS: High Frequency Structural Simulator. Available online: https://www.ansys.com/products/electronics/ansys-hfss (accessed on 27 January 2021).
37. Stein, F.M. *The Curve Parallel to a Parabola Is Not a Parabola: Parallel Curves*; Taylor & Francis, Ltd.: Abingdon, UK, 1980; Volume 11, pp. 239–246.
38. 3MTM FluorinertTM Electronic Liquid FC-70–3M Science Applied to Life. September 2019. Available online: https://multimedia.3m.com/mws/media/64891O/fluorinert-electronic-liquid-fc-70.pdf (accessed on 27 January 2021).

39. Vosoogh, A.; Kildal, P.-S. Simple Formula for Aperture Efficiency Reduction Due to Grating Lobes in Planar Phased Arrays. *IEEE Trans. Antennas Propag.* **2016**, *64*, 2263–2269. [CrossRef]
40. Bharambe, V.T.; Adams, J.J. Planar 2-D Beam Steering Antenna Using Liquid Metal Parasitics. *IEEE Trans. Antennas Propag.* **2020**, *68*, 7320–7327. [CrossRef]
41. Liu, Y.; Wang, Q.; Jia, Y.; Zhu, P. A Frequency- and Polarization-Reconfigurable Slot Antenna Using Liquid Metal. *IEEE Trans. Antennas Propag.* **2020**, *68*, 7630–7635. [CrossRef]
42. Singh, A.; Goode, I.; Saavedra, C.E. A Multistate Frequency Reconfigurable Monopole Antenna Using Fluidic Channels. *IEEE Antennas Wirel. Propag. Lett.* **2019**, *18*, 856–860. [CrossRef]

Article

Three-Dimensional Chipless RFID Tags: Fabrication through Additive Manufacturing

Sergio Terranova, Filippo Costa, Giuliano Manara and Simone Genovesi *

Dipartimento di Ingegneria dell'Informazione, Università di Pisa, 56122 Pisa, Italy; sergio.terranova@ing.unipi.it (S.T.); filippo.costa@unipi.it (F.C.); giuliano.manara@unipi.it (G.M.)
* Correspondence: simone.genovesi@unipi.it

Received: 20 July 2020; Accepted: 18 August 2020; Published: 21 August 2020

Abstract: A new class of Radio Frequency IDentification (RFID) tags, namely the three-dimensional (3D)-printed chipless RFID one, is proposed, and their performance is assessed. These tags can be realized by low-cost materials, inexpensive manufacturing processes and can be mounted on metallic surfaces. The tag consists of a solid dielectric cylinder, which externally appears as homogeneous. However, the information is hidden in the inner structure of the object, where voids are created to encrypt information in the object. The proposed chipless tag represents a promising solution for anti-counterfeiting or security applications, since it avoids an unwanted eavesdropping during the reading process or information retrieval from a visual inspection that may affect other chipless systems. The adopted data-encoding algorithm does not rely on On–Off or amplitude schemes that are commonly adopted in the chipless RFID implementations but it is based on the maximization of available states or the maximization of non-overlapping regions of uncertainty. The performance of such class of chipless RFID tags are finally assessed by measurements on real prototypes.

Keywords: 3D printing; additive manufacturing; Radio Frequency IDentification (RFID), chipless RFID; mounted on metal

---

1. Introduction

Additive Manufacturing (AM) is a rapid prototyping method of fabrication based on three-dimensional (3D) printing. Differently from the more traditional technologies based on a subtractive process (e.g., milling) or a formative one (e.g., forging), AM creates the desired shape starting from a three-dimensional Computer-Aided Design (CAD) model that is sliced into a set of layers, each one being a thin cross-section of the original model. The obtained object is a staggered approximation whose fidelity with the original one depends on the finite thickness of each layer, the properties of the employed material and the adopted AM technology [1]. The choice can be made among processes that exploit photopolymerization, such as the Stereolithography (SLA), material extrusion, as in the case of Fused Filament Fabrication (FFF), and power bed fusion, like Selective Laser Sintering (SLS), just to name a few that can print plastic-based materials, ceramics and even metal [2].

Two remarkable features provided by AM consist in providing the ability to build very complex geometries and in tailoring mechanical and electrical parameters of an object by using the capability of controlling the density of a printed object [3]. This unique combination has opened new and exciting paths in several areas that span from aerospace to medical, from transportation to energy [4]. Furthermore, there are numerous examples of 3D-printed devices suitable for chemical-, temperature- and pressure-sensing applications [5–7]. AM has also boosted new solutions in electromagnetics where it has been adopted for lenses [8], Frequency Selective Surfaces (FSS) [9], printed electronics [10], reflectarray [11], antennas for space [12,13] and several other devices [3,14–17]. Among the most interesting and profitable application areas, the Radio Frequency IDentification (RFID) is certainly one of the more appealing and fast-growing [18]. The most widespread identification device consists of an antenna connected to an Integrated Circuit (IC) that modulates the scattered field, commonly referred to as a RFID tag [19]. The interrogation process of the RFID tag differs depending on the frequency bandwidth employed, which, in turns, determines the read range. Low-Frequency (LF) tags work around 130 KHz and use near-field inductive coupling to obtain the necessary power to communicate their information to the nearby reader (typically 10–20 cm). High-frequency (HF) tags work at 13.56 MHz, exploit near-field coupling as well and achieve similar read range. The Ultra-High-Frequency (UHF) range is adopted by RFID tags that commonly resonate in a narrow band within the 860–960 MHz frequency interval. The interrogating signal is a circularly polarized plane wave, and this kind of tag guarantees a reliable communication range up to 10 m.

Recently, a class of RFID tags that do not rely on an IC for encoding the information, the chipless RFID tags, have been proposed for both identification and sensing duties [20–22]. The vast majority of chipless RFID designs assume a plane wave interrogation system that must provide the proper amount of energy to be scattered back by the tag and recollected to the reader into an intelligible form. The information is generally encoded in the Frequency Domain (FD) or in the Time Domain (TD), and particular attention has to be paid to isolate the meaningful electromagnetic field reflection of the tag from the undesired backscattering noise generated by clutter and antenna coupling [23–25].

The most common chipless tag operating in time domain is based on Surface Acoustic Waves (SAW), which encode the information exploiting the time of travel of surface waves to a set of suitably placed reflectors. However, this solution is not cheap if compared to UHF RFID tags and may require a not negligible footprint [26]. The most diffused FD-based chipless tags instead encode information into the spectrum using resonant structures. The encoding algorithm is based on a one-to-one association among an information bit and the presence or absence of a resonance peak in the spectral signature at a predetermined and predictable frequency. These tags are robust to interference, fully printable and have large data capacity, although they require a broad spectrum for data encoding and a wideband-dedicated RFID reader. Currently, there are also many hybrid approaches between FD and TD systems [27–30]. Recently, printed dielectric encoders based on linear chains of dielectric inclusions were presented in Reference [31], in which a line of equally spaced resonators was sequentially read by linearly moving them in front of a near-field reader.

The reading range of these solutions can be comparable with the UHF tag, although a large frequency bandwidth is generally required. In order to overcome this bandwidth requirement, new encoding algorithms have been presented in the literature, such as in Reference [32], where the information is encoded in the quantized values of the difference between the TE and TM phase response and a multifrequency reader can be adopted. Although it is possible to cope with the backscattering noise adopting various strategies [33–36], for some applications, it is not required to achieve read ranges greater than a few centimeters, and therefore, near-field reading schemes have been employed even for chipless RFIDs. This can be the case when sensitive information is transferred or for increased data capacity or monitoring purposes [37,38].

The degrees of freedom offered by AM offers the possibility to consider unconventional RFID tag designs that can be easily manufactured by using inexpensive materials and low-cost processes. In the proposed concept study, a cylindrical dielectric structure is manufactured with inner voids that are invisible from outside and with an exterior appearance of a solid bulk shape. The three-dimensional (3D) chipless tag is completely realized with cheap material, and it is easy to manufacture with low-cost 3D printers. The proposed design provides an interesting coding capacity and, at the same time, conceals the information in its inner structure that makes it suitable for envisioned physical-layer security tasks and also for anticounterfeiting purposes [39,40]. The 3D chipless RFID tag can be mounted on metal platforms, and it is easily scalable to other dimensions or frequency bands. The presented 3D chipless tag represents an example of a frequency-domain chipless encoder that fully takes advantage of the three-dimensional nature of the tag and, at the same time, conceals the encoded information. The manuscript is organized as follow. Section 2 describes the geometry of the tag and its working principle. Section 3 describes the encoding and decoding scheme. In Section 4, the tag performance is experimentally verified. Conclusions are drawn in Section 5.

2. 3D-Printed Chipless RFID Tag

The adoption of an additive manufacturing process for realizing a device capable of storing information has been stimulated by the high flexibility offered by 3D with respect to 2D printing. The idea at the basis of the 3D chipless RFID design is to choose a tag shape that can exhibit a multi-resonance response in the frequency domain that, in turn, is exploited to encode the information. The reading system should be able to read this frequency signature once placed in contact with the tag. Polylactic acid (PLA) has been chosen for the tag fabrication due to its low-cost and easiness of use, even with entry-level printers.

The theoretical framework of the idea at the basis of the investigated 3D chipless tag comes from the considerable existing literature on dielectric resonators [41] and Dielectric Resonator (DR) antennas [42–45], and, in particular, on dual frequency and wideband DR antennas [46–48]. Dielectric resonators are microwave devices with high-quality factor Q, and they are used as elements in microwave filter and oscillator designs since these are excellent substitutes for metal resonant cavities. In a DR, it is possible to store electromagnetic energy because, once a field is triggered inside the resonator, an electromagnetic wave propagates and bounces back and forth between the walls, giving rise to a standing wave. Generally, this type of behavior is obtained when the dielectric permittivity is very high (e.g., 40–50). This effect can be well explained by calculating the reflection coefficient at the dielectric–air interface.

$$\Gamma = \frac{\sqrt{\varepsilon_r} - 1}{\sqrt{\varepsilon_r} + 1} \tag{1}$$

As can be seen in Equation (1), the coefficient becomes equal to 1 when the dielectric constant becomes large. It is, therefore, possible to further approximate the air–dielectric interface according to the non-physical condition of Perfect Magnetic Conductor (PMC), which requires the tangential components of the magnetic field to vanish. Although the assumption of the PMC condition has been profitably exploited for the design of dielectric resonators, it has always been known that a portion of the electromagnetic field can leak from the resonator, leading to a decrease of the Q-factor. This important observation led to the first studies of the irradiation of a DR in Reference [49], and since then, several papers have been published on this subject. In order for the dielectric material to operate as an antenna, its permittivity must be moderate, approximately in the 5–30 range, so that the energy in the cavity can escape from the walls and be irradiated into the environment. It is therefore important to underline that the proposed 3D tag design can benefit from the dielectric resonator antenna theory, but in many ways, it differs because the theoretical models present in literature are valid only under the condition of high permittivity, while in the present case, the PLA printing filament is characterized by a low permittivity of 2.62.

In order to explain how these theoretical principles can be exploited for designing a 3D chipless tag, it is useful to briefly summarize the behavior of a cylindrical Dielectric Resonator Antenna (DRA) (Figure 1).

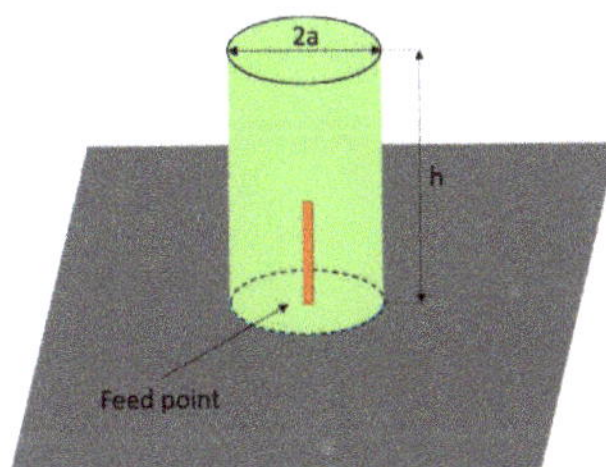

Figure 1. Three-dimensional (3D) view of the probe-fed cylindrical DRA.

Depending on the position of the probe, one of the principal radiating modes can beexcited [41,44,45]. The length of the coaxial probe, on the other hand, is sized to handle impedance matching. For a probe positioned in the center of the dielectric cylinder, as in our case, $TM_{01\delta}$ mode is excited. In addition, thanks to this type of coupling, a matching network is not required. Clearly, this length must be less than the height of the dielectric cylinder to avoid the probe radiation. It is interesting to notice that fundamental $TM_{01\delta}$ mode radiates like a short electric monopole. It should also be specified that this resonant mode coincides with that of the cylindrical DRA only in the semi-space $z > 0$. The distribution is in fact obtained by eliminating the ground plane and applying the method of images theory.

Unfortunately, there are no simple relationships to dimension the probe height once the dielectric cylinder dimensions are fixed. However, through an intensive campaign of experimental investigations and curve fitting, useful relationships have been obtained to calculate the resonance frequency and the merit factor associated with a particular resonant mode [45]. The approximated expression used to calculate the resonance frequency of the $TM_{01\delta}$ is reported in Equation (2), which is also suitable for low permittivity values.

$$f_R = \frac{c}{2\pi a\sqrt{\varepsilon_r + 2}}\sqrt{3.83^2 + \left(\frac{\pi}{2}\right)^2\left(\frac{a}{h}\right)^2} \qquad 0.125 \le \frac{a}{h} \le 5 \tag{2}$$

The first iteration of the 3D tag design process is shown in Figure 2, where a structure consisting of a solid cylinder of PLA ($\varepsilon_r = 2.62 - j0.05$), with metallized top and bottom face, is considered. This tag is probed by using a standard coaxial connector introduced at the tag center. The dimensions of the tag (h_d, r_d) are set by considering the cutoff frequency of the fundamental $TM_{01\delta}$ mode of the dielectric structure. Since the predefined frequency bandwidth has been set within 1.0 and 5.0 GHz, the radius, r_d, is set equal to 35 mm. The coaxial probe (l = 18 mm) is employed as a tag reader, and the collected measured impedance provides the input data for the data-encoding process. The effects of the coupling between the probe and the tag are described in Figure 3, where the real and imaginary part of the tag input impedance (Z_{in}) are reported. The case of the probe placed on the top metal cap and radiating in free space is obviously similar to that of a monopole with a finite ground plane. There is a first resonance around 3 GHz in correspondence of a low value of the real part of the impedance and then an anti-resonance around 3.8 GHz, with a higher real component. The loading effect of placing the 3D structure dielectric structure around the probe determines the downshift around 2 GHz of the resonance and 2.8 GHz of the antiresonance, and an increase of the real part. A hint of another couple, resonance/antiresonance, starts to be visible within the 3.5–4 GHz band.

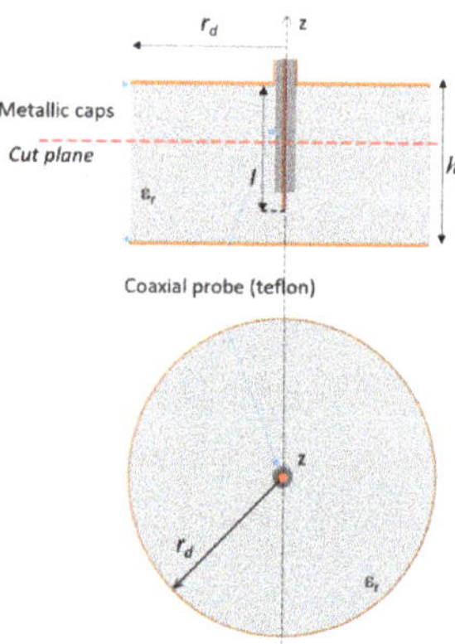

Figure 2. Lateral and top view of the first iteration for the tag design process.

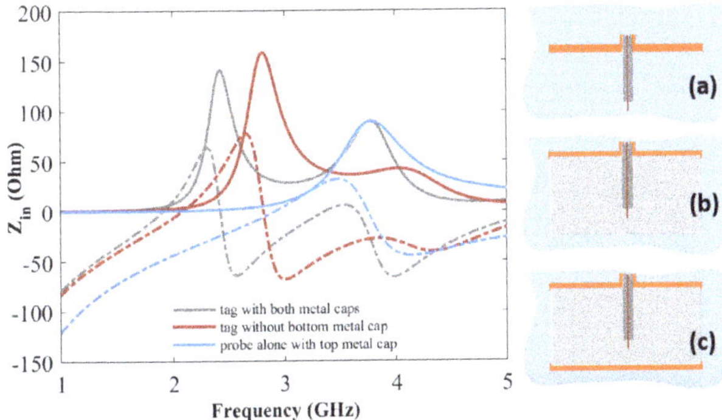

Figure 3. Real (continuous) and imaginary (dashed) part of the input impedance as a function of frequency considering the probe alone placed on the top metal cap (**a**), the dielectric structure without the metal bottom cap (**b**) and with both metal caps (**c**).

In order to explain the chosen geometric parameters of the 3D tag, the above-mentioned expression (2) can be adopted. By using the values of height, h_d = 22 mm, radius, a = 35 mm and ε_r = 2.62, a resonance frequency equal to f_R = 2.902 GHz is obtained. This resonance frequency deviates by about 100 MHz from the simulated resonance, as can be seen from the red curve in Figure 3B, where the imaginary part of the impedance crosses the zero level at f_R = 2.801 GHz. Once this first project was done, it was thought to add a metal cap in order to avoid an antenna behavior and make this tag similar to a dielectric resonator, despite its low permittivity ε_r.

A second metallic cap is then added to the bottom surface of the structure to provide an isolation from objects on which the tag could be placed, even metallic ones. This tag exhibits a resonance/antiresonance at lower frequency but also a more pronounced sign of another one around 3.5 GHz. It could be useful for encoding purposes to fully excite this latter one as well in order to possibly exploit it for encoding purposes. Therefore, the next step is to find a way for enhancing the resonance/antiresonance couples and be able to vary their position. To this aim, in view of the degrees of freedom offered by the additive manufacturing process, an empty annular sector has been included inside the original dielectric structure (Figure 4).

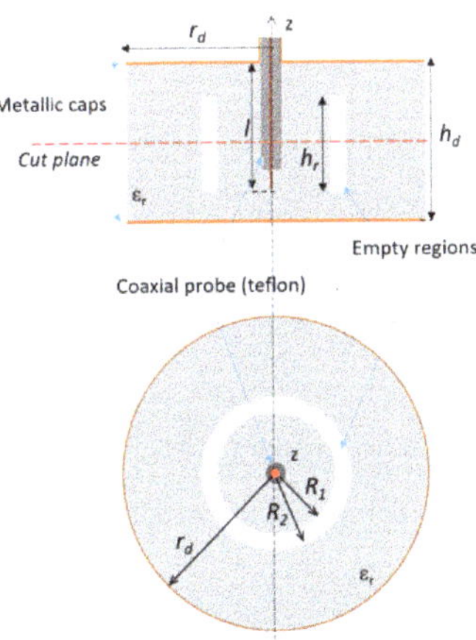

Figure 4. Lateral and top view of the second iteration of the design process in which an empty annular sector is introduced.

The effects of changing the dimensions of this ring-shaped void are reported in Figure 5, for the case of a fixed thickness (i.e., R_2-R_1 = constant) and for a variable thickness when R_2 is assigned. As it can be seen from Figure 5a, by fixing the thickness and the height of the empty ring-shaped void to 3 and 14.5 mm respectively, the position of the second couple of resonance/antiresonance can be changed by varying R_1 and R_2. The resonance shift is also obtained by varying the thickness of the vacuum inclusion and fixing one of the two radii, for example R_2 (Figure 5b). These frequency shifts of the second resonance or antiresonance can be exploited for encoding purposes, although other variable zero-crossings of the imaginary part of the impedance are needed for increasing the number of codified states.

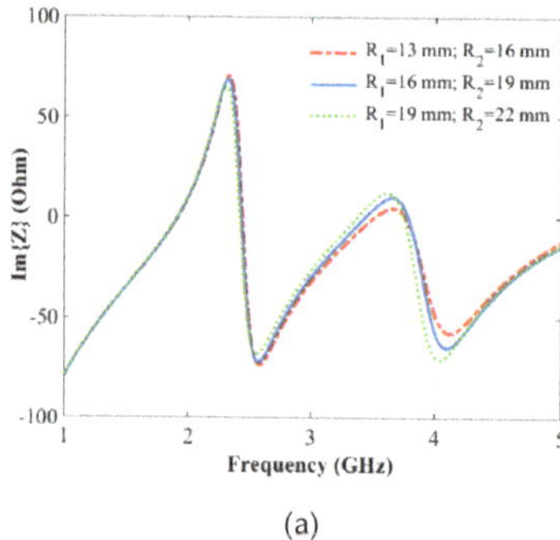

(a)

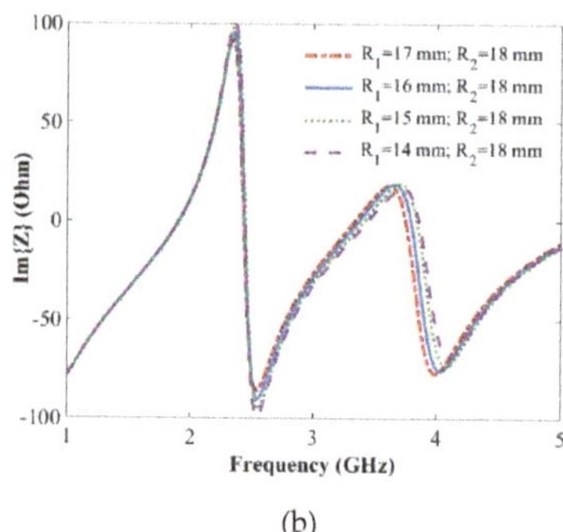

(b)

Figure 5. Imaginary part of the input impedance as a function of frequency in case the width and the height of the empty ring are fixed (**a**), and by varying the width of the empty ring for a fixed R_2 (**b**).

The previous analysis has suggested that a void inclusion is helpful in shifting the resonances. Therefore, an additional void inclusion is added (Figure 6).

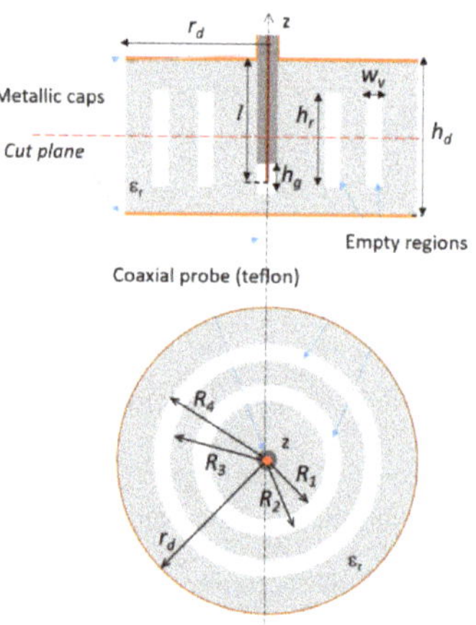

Figure 6. Lateral and top view of the third iteration of the design process in which an empty annular sector is introduced.

An exhaustive study of the possible configurations of the structure presented was carried out using Ansys High-frequency structure simulator (HFSS) [50] to investigate the behavior of the two pairs of resonances as a function of the void inclusion radius and reciprocal position. In more detail, with R_1 and R_2 being the inner and outer radius of the first inclusion and R_3 and R_4 those of the second one, a set of configurations has been considered in which the dimension and position of the first inclusion are fixed (i.e., R_1 and R_2 are constant) while the position of the second ring is changed (i.e., R_3 and R_4 are variable) and only the thickness is kept constant (i.e., R_4 and R_3 are constant). It is apparent that the first resonance is not affected by this change, whereas the first antiresonance and, more significantly, the second couple, resonance/antiresonance, are shifted (Figure 7a). On the contrary, fixing the dimensions of the second ring (R_3, R_4) and varying the position of the first ring does not affect the second resonance but alters all the other zero-crossing of the impedance imaginary part (Figure 7b). It can be concluded that the resonance/antiresonance related to each void inclusion cannot be controlled independently and that a change in one inclusion alters the whole frequency response. It is, therefore, necessary to map the resonance/antiresonance couples of all the allowed configurations to introduce a suitable coding of the electromagnetic signature. In particular, the height of the rings is set equal to h_r = 14.5 mm and the width of the voids. w_v = 3 mm (Figure 7). By considering the dimension of the tag and the manufacturing process, it has been considered to vary R_1 within (4–15) mm, R_2 in (7–18) mm or up to half of the radius of the dielectric cylinder and R_3 within (19–31) mm and R_4 in (22–34) mm, or at 1 mm from the external surface of the cylinder.

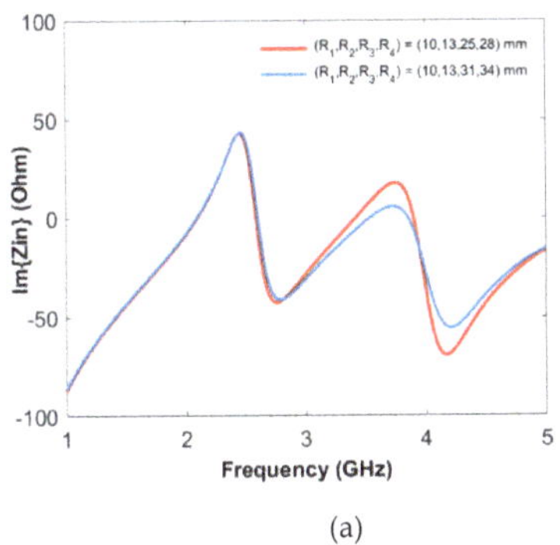

(a)

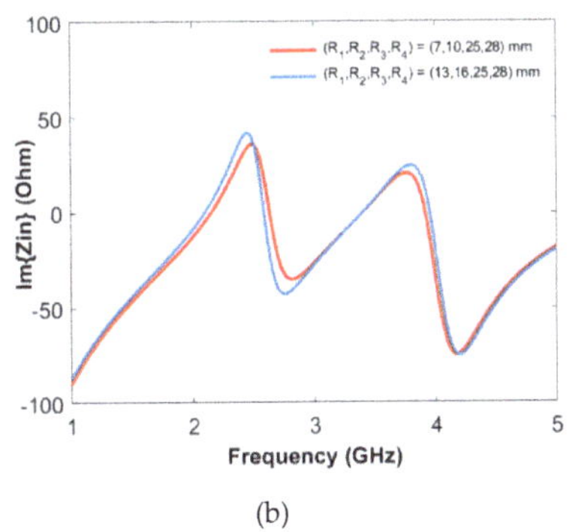

(b)

Figure 7. Imaginary part of the input impedance as a function of frequency in case the first empty ring is fixed and the position of the second one is changed (**a**), and by fixing the second empty ring and changing the position of the first one (**b**).

In view of the maximization of the number of the states, it is important that the imaginary part of the impedance has a large dynamic so as to have unambiguous and non-intersecting areas associated to the encoding based on resonance frequencies. To achieve this goal, an air gap, h_g, has been introduced in the design (Figure 8).

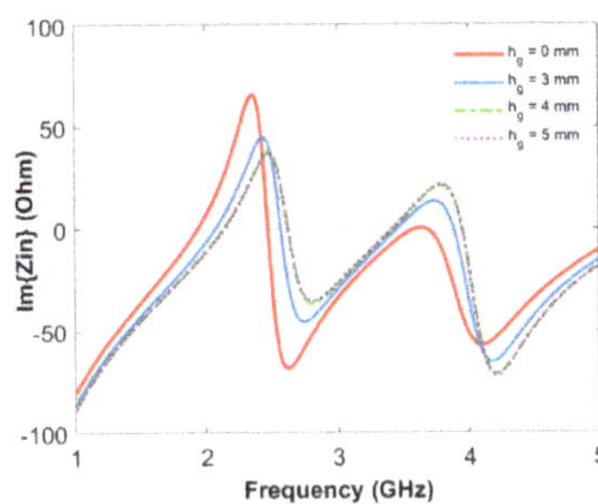

Figure 8. Effect of the presence of an air gap around the connector tip and the dielectric material (see Figure 6) on the imaginary part of impedance Im(z). All curves refer to a tag with total height h_d = 22 mm, height of void inclusions h_r = 14.5 mm and void inclusions radii R_1, R_2, R_3, R_4 = 10, 13, 25, 28 mm, respectively.

In fact, by varying this latter parameter from 3 to 5 mm, it is possible to observe a significant shift of the resonance frequencies (Figure 8) and therefore, an increase of the potential number of states. It is interesting to note that in the absence of an air gap, the second resonance around 3.6 GHz is hardly detectable. By increasing the value of h_g, the zero-cross is much more easily detectable. When the height of the air gap is 3 mm, it reaches the tip of the probe of length l. By further increasing the value of h_g from 4 to 5 mm, the tip of the probe is no longer in contact with the dielectric material, and it can be seen that then, the curves are practically superimposed. Looking at the reported trend, a value of h_g = 4 mm has been chosen since a further increment does not provide significant impedance variations.

3. Encoding and Decoding Scheme

Most of the adopted encoding algorithms for chipless RFID tags are based on ON/OFF schemes applied to the amplitude of resonance peaks in the electromagnetic response of the device. This approach is also encouraged by the fact that each peak is associated with a resonator, sometimes referred to as a resonating particle, that can be activated (or removed) in order to apply the coding. Since the proposed resonator does not rely on any single independent particle but on the overall interactions among the voids inside the dielectric structure, it has been of fundamental relevance to conceive a coding strategy that manages to extract useful information from the spectral signatures of a 3D chipless tag. This task has been accomplished by exploiting the correlations that exist between the variations of the geometric parameters of the tag, such as the positions and dimensions of the voids, that have a non-predictable effect on the resonances.

The algorithm is based on a positional resonance encoding, so each pair of resonances that can be embedded in the spectral signature can be represented as a point on a two-dimensional (2D) map, whose axes identify the possible values that each resonance can assume. The main purpose of the algorithm is then to unambiguously distinguish two points on this map that represent the coded information, considering, at the same time, that they can slightly change because of imperfections and errors due to the reading or manufacturing. It is possible to model this range of indecision as a certain area (coding zone) centered around the ideal coded position. So, the algorithm aims to find the largest non-intersecting subset of coding zones in the plane. Usually, the problem is not easy to solve because it belongs to the class of Non-determistic Polynomial-time hard problems (NP-hard problems). Algorithms of this type take the name of MISR (Maximum Independent Set of Rectangles) and can be solved, although in an approximate way when the considered coding zone has the shape of a rectangle.

At each allowed configuration of the resonator, that is a quadruple (R_1, R_2, R_3, R_4), a state is associated, which is defined by using the first two resonance frequencies (f_{Res1}, f_{Res2}). In order to define an unambiguous encoding scheme, each state is mapped into a point P of coordinate (f_{Res1}, f_{Res2}) in the two-dimensional space defined by the resonance frequencies exhibited by the tag, under the assumed geometrical constraints. In order to elaborate a robust encoding scheme, it is also considered that measurements on the real device may differ from the expected ones due to, for example, small imperfections in the realization of the prototype or the deterioration of the employed material. For this reason, a certain percentage tolerance in the mapping of each couple of frequencies into a codified state has been considered. This implies that the state is not codified by a single couple of resonances (i.e., point P), but by an area around the point P representing the state. For the sake of simplicity, in the hypothesis of an error Δ, it may consist of a certain percentual change of each resonance frequency or of a fixed quantity not directly related to the resonant frequencies. In the former case, the error can be associated to a surface which has the form of a rectangle, whereas in the latter case, it is represented by a square. An example of this mapping is reported in Figure 9 by square regions (Δ = 10 MHz). This choice of considering a certain error clearly has an impact on the encoding phase, where it will be necessary to adopt a criterium for maximizing the states that can actually be distinguished, that means selecting the maximum number of regions that do not intersect.

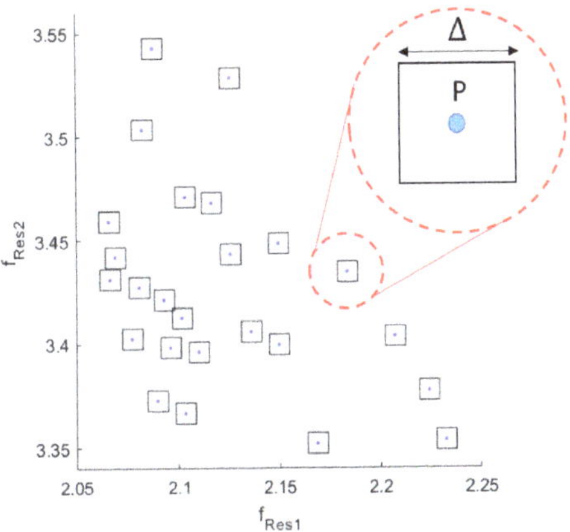

Figure 9. Representation of the tolerated uncertainty regions obtained by considering the first and Table 3. mm, tag height is 22 mm and the height of the rings is 14.5 mm.

It is, therefore, possible, in principle, to apply the encoding algorithm to a state comprising any pair of these zeros. Given a family of Im$\{Z_{in}\}$, there are 6 possible pairs of frequencies (f_i,f_j, with i,j = 1, 2, 3, 4, $i \neq j$) that can be considered. Once all the possible combinations have been tested, it will be possible to know which is the best pair of zeros of Im$\{Z_{in}\}$ that will provide the maximum number of distinguishable states suitable for decoding. In order to maximize the final number of states, a further set of device configurations has been considered, in which, other than the position of the empty annular regions, their thickness and height are varied. All the consequent resonance pairs (f_{Res1}, f_{Res2}) are shown in Figure 10. In particular, the configurations having different thickness and height are highlighted with a different marker. The initial step of the encoding algorithm is to create a data structure that stores the number and coordinates of the neighboring points for each resonance frequency pair (f_{Res1}, f_{Res2}), together with the information on those neighbors that intersect the area of the considered pair. In this way, it is possible to build an organized map of all disjointed points, highlighting points that have a number of neighbors equal to 1, 2, 3 and so on. In Figure 11, these rectangular regions are shown with different colors, depending on the number of neighbor points. For simplicity, only resonance frequency pairs that have up to 5 neighbors are represented, although the maximum order of intersections can be much higher. Next, the indices of the states that have the same number of neighbors are stored in separate vectors in order to consider sub-sets of the more general initial structure. The points that are classified as states with 0 neighbors are clearly disjointed and unambiguously identifiable, and they are therefore considered as resolved values. After this first selecting step, the remaining states are ranked in ascending order depending on the number of their neighbors. The group of configurations with only one neighbor (order 1) are the one considered at this stage. To avoid any possible ambiguity, the algorithm removes the neighbor state from those allowed and updates the entire data structure so that this deleted pair (which may have more than one neighbor) is also deleted from the list of other states. By having eliminated the only nearby node, the states that were ranked first now have no neighbors and so they are aggregated to the set of disjointed points. Once all nodes of order 1 have been resolved, the next order set, e.g., order 2, is considered. In this case, the list of states with 2 neighbors is scrolled, and for each of these, only one of the two neighbors is eliminated. In particular, the neighbor that has the maximum number of neighbors is deleted and also, this time, the whole data structure is updated and nodes of order 1 will be revealed. The above-stated criterion of eliminating the neighbor that has the largest number of neighbors is also valid if the order of the set is greater than 2. Applying this procedure iteratively always to the set having the lowest number of neighbors will finally lead to only states (regions) disjointed in frequency and therefore, distinguishable during the decoding phase. A comprehensive flow chart of this process is illustrated in Figure 12.

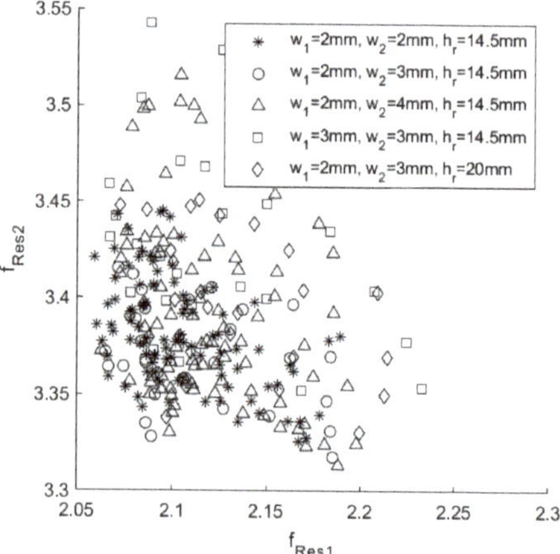

Figure 10. Representation of the couples of resonance frequencies of different tag families characterized by different thicknesses and heights. w_1 is the thickness of the first empty ring, w_2 is the thickness of the second one, and h_r is the height of both rings.

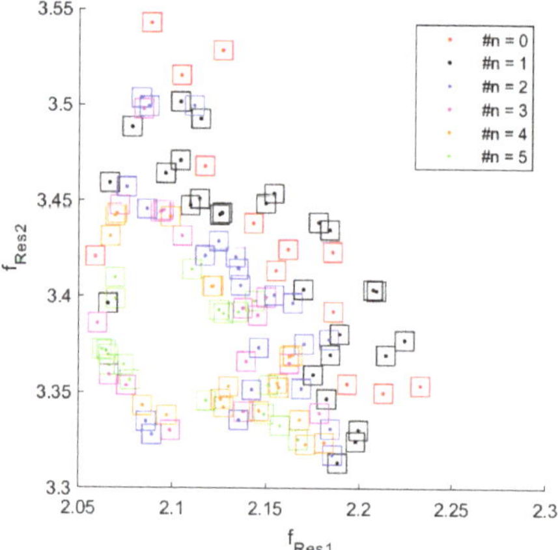

Figure 11. Regions of uncertainty classified according to the number of their intersections marked by the symbol # in the figure.

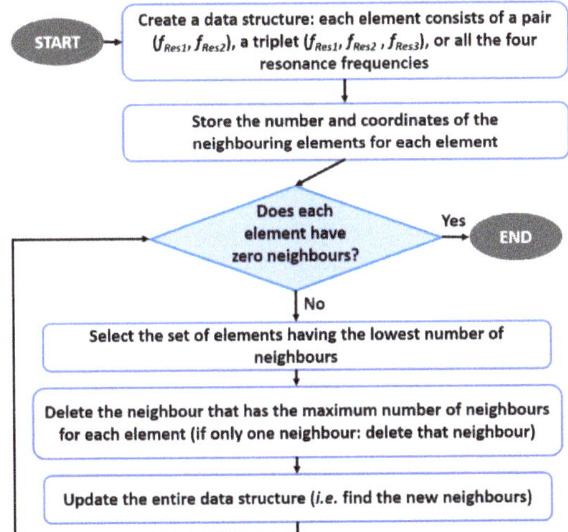

Figure 12. Flow chart of the algorithm to select the maximum number of unambiguous identification states.

It is interesting to notice that the initial choice of the Im{Z_{in}} zeros can lead to a different number of disjointed final states, as illustrated in Table 1. It is shown that by setting = 10 MHz, the best result is obtained when the third and fourth zero are chosen, which are the second resonance and second antiresonance, respectively. In this case, there are 105 distinguishable final states which correspond to 6.71 information bits. At the end of the encoding phase, all the possible separate regions associated with the couple of frequencies (f_{Res1}, f_{Res2}) have been selected. During the decoding phase, a search is performed through all the couples and the corresponding codified state is associated to the spectral signature. In more detail, it is checked if the couple of frequencies (f_{1^*}, f_{2^*}) related to the considered tag belongs to the intersection of the interval $f_{Res1}-\Delta/2 \leq f_{1^*} \leq f_{Res1} + \Delta/2$ and $f_{Res2}-\Delta/2 \leq f_{2^*} \leq f_{Res2}+\Delta/2$. If the condition is never verified, then the error made during the reading exceeds the estimated one and no decoding is possible.

Table 1. Number (#) of encoded states and corresponding number of bits obtained by setting Δ = 10 MHz.

f_{Res1}	f_{Res2}	# of Values	# of Bits
1	2	49	5.61
1	3	84	6.39
1	4	102	6.67
2	3	73	6.19
2	4	90	6.49
3	4	105	6.71

4. Prototype Realization and Measurements

In order to evaluate the performance of the proposed detection system, several prototypes of the 3D chipless RFID tag have been fabricated by using a 3D PowerWasp EVO printer [51]. A vector network analyzer (Keysight E5071C, Keysight, Santa Rosa, California, U.S.) has been used to measure the spectral signature of the tag in a non-anechoic environment.

The g-code file for the printing process has been generated by using the Ultimaker Cura software [52]. The diameter of the adopted PLA filament is 1.8 mm, and the selected extruder temperature is 185°. The print bed was not heated, so it was necessary to spread glue on it to stabilize the tag during the filament deposition process. To further improve adhesion during printing, a filament mesh was created before printing the piece. This substrate has a thickness of 0.3 mm, and it is printed by stacking 3 layers, each 0.1 mm high. The printing speed adopted was 70 mm/s in order to guarantee a good tradeoff between a good surface finish and not too long manufacturing times. The height of the inner layers, which determines the vertical resolution, has been set to 0.1 mm, a sufficient value to guarantee a good resolution given the cylindrical symmetry of the object. These layers have been printed with an infill percentage of 80%, in order to achieve a good mechanical stiffness. Moreover, a sufficiently high prototype density is important to ensure the desired dielectric properties. Among the different printing patterns available (e.g., rectilinear, honeycomb, grid), the one used for the internal layers was the grid type. For the external layers, i.e., top and bottom surfaces, a rectilinear pattern and an infill percentage of 100% has been used. The adopted line width was 0.4 mm, which is equal to the nozzle size. The wall thickness determines the width of the vertical walls of the tag, and it must be a multiple of the nozzle size. However, increasing the thickness of the shell determines a significant increase in the printing time and, at the same time, a greater mechanical stiffness. Generally, a value of 0.8 mm is a good design choice, and this was the set value. Once the PLA part of the device has been printed, a 0.5 mm thick copper metal disc was applied to the top and bottom layer using a thin coat of glue.

The 3D tag is shown in Figure 13, during the realization phase. It is worth noticing the presence of the two empty annular regions and the central hole in which the SMA connector will be inserted. It is also important to highlight that the additive manufacturing process allows a simple and seamless implementation of this prototype, whereas traditional techniques (i.e., subtractive ones) would have required the fabrication of the bulk cylinder, the drilling of the voids and a subsequent gluing to cover the inner structure. As a result, from the outside, it is not possible to infer the dimension of the inner voids (Figure 14).

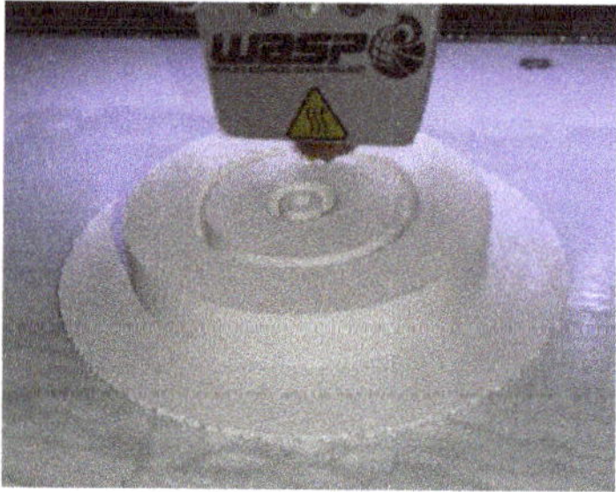

Figure 13. Three-dimensional (3D)-printed tag prototype during the realization phase.

A representative measure of the input impedance exhibited by one of the tags is compared with the corresponding simulation result in Figure 15. The geometric parameters of the manufactured prototype are: h_d = 22 mm, h_r = 14.5 mm, R_1, R_2, R_3, R_4 = 4, 7, 19, 22 mm respectively, r_d = 35 mm and h_g = 4 mm. The agreement is quite good for all the four resonance frequencies.

Figure 14. Final assembled prototype of the realized 3D chipless tag.

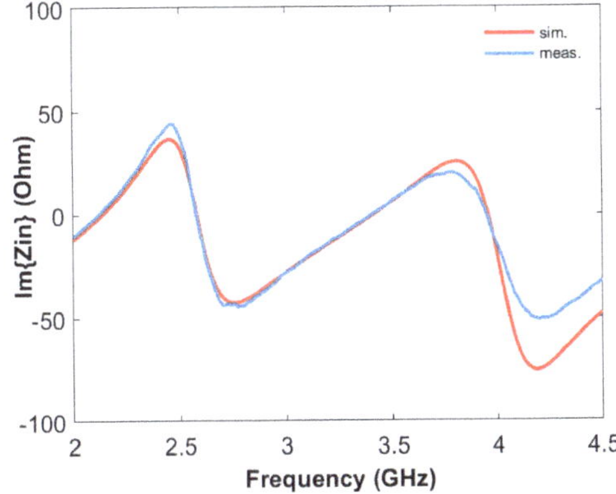

Figure 15. Comparison between simulated and measured results for the manufactured prototype (h_d = 22 mm, h_r =14.5 mm, R_1, R_2, R_3, R_4 = 4, 7, 19, 22 mm respectively, r_d = 35 mm and h_g = 4 mm).

Different encoding sequences can be used depending on the number of considered frequencies (i.e., 2, 3 or 4) and the selected ones (i.e., f_{RES1}, f_{RES2}, f_{RES3}, f_{RES4}). Table 2 shows the range of variations associated with each resonance frequency. The number of available bits (*#bit*) has been calculated for all the possible options that can be exploited by using different combinations of the resonance frequencies for several values of error expressed in MHz, Δ (Table 3).

Table 2. Range of values for each resonance frequency (MHz).

	Lower Value	Upper Value
f_{RES1}	2058	2232
f_{RES2}	2545	2660
f_{RES3}	3313	3542
f_{RES4}	3811	4060

Table 3. Number of available encoded states and bits for different combination of resonance frequencies and values of Δ, the error expressed in MHz.

f_{RES1}	f_{RES2}	f_{RES3}	f_{RES4}	#enc Δ = 10	#bit	#enc Δ = 20	#bit	#enc Δ = 30	#bit	#enc Δ = 40	#bit
1	2			49	5,6	17	4,1	10	3,3	7	2,8
1	3			84	6,4	37	5,2	22	4,5	15	3,9
1	4			102	6,7	45	5,5	25	4,6	17	4,1
2	3			73	6,2	32	5,0	17	4,1	12	3,6
2	4			90	6,5	37	5,2	20	4,3	13	3,7
3	4			105	6,7	45	5,5	28	4,8	19	4,2
1	2	3		117	6,9	44	5,5	24	4,6	14	3,8
1	2	4		145	7,2	61	5,9	29	4,9	18	4,2
1	3	4		190	7,6	89	6,5	51	5,7	32	5,0
2	3	4		177	7,5	80	6,3	44	5,5	27	4,8
1	2	3	4	197	7,6	96	6,6	52	5,7	33	5,0

On the basis of the manufactured prototypes, it has been found that a value of Δ = 20 MHz can provide a reliable measurement and therefore an intelligible and unambiguous recovery of the encoded data. This means that in case that all four resonance frequencies are exploited, the proposed 3D chipless RFID can offer 6.6 bits.

Since the number of encoded states is strictly related to the number of resonances, it is apparent that increasing their number can enhance the number of bits. This can be obtained, for example, by using a material with higher dielectric permittivity or by exploring the possibility offered by multi-material 3D tags.

It is interesting to note that the presented structure is not based on a unique and irreproducible signature, such as chaosmetry-based methods. The aim is to maximize the information regardless of the correlation that exists between the resonance/antiresonance pairs of the tag. In this case, the superposition principle of effects cannot be applied, as it would happen for a tag made up of several resonators in which a sharp and predictable resonance occurs due to the presence/absence of a specific resonator, such as in the most common chipless RFID tag implementations. It is interesting to note that there are different geometric configurations (thickness of annular voids, radial distance of voids from the axis of the cylindrical structure, height of voids, etc.) that give rise to the same zeros of the resonances, although the entire signatures are different. It is, therefore, possible to exploit this simple fact in order to introduce a degree of unpredictability for anti-counterfeiting purposes, since only those who know the encoding scheme know which bit pattern corresponds to a particular vector of zeros, extracted from the imaginary part of the input impedance.

5. Conclusions

A novel Chipless 3D-printed tag, fabricated by using PLA filament and a 3D low-cost printer, has been proposed in this paper. The tag has been metallized on the top and bottom surface in order to obtain a resonant-type structure by confining the electromagnetic fields within it. In addition, thanks to the metallization, the tag can be easily mounted on metal surfaces. The reading process is performed by inserting a probe inside a small hole on the bottom surface of the tag. The proposed reading scheme is particularly recommended for anti-counterfeiting applications for avoiding unwanted leakage of information and ensuring a closed reading system. The information has been encoded in the imaginary part of the tag impedance and measured through a vector network analyzer. The employed design ensures good coding capacity. The fabricated tags appear identical through a visual inspection, with an external appearance of a solid cylinder, as the geometry of the specific tag configuration depends on the internal features. A data-encoding algorithm based on the maximization of available states or the maximization of non-overlapping regions of uncertainty has been presented. The design presented in this work can be extended to other geometric shapes as well as other alterations within the tag during the printing process. It is also possible to adopt other printing filaments, characterized by different dielectric properties, to adequately modify the electromagnetic response of the tag and further increase the coding capacity. An improvement in coding density can be achieved by adopting a high permittivity filament and including appropriate modifications, such as voids, inclusions with dielectrics of different permittivity or doped with metal powders in order to mimic a metal inclusion. In this way, the spectral signature of the tag, e.g., the imaginary part of the impedance or the phase of the reflection coefficient, may contain dozens of resonances and the latter ones can be exploited to greatly increase the storable information [53,54]. It is interesting to notice that by adopting a higher permittivity printing filament (e.g., $\varepsilon_r = 10$ or more [53]), the size of the tag and its weight can drop significantly and, although the cost of the material increases, the final cost of the tag remains almost unaffected.

Author Contributions: Conceptualization, S.T. and S.G.; methodology, S.T., F.C., G.M. and S.G.; software, S.T. and S.G.; validation, S.T., F.C., G.M. and S.G.; formal analysis, S.T. and S.G.; investigation, S.T., F.C. and S.G.; writing—original draft preparation, S.T. and S.G.; writing—review and editing, S.T., F.C., G.M. and S.G. All authors have read and agreed to the published version of the manuscript.

Funding: This work was partially supported by the Italian Ministry of Education and Research (MIUR) in the framework of the CrossLab project (Departments of Excellence) and by TT Tecnosistemi S.p.A., Prato, Italy.

Conflicts of Interest: The authors declare no conflict of interest.

References

1. Gibson, I.; Rosen, D.W.; Stucker, B. *Additive Manufacturing Technologies: 3D Printing, Rapid Prototyping and Direct Digital Manufacturing*, 2nd ed.; Springer: New York, NY, USA, 2015.
2. Yang, L.; Hsu, K.; Baughman, B.; Godfrey, D.; Medina, F.; Menon, M.; Wiener, S. *Additive Manufacturing of Metals: The Technology, Materials, Design and Production*; Springer International Publishing: Cham, Switzerland, 2017.
3. Zhang, S.; Cadman, D.; Whittow, W.; Vardaxoglou, J.C. Enabling extrusion based additive manufacturing for RF applications: Challenges and opportunities. In Proceedings of the 12th European Conference on Antennas and Propagation, London, UK, 9–13 April 2018; pp. 1–5.
4. Ngo, T.D.; Kashani, A.; Imbalzano, G.; Nguyen, K.T.Q.; Hui, D. Additive manufacturing (3D printing): A review of materials, methods, applications and challenges. *Compos. Part B Eng.* **2018**, *143*, 172–196. [CrossRef]
5. Salim, A.; Ghosh, S.; Lim, S. Low-Cost and Lightweight 3D-Printed Split-Ring Resonator for Chemical Sensing Applications. *Sensors* **2018**, *18*, 3049. [CrossRef] [PubMed]
6. Herter, J.; Wunderlich, V.; Janeczka, C.; Zamora, V. Experimental Demonstration of Temperature Sensing with Packaged Glass Bottle Microresonators. *Sensors* **2018**, *18*, 4321. [CrossRef] [PubMed]
7. Emon, M.O.F.; Choi, J.W. Flexible Piezoresistive Sensors Embedded in 3D Printed Tires. *Sensors* **2017**, *17*, 656. [CrossRef] [PubMed]
8. Manafi, S.; González, J.M.F.; Filipovic, D.S. Design of a Perforated Flat Luneburg Lens Antenna Array for Wideband Millimeter-Wave Applications. In Proceedings of the 2019 13th European Conference on Antennas and Propagation, Małopolskie, Poland, 31 March–5 April 2019; pp. 1–5.
9. Sanz-Izquierdo, B.; Parker, E.A. 3D printed FSS arrays for long wavelength applications. In Proceedings of the 8th European Conference on Antennas and Propagation, The Hague, The Netherlands, 6–11 April 2014; pp. 2382–2386.
10. Espalin, D.; Muse, D.W.; MacDonald, E.; Wicker, R.B. 3D Printing multifunctionality: Structures with electronics. *Int. J. Adv. Manuf. Technol.* **2014**, *72*, 963–978. [CrossRef]
11. Nayeri, P.; Liang, M.; Sabory-Garcı, R.A.; Tuo, M.; Yang, F.; Gehm, M.; Xin, H.; Elsherbeni, A.Z. 3D Printed Dielectric Reflectarrays: Low-Cost High-Gain Antennas at Sub-Millimeter Waves. *IEEE Trans. Antennas Propag.* **2014**, *62*, 2000–2008. [CrossRef]
12. Teniente, J.; Iriarte, J.C.; Caballero, R.; Valcázar, D.; Goni, M.; Martínez, A. 3-D Printed Horn Antennas and Components Performance for Space and Telecommunications. *IEEE Antennas Wirel. Propag. Lett.* **2018**, *17*, 2070–2074. [CrossRef]
13. BYRNE, B.; CAPET, N. Compact 3D Printed Antenna Technology for Nanosat/Cubesat Applications. In Proceedings of the 2019 13th European Conference on Antennas and Propagation, Małopolskie, Poland, 31 March–5 April 2019; pp. 1–3.
14. Ramirez, R.A.; Rojas-Nastrucci, E.A.; Weller, T.M. UHF RFID Tags for On-/Off-Metal Applications Fabricated Using Additive Manufacturing. *IEEE Antennas Wirel. Propag. Lett.* **2017**, *16*, 1635–1638. [CrossRef]
15. Moscato, S.; Bahr, R.; Le, T.; Pasian, M.; Bozzi, M.; Perregrini, L.; Tentzeris, M.M. Infill-Dependent 3-D-Printed Material Based on NinjaFlex Filament for Antenna Applications. *IEEE Antennas Wirel. Propag. Lett.* **2016**, *15*, 1506–1509. [CrossRef]
16. Chieh, J.C.S.; Dick, B.; Loui, S.; Rockway, J.D. Development of a Ku-Band Corrugated Conical Horn Using 3-D Print Technology. *IEEE Antennas Wirel. Propag. Lett.* **2014**, *13*, 201–204. [CrossRef]
17. Zhang, B.; Guo, Y.X.; Zirath, H.; Zhang, Y.P. Investigation on 3-D-Printing Technologies for Millimeter- Wave and Terahertz Applications. *Proc. IEEE* **2017**, *105*, 723–736. [CrossRef]
18. Bolic, M.; Simplot-Ryl, D.; Stojmenovic, I. *RFID Systems: Research Trends and Challenges*; John Wiley & Sons: Hoboken, NJ, USA, 2010.
19. Rao, K.S.; Nikitin, P.V.; Lam, S.F. Antenna design for UHF RFID tags: A review and a practical application. *IEEE Trans. Antennas Propag.* **2005**, *53*, 3870–3876. [CrossRef]
20. Genovesi, S.; Costa, F.; Borgese, M.; Dicandia, F.A.; Monorchio, A.; Manara, G. Chipless RFID sensor for rotation monitoring. In Proceedings of the 2017 IEEE International Conference on RFID Technology Application, Warsaw, Poland, 20–22 September 2017; pp. 233–236.

21. Costa, F.; Gentile, A.; Genovesi, S.; Buoncristiani, L.; Lazaro, A.; Villarino, R.; Girbau, D. A Depolarizing Chipless RF Label for Dielectric Permittivity Sensing, *IEEE Microw. Wirel. Compon. Lett.* **2018**, *28*, 371–373. [CrossRef]
22. Lazaro, A.; Villarino, R.; Costa, F.; Genovesi, S.; Gentile, A.; Buoncristiani, L.; Girbau, D. Chipless Dielectric Constant Sensor for Structural Health Testing. *IEEE Sens. J.* **2018**, *18*, 5576–5585. [CrossRef]
23. Costa, F.; Genovesi, S.; Monorchio, A. Normalization-Free Chipless RFIDs by Using Dual-Polarized Interrogation. *IEEE Trans. Microw. Theory Tech.* **2016**, *64*, 310–318. [CrossRef]
24. Rezaiesarlak, R.; Manteghi, M. Complex-Natural-Resonance-Based Design of Chipless RFID Tag for High-Density Data. *IEEE Trans. Antennas Propag.* **2014**, *62*, 898–904. [CrossRef]
25. Ramos, A.; Perret, E.; Rance, O.; Tedjini, S.; Lázaro, A.; Girbau, D. Temporal Separation Detection for Chipless Depolarizing Frequency-Coded RFID. *IEEE Trans. Microw. Theory Tech.* **2016**, *64*, 2326–2337. [CrossRef]
26. Plessky, V.P.; Reindl, L.M. Review on SAW RFID tags. *IEEE Trans. Ultrason. Ferroelectr. Freq. Control.* **2010**, *57*, 654–668. [CrossRef]
27. Vena, A.; Babar, A.A.; Sydänheimo, L.; Tentzeris, M.M.; Ukkonen, L. A Novel Near-Transparent ASK-Reconfigurable Inkjet-Printed Chipless RFID Tag. *IEEE Antennas Wirel. Propag. Lett.* **2013**, *12*, 753–756. [CrossRef]
28. Vena, A.; Perret, E.; Tedjini, S. Chipless RFID Tag Using Hybrid Coding Technique. *IEEE Trans. Microw. Theory Tech.* **2011**, *59*, 3356–3364. [CrossRef]
29. Rance, O.; Siragusa, R.; Lemaitre-Auger, P.; Perret, E. RCS magnitude coding for chipless RFID based on depolarizing tag. In Proceedings of the 2015 IEEE MTT-S International Microwave Symposium, Phoenix, AZ, USA, 17–22 May 2015; pp. 1–4.
30. Herrojo, C.; Naqui, J.; Paredes, F.; Martín, F. Spectral signature barcodes implemented by multi-state multi-resonator circuits for chipless RFID tags. In Proceedings of the 2016 IEEE MTT-S International Microwave Symposium, San Francisco, CA, USA, 22–25 May 2016; pp. 1–4.
31. Herrojo, C.; Paredes, F.; Bonache, J.; Martín, F. 3-D-Printed High Data-Density Electromagnetic Encoders Based on Permittivity Contrast for Motion Control and Chipless-RFID. *IEEE Trans. Microw. Theory Tech.* **2020**, *68*, 1839–1850. [CrossRef]
32. Genovesi, S.; Costa, F.; Monorchio, A.; Manara, G. Chipless RFID Tag Exploiting Multifrequency Delta-Phase Quantization Encoding. *IEEE Antennas Wirel. Propag. Lett.* **2016**, *15*, 738–741. [CrossRef]
33. Genovesi, S.; Costa, F.; Dicandia, F.A.; Borgese, M.; Manara, G. Orientation-Insensitive and Normalization Free Reading Chipless RFID system based on Circular Polarization Interrogation. *IEEE Trans. Antennas Propag.* **2019**, *68*, 2370–2378. [CrossRef]
34. Kracek, J.; Svanda, M.; Hoffmann, K. Scalar Method for Reading of Chipless RFID Tags Based on Limited Ground Plane Backed Dipole Resonator Array. *IEEE Trans. Microw. Theory Tech.* **2019**, *67*, 4547–4558. [CrossRef]
35. Costa, F.; Borgese, M.; Gentile, A.; Buoncristiani, L.; Genovesi, S.; Dicandia, F.A.; Bianchi, D.; Monorchio, A.; Manara, G. Robust Reading Approach for Moving Chipless RFID Tags by Using ISAR Processing. *IEEE Trans. Microw. Theory Tech.* **2018**, *66*, 2442–2451. [CrossRef]
36. Vena, A.; Perret, E.; Tedjni, S. A Depolarizing Chipless RFID Tag for Robust Detection and Its FCC Compliant UWB Reading System. *IEEE Trans. Microw. Theory Tech.* **2013**, *61*, 2982–2994. [CrossRef]
37. Mukherjee, S. Chipless near field resistive element sensor using phase processing. In Proceedings of the 2016 IEEE International Conference on RFID (RFID), Orlando, FL, USA, 3–5 May 2016; pp. 1–5.
38. Herrojo, C.; Mata-Contreras, J.; Nunez, A.; Paredes, F.; Ramon, E.; Martin, F. Near-Field Chipless-RFID System with High Data Capacity for Security and Authentication Applications. *IEEE Trans. Microw. Theory Tech.* **2017**, *65*, 5298–5308. [CrossRef]
39. Nguyen, H.P.; Retraint, F.; Morain-Nicolier, F.; Delahaies, A. A Watermarking Technique to Secure Printed Matrix Barcode—Application for Anti-Counterfeit Packaging. *IEEE Access.* **2019**, *7*, 131839–131850. [CrossRef]
40. Miao, J.; Ding, X.; Zhou, S.; Gui, C. Fabrication of Dynamic Holograms on Polymer Surface by Direct Laser Writing for High-Security Anti-Counterfeit Applications. *IEEE Access.* **2019**, *7*, 142926–142933. [CrossRef]
41. Kajfez, D.; Guillon, P. *Dielectric Resonators*, 2nd ed.; SciTech Publishing, Incorporated: Raleigh, NC, USA, 1998.
42. Keyrouz, S.; Caratelli, D. Dielectric Resonator Antennas: Basic Concepts, Design Guidelines, and Recent Developments at Millimeter-Wave Frequencies. *Int. J. Antennas Propag.* **2016**, *2016*, 1–20. [CrossRef]

43. Huitema, L.; Monédière, T. *Dielectric Material*; InTech: Rijeka, Croatia, 2012.
44. Luk, K.M.; Leung, K.W. *Dielectric Resonator Antennas*; Research Studies Press: Baldock, UK, 2003.
45. Petosa, A. *Dielectric Resonator Antenna Handbook*; Artech House: Boston, MA, USA, 2007.
46. Leung, K.W.; Lim, E.H.; Fang, X.S. Dielectric Resonator Antennas: From the Basic to the Aesthetic. *Proc. IEEE* **2012**, *100*, 2181–2193. [CrossRef]
47. Fang, X.S.; Leung, K.W. Designs of Single-, Dual-, Wide-Band Rectangular Dielectric Resonator Antennas. *IEEE Trans. Antennas Propag.* **2011**, *59*, 2409–2414. [CrossRef]
48. Fang, X.S.; Leung, K.W. Linear-/Circular-Polarization Designs of Dual-/Wide-Band Cylindrical Dielectric Resonator Antennas. *IEEE Trans. Antennas Propag.* **2012**, *60*, 2662–2671. [CrossRef]
49. Long, S.; McAllister, M.; Shen, L. The resonant cylindrical dielectric cavity antenna. *IEEE Trans. Antennas Propag.* **1983**, *31*, 406–412. [CrossRef]
50. Available online: https://ansys.com (accessed on 16 July 2020).
51. Available online: https://www.3dwasp.com/power-wasp-evo/ (accessed on 16 July 2020).
52. Available online: https://ultimaker.com/software/ultimaker-cura (accessed on 16 July 2020).
53. PREPERM® 3D ABS1000 filament 1.75mm 750g. Available online: https://www.preperm.com/webshop/product/preperm-3d-abs-%C9%9Br-10-0-filament-1-75mm/ (accessed on 14 June 2020).
54. RS PRO 1.75mm Copper MT-COPPER 3D Printer Filament, 750g. Available online: https://uk.rs-online.com/web/p/3d-printing-materials/1254347/ (accessed on 10 August 2020).

MDPI
St. Alban-Anlage 66
4052 Basel
Switzerland
Tel. +41 61 683 77 34
Fax +41 61 302 89 18
www.mdpi.com

Sensors Editorial Office
E-mail: sensors@mdpi.com
www.mdpi.com/journal/sensors

www.ingramcontent.com/pod-product-compliance
Lightning Source LLC
LaVergne TN
LVHW071254170726
843515LV00006BA/1400

* 9 7 8 3 0 3 6 5 4 4 8 8 5 *